THE 8051 MICROCONTROLLER

Third Edition

THE 8051 MICROCONTROLLER

Third Edition

KENNETH AYALA

THOMSON

DELMAR LEARNING™

Australia Canada Mexico Singapore Spain United Kingdom United States

The 8051 Microcontroller, 3E

Kenneth Ayala

Vice President, Technology and Trades SBU:
Alar Elken

Editorial Director:
Sandy Clark

Senior Acquisitions Editor:
Steve Helba

Development:
Dawn Daugherty

Marketing Director:
Dave Garza

Channel Manager:
Fair Huntoon

Marketing Coordinator:
Casey Bruno

Production Director:
Mary Ellen Black

Production Manager:
Andrew Crouth

Production Editor:
Stacy Masucci

Printed in Canada

1 2 3 4 5 XX 06 05 04

For more information contact
Delmar Learning
Executive Woods
5 Maxwell Drive, PO Box 8007,
Clifton Park, NY 12065-8007

Or find us on the World Wide Web at
www.delmarlearning.com

Library of Congress Cataloging-in-Publication Data:

Ayala, Kenneth J.
The 8051 microcontroller / by Kenneth J. Ayala. — 3rd ed.
p. cm.
ISBN 1-4018-6158-X (pbk. : alk. paper)
1. INTEL 8051 (Computer)
2. Digital control systems. I. Title.
QA76.8.I27A92 2004
004.165—dc22

2004046070

NOTICE TO THE READER

To John Jamison of VMI and
John Peatman of Georgia Tech,
both of whom made this book possible.

Contents

Preface

The microprocessor has been with us for some 30 years now, growing from an awkward 4-bit child to a robust 64-bit adult. Soon, 128-bit wizards will appear to crunch numbers, spreadsheets, and CAD CAM. The engineering community became aware of, and enamored with, the 8-bit microprocessors of the middle to late 1970s. The bit size, cost, and power of these early CPUs were particularly useful for specific tasks involving data gathering, machine control, human interaction, and many other applications that granted a limited intelligence to machines and appliances.

The personal computer that was spawned by the 8-bit units predictably became faster by increasing register word size and more complex by the addition of operating system hardware. This process evolved complex CPUs that are poorly suited to dedicated applications and more applicable to the generic realm of the computer scientist and system programmer. Engineering applications, however, did not change; these applications continue to be well served by 8-bit CPUs with limited memory size and I/O power. Cost per unit also continues to dominate manufacturing considerations. Using an expensive 32-bit microprocessor to perform functions that can be as efficiently served by an inexpensive 8-bit microcontroller will doom the 32-bit product to failure in any competitive marketplace.

The early 1980s produced a new type of small computer — one with not only the CPU on the chip, but RAM, ROM, timers, UARTS, ports, and other common peripheral I/O functions also. The 8-bit microprocessor became the microcontroller.

Some manufacturers, hoping to capitalize on our software investment, have brought out families of microcontrollers that are software compatible with older microprocessors. Others, wishing to optimize the instruction set and architecture to improve speed and reduce code size, produced totally new designs that had little in common with their earlier microprocessors. Both of these trends continue.

This book has been written for a diverse audience. It is meant for use primarily by those who work in the area of the electronic design and assembly language programming of small, dedicated computers.

An extensive knowledge of electronics is not required to program a microcontroller. Many practitioners in disciplines not normally associated with computer electronics — transportation, HVAC, mechanisms, medicine, and manufacturing processes of all types—can benefit from a knowledge of how these "smart chips" work and how they can be used to improve their particular product.

Persons quite skilled in the application of classical microprocessors, as well as novice users who have no understanding of computer operation, should all find this book useful. The seasoned professional can read Chapter 3 with some care, glance at the mnemonics in Chapters 5 through 8, and inspect the applications in Chapters 9, 10, and 11. The student may wish to read Chapters 3 and 4, study the mnemonics and program examples carefully in Chapters 5 through 8, and then exercise the example programs in Chapters 9, 10, and 11 to see how it all works.

The text is suitable for a one- or two-semester course in microcontrollers. A two-semester course sequence could involve the study of Chapters 1 to 8 in the first semester and Chapters 9, 10, and 11 in the second semester in conjunction with several involved student programs. A one-semester course might stop with Chapter 9 and use many short student assignments drawn from the problems at the end of each chapter. The only prerequisite would be introductory topics concerning combinational and sequential logic operation and a working knowledge of using a PC-compatible personal computer.

No matter what the interest level, I hope all groups will enjoy using the software that has been included as part of the text. It is my belief that one should not have to buy unique hardware evaluation boards, or other hardware-specific items, in order to "try out" a new microcontroller. I also believe that it is important to get to the job of writing code as easily and as quickly as possible. The time spent learning to use the hardware board, board monitor, board communication software, and other boring overhead is time taken from learning to write code for the microcontroller.

The program included on the disk is a Keil, Inc., IDE. Keil has provided us all with a software development environment that is not only easy to use but one that we can uniquely configure for our own special purposes. Details on the assembler and simulator are provided in the appropriate appendices; use them as early as possible in your studies. Many points that are awkward to explain verbally become clear when you see them work in the simulator windows!

I have purposefully not included a great deal of hardware-specific information with the text. If your studies include building working systems that interface digital logic to the microcontroller, you will become very aware of the need for precise understanding of the electrical loading and timing requirements of an operating microcontroller. These details are best discussed in the manufacturer's data book(s) for the microcontroller and any associated memories and interface logic. Timing and loading considerations are not trivial; an experienced designer is required to configure a system that will work reliably. Hopefully, many readers will be from outside the area of electronic design and are mainly concerned with the essentials of programming and interfacing a microcontroller. For these users, I would recommend the purchase of complete boards that have the electrical design completed and clear directions as to how to interface common I/O circuits.

Anyone who writes a book soon becomes aware of a hidden "army" of others who help make it possible. At the risk of forgetting to mention some names, I would like to thank the following people:

Hesham Shaalan, Texas A&M University, Corpus Christi, Texas.

Max Rabiee, University of Cincinnati, Cincinnati, Ohio.

Gang Feng, University of Wisconsin, Platteville, Wisconsin.

Ed Kuhlman, sales manager at Modular Micro Controls, Norfield, Minnesota, who has brought to market an excellent student trainer built around hardware examples from this book.

The engineers, and lately the professors, who have bought or adopted this book.

Finally, let me thank you, the reader. I would be very grateful if any errors of commission or omission are gently pointed out to me at kenayala@aol.com.

Kenneth J. Ayala

1

Microprocessors and Microcontrollers

Chapter Outline

CHAPTER OBJECTIVES

On successful completion of this chapter you will be able to:

- List the differences between microcontrollers and microprocessors.
- Describe the prominant standard features of a typical microcontroller.
- Name several contemporary microcontroller manufacturers and notable features of their products.
- Identify the major components of a microcontroller development system.

1.0 Introduction

The past three decades have seen the introduction of a technology that has radically changed the way in which we analyze and control the world around us. Born of parallel developments in computer architecture and integrated circuit fabrication, the microprocessor, or "computer on a chip," first became a commercial reality in 1971 with the introduction of the 4-bit 4004 by a small, unknown company by the name of Intel Corporation. Other, more well-established, semiconductor firms soon followed Intel's pioneering technology so that by the late 1970s we could choose from a half dozen or so microprocessor types.

The 1970s also saw the growth of the number of personal computer users from a handful of hobbyists and "hackers" to millions of business, industrial, governmental, defense, educational, and private users now enjoying the advantages of inexpensive computing.

A by-product of microprocessor development was the microcontroller. The same fabrication techniques and programming concepts that make possible the general-purpose microprocessor also yielded the microcontroller.

Microcontrollers are not as well known to the general public, or to many in the technical community, as are the more glamorous microprocessors. The public is, however, very well aware that "something" is responsible for all of the smart DVDs, clock radios, washers and dryers, video games, telephones, microwaves, TVs, automobiles, toys, vending machines, copiers, elevators, irons, and a myriad of other articles that are intelligent and "programmable." Companies are also aware that being competitive in this age of the microchip requires their products, or the machinery they use to make those products, to have some "smarts."

The purpose of this chapter is to introduce the concept of a microcontroller and discuss current trends. The remainder of the book will study one of the most popular types, the 8051, in detail.

1.1 Microprocessors and Microcontrollers

Readers who have no prior concepts of computer operation should read Chapter 4 now.

Microprocessors and microcontrollers stem from the same basic idea, are made by the same people, and are sold to the same types of system designers and programmers. What is the difference between the two?

Microprocessors

A *microprocessor,* as the term has come to be known, is a general-purpose digital computer central processing unit (CPU). Although popularly known as a "computer on a chip," the microprocessor is in no sense a complete digital computer.

Figure 1.1 shows a block diagram of a microprocessor CPU, which contains an arithmetic and logic unit (ALU), a program counter (PC), a stack pointer (SP), some working registers, a clock timing circuit, and interrupt circuits.

To make a complete microcomputer, one must add memory, usually read-only program memory (ROM) and random-access data memory (RAM), memory decoders, an oscillator, and a number of input/output (I/O) devices, such as parallel and serial data ports. In addition, special-purpose devices, such as

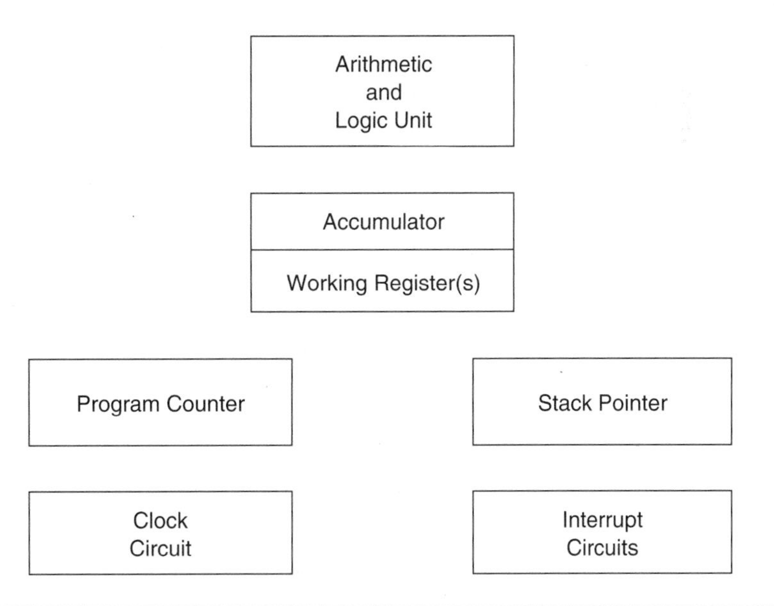

FIGURE 1.1 ◆ A Block Diagram of a Microprocessor

interrupt handlers and counters, may be added to relieve the CPU from time-consuming counting or timing chores. Equipping the microcomputer with mass storage devices, commonly a floppy and hard disk drives, and I/O peripherals, such as a keyboard and a CRT display, yields a small computer that can be applied to a range of general-purpose software applications.

The key term in describing the design of the microprocessor is *general-purpose.* The hardware design of a microprocessor CPU is arranged so that a small or very large system can be configured around the CPU as the application demands. The internal CPU architecture, as well as the resultant machine-level code that operates that architecture, is comprehensive but as flexible as possible.

The prime use of a microprocessor is to read data, perform extensive calculations on that data, and store those calculations in a mass storage device or display the results for human use. The programs used by the microprocessor are stored in the mass storage device and loaded into RAM as the user directs. A few microprocessor programs are stored in ROM. The ROM-based programs are primarily small fixed programs that operate peripherals and other fixed devices that are connected to the system. The design of the microprocessor is driven by the desire to make it as expandable and flexible as possible, in the expectation of commercial success in the marketplace.

Microcontrollers

Figure 1.2 shows the block diagram of a typical *microcontroller,* which is a true computer on a chip. The design incorporates all of the features found in a microprocessor CPU: ALU, PC, SP, and registers. It also has added the other features needed to make a complete computer: ROM, RAM, parallel I/O, serial I/O, counters, and a clock circuit.

Like the microprocessor, a microcontroller is a general-purpose device, but one that is meant to read data, perform limited calculations on that data, and control its environment based on those calculations. The prime use of a microcontroller is to control the operation of a machine using a fixed program that is stored in ROM and that does not change over the lifetime of the system.

The design approach of the microcontroller mirrors that of the microprocessor: make a single design that can be used in as many applications as possible in order to sell, hopefully, as many as possible. The microprocessor design accomplishes this goal by having a very flexible and extensive repertoire of multi-byte instructions. These instructions work in a hardware configuration that enables large amounts of memory and I/O to be connected to address and data bus pins on the integrated circuit package. Much of the activity in the microprocessor has to do with moving code and data to and from *external* memory to the CPU. The architecture features working registers that can be programmed to take part in the memory access process, and the instruction set is

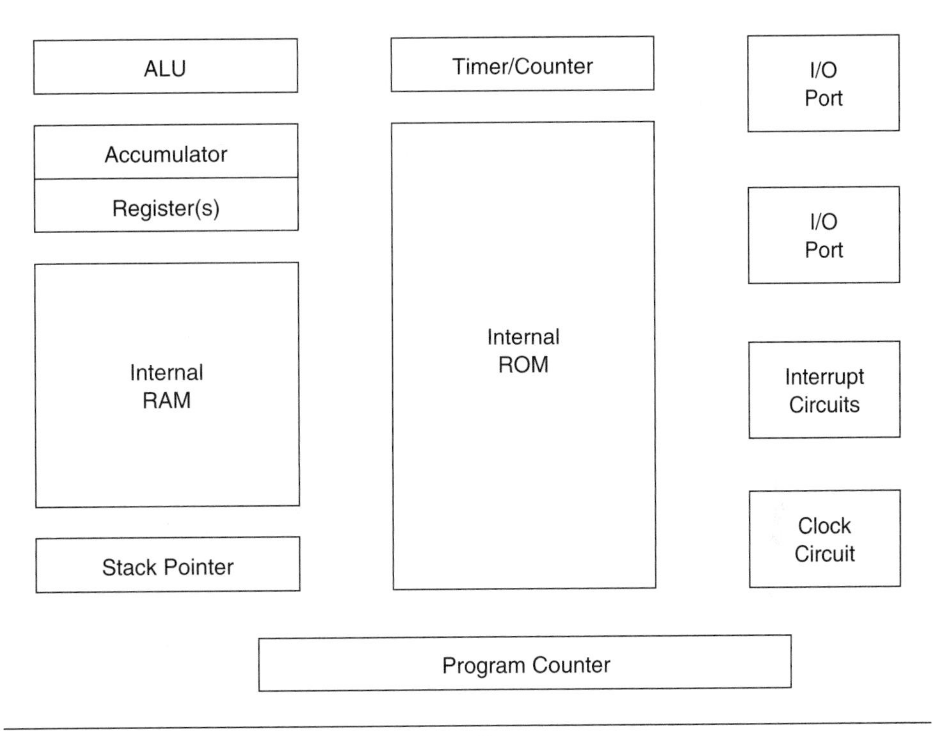

FIGURE 1.2 ◆ A Block Diagram of a Microcontroller

aimed at expediting this activity in order to improve throughput. The pins that connect the microprocessor to external memory are unique, each having a single function. Data is handled in word, double-word, or larger, sizes.

The microcontroller design uses a much more limited set of instructions that are used to move code and data from *internal* memory to the ALU. Many instructions are coupled with pins on the integrated circuit package; the pins are "programmable"—that is, capable of having several different functions depending on the wishes of the programmer.

The microcontroller is concerned with getting data from and to its own pins; the architecture and instruction set are optimized to handle data in bit, byte, and word size.

Figure 1.3 is a photograph of several members of the 8051 family.

Comparing Microprocessors and Microcontrollers

The contrast between a microcontroller and a microprocessor is best exemplified by the fact that most microprocessors have many operational codes (opcodes) for moving data from external memory to the CPU; microcontrollers may

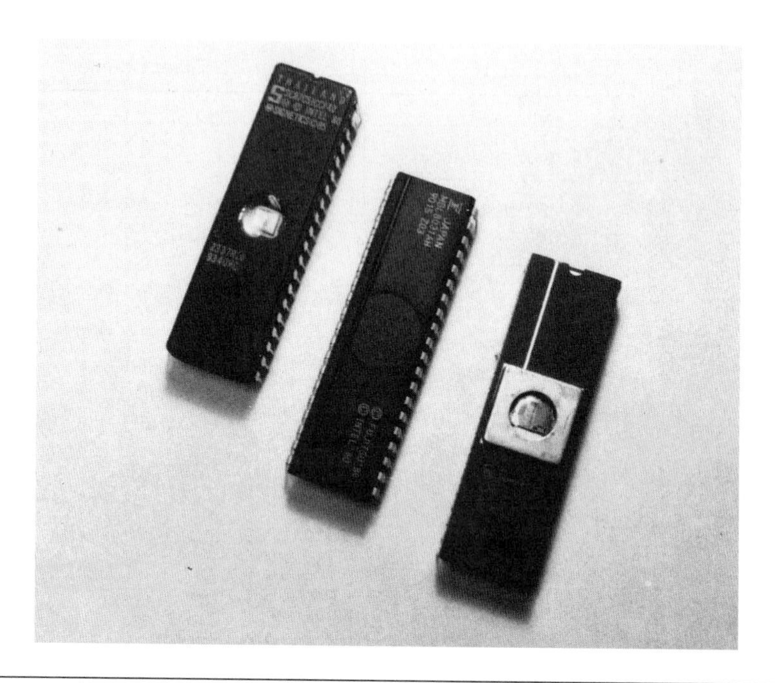

FIGURE 1.3 ◆ The 8051 Microprocessor

have one or two. Microprocessors may have one or two types of bit-handling instructions; microcontrollers will have many.

To summarize, the microprocessor is concerned with rapid movement of code and data from external addresses to the chip; the microcontroller is concerned with rapid movement of bits within the chip. The microcontroller can function as a computer with the addition of no external digital parts; the microprocessor must have many additional parts to be operational.

1.2 The Z80 and the 8051

To see the differences in concept between a microprocessor and a microcontroller, in Table 1.1 we will examine the pin configurations, architecture, and instruction sets for a 1980s-era 8-bit microprocessor, the Zilog Z80, and a microcontroller, the 8-bit Intel 8051.

Note that the point here is not to show that one design is "better" than the other; the two designs are intended to be used for different purposes and in different ways. For example, the Z80 has a very rich instruction set. The penalty

TABLE 1.1

	Z80	8051
Pin Configurations		
Total pins	40	40
Address pins	16 (fixed)	16
Data pins	8 (fixed)	8
Interrupt pins	2 (fixed)	2
I/O pins	0	32
Architecture		
8-bit registers	20	34
16-bit registers	4	2
Stack size	64K	128
Internal ROM	0	4K
Internal RAM	0	128
External memory	64K	128K
Flags	6	4
Timers	0	2
Parallel port	0	4
Serial port	0	1
Instruction Sets (types/variations)		
External moves	4/14	2/6
Block moves	2/4	0
Bit manipulate	4/4	12/12
Jump on bit	0	3/3
Stack	3/15	2/2
Single byte	203	49
Multi-byte	490	62

that is paid for this abundance is the number of multi-byte instructions needed, some 71% of the total number. Each byte of a multi-byte instruction must be fetched from program memory, and each fetch takes time; this results in longer program byte counts and slower execution time versus single-byte instructions. The 8051 has a 62% multi-byte instruction content; the 8051 program is more compact and will run faster to accomplish similar tasks.

The disadvantage of using a *lean* instruction set as in the 8051 is increased programmer effort (expense) to write code; this disadvantage can be overcome when writing large programs by the use of high-level languages such as BASIC and C, both of which are popular with 8051 system developers. The price paid for reducing programmer time (there is *always* a price) is the size of the program generated.

1.3 Four-Bit to Thirty-Two-Bit Microcontrollers

Every application demands a microcontroller that offers the right amount of functionality at the minimum cost. Applications vary from controlling an appliance to controlling an automobile. No single microcontroller design can economically meet these demands, so semiconductor manufacturers offer an array of microcontrollers designed to handle data in 4-, 8-, 16- and 32-bit words.

Bits are not the only thing that increase as word length grows. Additional functions are also added to meet market needs. Some of the more popular additional functions are

Analog-to-digital (A/D) converters, which change external analog signals to digital bits

Counter arrays, used to count and generate pulses

Watchdog timers (WDTs), which reset the controller if program execution hangs up

Serial data, both asynchronous (UART) and synchronous

Pulse width modulation (PWM), used, among other things, for motor speed control

Phase-locked-loops (PLLs), used for synchronous communications

External bus controllers for static (RAM/ROM) and dynamic (SDRAM) memories

If the market is large enough, say for automotive dashboard control and display, a high-end microcontroller will be designed and marketed by numerous companies just to meet that single application.

Four-Bit Microcontrollers

In terms of sheer volume, 4-bit microcontrollers may be used more than any other type. For a commodity microcontroller, cost depends as much on the volume of the package and the number of pins as on the amount of silicon inside. Pin count, in turn, depends on the number of data bits commonly handled by the microcontroller, and its I/O capability. Using 4 bits reduces package cost and pin count to a minimum. Employed in applications ranging from alphanumeric LED/LCD display drivers to portable battery chargers, these are the least expensive "smart chips" available. An example of a contemporary 4-bit microcontroller is the Renasas M34501, in a 20-pin DIP:

4-bit RAM	10-bit ROM	Counters	I/O Pins	Other Features
256	4K	2	14	A/D, WDT

Eight-Bit Microcontrollers

Eight-bit microcontrollers are perhaps the most popular microcontrollers in use today, judging from the number of semiconductor companies making them.

Eight bits have proven to be a very useful word size for everyday controller tasks. Capable of 256 decimal values, or quarter-percent resolution, the 1-byte data word is adequate for many control and monitoring applications. Serial ASCII code is also byte size, making 8 bits the natural choice for data communication applications. In addition, most low-cost RAM and ROM memories store 1 byte per memory location for easy interfacing to an 8-bit microcontroller.

One indication of the popularity of 8-bit microcontrollers is the fact that some 44 manufacturers produce over 600 models based on the 8051 architecture alone! Other popular microcontrollers, such as those designed by Microchip, Motorola, and Zilog, add hundreds of additional choices to the 8-bit menu. Here is what the 8051 offers when mounted in a 40-pin DIP:

ROM	ROM	Counters	I/O Pins	Other Features
128 bytes	4K	2	32	UART

Variations on a design occur when manufacturers offer models that include application-specific extras such as A/D and D/A converters, counter arrays, UARTs, WDT, and different memory configurations. On-chip ROM memory size may be increased, or ROMless models made that use off chip EPROMs for prototyping purposes. Flash and EEPROM memories may also be incorporated into the design.

Sixteen-Bit Microcontrollers

Sixteen-bit microcontrollers offer much of the generality of 8-bit models, but with greatly increased memory size and speed. Sixteen-bit microcontrollers are also much better suited for programming in high-level languages, such as C.

Applications for 16-bit microcontrollers are calculation and data intensive and include disk drives, modems, printers, scanners, pattern recognition, and automotive and servomotor control. A typical 16-bit microcontroller, the Motorola 68HC16Z3, has these attributes when mounted in a 144-lead LQFP package:

RAM	ROM	Counters	I/O Pins	Other Features
4K	8K	2	24	Counter Array, UART, A/D, WDT

Inspection of this summary shows that very few (24) of the 144 pins are used for general-purpose I/O. In fact, most of the pins are dedicated to external memory addressing, A/D input channels, counting/timing, serial data, pulse width modulation, phase-locked loops, and external memory bus control.

Thirty-Two-Bit Microcontrollers

Thirty-two-bit microcontrollers are currently evolving away from general-purpose applications to targeted markets such as PDAs, GPS, automotive control, communication networks, robotics, entertainment/game boxes, digital cameras, cell phones, and similar high-end uses. As an example, the Sharp LH79520 housed in a 176-pin LQFP package offers the following features, most of which could be used to implement a notebook computer: 32K RAM, color LCD controller, three UARTs, synchronous serial, two PWMs, 64 I/O pins, four counters, WDT, real-time clock, PLL, and direct memory access (DMA). Clearly, the development of 32-bit microcontrollers is driven by large, well-defined markets, and must be studied on an individual basis.

1.4 Development Systems for Microcontrollers

What is needed to be able to apply a microcontroller to your product? That is, what package of hardware and software will allow the microcontroller to be programmed and connected to your application? A package commonly called a *development system* is required.

First, trained personnel must be available either on your technical staff or as consultants. One person who is versed in digital hardware and computer software is the minimum number.

Second, a device capable of programming EPROMs must be available to test the prototype device. Many of the microcontroller families discussed have a ROMless version, an EPROM version, or an electrically erasable and programmable read only memory (EEPROM) version that lets the designer debug the hardware and software prototype before committing to full-scale production. Many inexpensive EPROM programmers are sold that plug into a port of most popular personal computers. More expensive, and more versatile, dedicated programmers are also available. An alternative to EPROMs are vendor-supplied prototype cards that allow code to be downloaded from a host computer and the program run from RAM for debugging purposes. An EPROM will eventually have to be programmed for the production version of the microcontroller.

Finally, software is needed, along with a personal computer to host it. The minimum software package consists of a machine-language assembler, which can be supplied by the microcontroller vendor or bought from independent developers. More expensive software, mainly consisting of high-level language compilers and debuggers, is also available.

A minimum development system, then, consists of a personal computer, a plug-in EPROM programmer, and a public-brand assembler. A more extensive system would consist of vendor-supplied dedicated computer systems with attendant high-level software packages and in-circuit emulators for hardware and software debugging. In 2003 dollars, the cost for the range of solutions outlined here is from $500 to $5,000.

1.5 Summary

The fundamental differences between microprocessors and microcontrollers are these:

- Microprocessors are intended to be general-purpose digital computers whereas microcontrollers are intended to be special-purpose digital controllers.
- Microprocessors contain a CPU, memory addressing circuits, and interrupt handling circuits. Microcontrollers have these features as well as timers, parallel and serial I/O, and internal RAM and ROM.
- Microcontroller models vary in data size from 4 to 32 bits. Four-bit units are produced in huge volumes for very simple applications, and 8-bit units are the most versatile. Sixteen- and 32-bit units are used in high-speed control and signal processing applications.
- Many models feature programmable pins that allow external memory to be added with the loss of I/O capability.

1.6 Questions

1. Name four major differences between a microprocessor and a microcontroller.
2. The 8051 has 40 pins on a dual inline package (DIP), yet the comparison with the Z80 microprocessor shows the 8051 has 58 pin functions. Explain this difference.
3. Name 20 items that have a built-in microcontroller.
4. Name 10 items that should have a built-in microcontroller.
5. Name the most unusual application of a microcontroller that you have seen actually for sale.

6. Name the most likely bit size for each of the following products:
   Modem
   Printer
   Toaster
   Automobile engine control
   Robot arm
   Small ASCII data terminal
   Chess player
   House thermostat
7. Explain why ROMless versions of microcontrollers exist.
8. List three essential items needed to make up a development system for programming microcontrollers.
9. Search the Internet and determine whether any manufacturer has announced a 64-bit microcontroller.

2

Numbering Systems and Binary Arithmetic

Chapter Outline

CHAPTER OBJECTIVES

On successful completion of this chapter you will be able to:

- Describe the differences between symbolic and positional number systems.
- Understand how positional number magnitudes are formed.
- Calculate the decimal equivalent of a nondecimal base number.
- Convert numbers from decimal to binary and hexadecimal, and the reverse.
- Understand the concept of signed complementary number systems.
- Structure a complementary signed number system of any size.
- Perform addition and subtraction operations using signed complementary numbers.
- Understand the meaning of an overflow in signed number mathematics.
- Perform multiplication and division of unsigned binary numbers.
- Describe some of the uses for the ASCII, BCD, Gray, and Parity error binary codes.

2◆0 Introduction

The human need to count things goes back to the dawn of civilization. How many sheep did one possess? More ominously, how much land was owned, and how much tax was due? Most urgently, what was the size of the opposing army? To answer the questions of "How much" or "How many," people invented number *systems.* A number system is any scheme used to count things. There are two general types of number systems, *symbolic* and *positional.* Positional number systems are the most popular number systems in use today and have been for hundreds of years.

2◆1 Symbolic Number Systems

The symbolic number system uses *distinct* symbols, called *numerals,* to construct larger numbers. The most famous symbolic number system, in the Western world, is the Roman numeral system. The Roman system has seven distinct numerals from which all larger numbers can be constructed. These and their decimal number equivalents are shown in Table 2.1.

A symbolic number system, such as the Roman, expresses large numbers by adding together a string of numerals. For instance, to express the decimal number 27 in Roman numerals results in the Roman number XXVII, or X + X + V + I + I. The Roman number for decimal 6,000 is obtained by placing together six Roman symbols for 1,000, MMMMMM, or M + M + M + M + M + M.

TABLE 2.1

Roman Number	Decimal Number
I	1
V	5
X	10
L	50
C	100
D	500
M	1,000

The Roman system forms large numbers from seven numerals by placing numerals of the same or increasing size to the left of each other, as in our previous examples. If a smaller number is placed to the left of a larger number, however, then the smaller is subtracted from the larger. For instance, the Roman number for decimal 9 is IX (10 − 1), not VIIII (5 + 1 + 1 + 1 + 1).

Clearly, the Roman numbering system is not convenient for expressing large numbers and certainly must have been a nightmare for students learning the Roman numeral multiplication tables. Moreover, there was no Roman symbol for zero, a much later invention of clever Arabic mathematicians. The Arabs, in turn, had gotten a very innovative numbering system from India that uses the *position* of a numeral to express its *size.* The Indians obtained the idea of a positional numbering system from the Mesopotamians, who had also decided to use *sixty* numerals! The Indians pared the number of numerals down to nine, the Arabs added the symbol for zero, and the positional decimal number system was introduced to Western Europe, via Moorish Spain, around the 11th century. The Arabic system became the *decimal* number system, from the Latin name for *tenth,* "decima."

The computational advantages of a decimal positional number system slowly gained acceptance in the Western world, starting with commercial companies located in Italy and Germany. The positional decimal number system has now entirely supplanted the symbolic Roman numerals in our society except for use in books, buildings, and the Superbowl.

2.2 Positional Number Systems

A positional number system is one in which a set of numerals form the basic counting symbols, or *digits.* The decimal number system, for instance, has 10 numerals: 0, 1, 2, 3, 4, 5, 6, 7, 8, and 9. Numbers larger than a single digit are expressed by arranging the numerals from right to left in a row. The position of a digit, in the number, is counted from right to left with the first position counted as position zero.

The total number of numerals used for a given positional number system is the *radix* (r) of the system. The radix of the decimal numbering system is 10, the total number of decimal numerals. The radix is used to make up numbers in the system that are greater than a single digit. The total value of a positional number is found as follows:

Count the position of a digit, starting at the right and numbering each position toward the left as position number 0, 1, 2, 3, and so on.

Raise the radix of the system to a power equal to each position number, as r^3, r^2, r^1, r^0, and so on.

Multiply the digit found at each position by its associated raised radix.

Add the resulting numbers together.

For instance, let us investigate the positional decimal numbering system, which has 10 numerals. Ten seems like a goodly number of digits, because humans are equipped with 10 fingers, and 10 numeric symbols are easy to learn. The 10 numeric symbols chosen for the Arab-Indian system are, in increasing order of size, 0, 1, 2, 3, 4, 5, 6, 7, 8, and 9. Ten digit symbols results in a radix of 10, which is a number made up of numerals 1 and 0. There is *no* numeral *10* in our system. Now, we construct a small table of the radix 10 raised to positional powers, as shown in Table 2.2.

Note that the first position is position number 0, *not* position number 1. Any number (except 0) raised to the power of 0 is a 1. Many grammar school children learn to express large decimal numbers by memorizing a few of the powers of 10 from Table 2.2, such as the "1's place," "10's place," "100's place," and so on.

We may now construct numbers in the decimal system using the 10 decimal numerals and Table 2.2. For instance, the decimal number 654,321 is constructed as shown in Table 2.3.

TABLE 2.2

Position	Power of 10
0	1
1	10
2	100
3	1,000
4	10,000
5	100,000
6	1,000,000
7	10,000,000
8	100,000,000

TABLE 2.3

Numeral	Position	Numeral $\times$ Radix^{Position}
1	0	$1 \times 10^0 = 000001$
		+
2	1	$2 \times 10^1 = 000020$
		+
3	2	$3 \times 10^2 = 000300$
		+
4	3	$4 \times 10^3 = 004000$
		+
5	4	$5 \times 10^4 = 050000$
		+
6	5	$6 \times 10^5 = \underline{600000}$
		654,321

The decimal number system has succeeded because very large numbers can be expressed using relatively short series of easily memorized numerals. Systems with a radix larger than 10 require additional numeric symbols but do result in more compact numbers than the decimal system. Systems with a radix smaller than 10 do not require as many numerals but result in longer numbers. Humankind also uses number systems based on a radix of 20 ("Four score and seven years ago . . ."), a radix of 60 (60 seconds in a minute, 60 minutes in an hour), and, since about 1940, a radix of 2.

Positional numbers permit the number system *inventor* to choose any radix and create any desired *symbol* for each numeral in the system. It is usually most convenient to pick from the decimal numeral symbols for number systems with a radix up to 10. For systems with a radix larger than 10, new symbols have to be invented, the most convenient being letters of the alphabet.

For instance, suppose that a system with a radix of 5 is chosen. A radix of 5 means that there must be five digits in the system, equivalent to decimal 0 to decimal 4. If we use decimal digits for our radix-5 system, then the number 12344 radix 5 has the decimal number system equivalent shown in Table 2.4.

To convert from nondecimal radix number A to decimal radix number B, multiply each digit of number A by its radix raised to the appropriate positional power. The product of each multiplication is expressed as a *decimal* number, and the resulting individual decimal numbers are added together to form the final, equivalent decimal number.

The reverse process, going from a decimal radix number back to a radix 5 number, involves division. From the previous example, 974 decimal is converted back to a 12344 radix 5 as shown in Table 2.5.

TABLE 2.4

Position	Numeral		$5^{Position}$		Decimal Equivalent
4	1	×	5^4	=	1 × 625 = 625
					+
3	2	×	5^3	=	2 × 125 = 250
					+
2	3	×	5^2	=	3 × 25 = 75
					+
1	4	×	5^1	=	4 × 5 = 20
					+
0	4	×	5^0	=	4 × 1 = 4
					974 radix 10

The process of converting decimal radix number B to a different radix number A is to divide decimal number B by the radix of number A and keep the remainder as a digit of the new number. The quotient from the first division is divided by the A radix, and the process repeated until the quotient is zero. The first division produces the *least significant digit* (LSD) of number A, and the last remainder is the *most significant digit* (MSD) of number A.

TABLE 2.5

Dividend ÷ 5		Quotient	Remainder	
$\frac{974}{5}$	=	194	Remainder of 4	(least significant digit)
$\frac{194}{5}$	=	38	Remainder of 4	
$\frac{38}{5}$	=	7	Remainder of 3	
$\frac{7}{5}$	=	1	Remainder of 2	
$\frac{1}{5}$	=	0	Remainder of 1	(most significant digit)
			12344	

2.3 Integer Binary Numbers

Sixty years ago, the practice of performing mathematical operations using binary numbers would have been thought absurd. To be sure, the mathematics of positional number systems had been well developed since the 9th century, but a positional number system based on a radix of 2 would likely have been thought an odd curiosity 60 years ago.

There is nothing new, or "modern," about binary numbers or binary mathematics. Binary numbers follow the same rules for arithmetic as does the more familiar decimal number system. Performing numerical calculations using binary numbers is no more difficult than performing calculations using decimal numbers. The challenge of binary arithmetic is to become comfortable with a system that has only two numerals—and to become comfortable with another system, called *hexadecimal,* that has 16 numerals.

Binary Digital Computers

So, the question might be, what happened 60 years ago that requires us to deal with binary numbers? The answers to that question are: the digital computer and the integrated circuit. Binary numbers and binary math are used today because humans have invented the digital computer and developed very *reliable* semiconductor electronic switches with which to build digital computers.

To date, reliable semiconductor switches must operate either turned on or turned off; they are inherently binary devices. Binary has evolved as the electronic digital system of choice because it is fairly easy to build reliable electronic switches that are either turned fully on or turned fully off. It is very difficult to build *reliable* electronic switches that can be on, half-on, and off, for instance. Electronic technology drives digital computer development, and electronic technology has progressed to the point at which extremely reliable and inexpensive binary electronic solid-state switches can be made.

The ability to make low-cost, reliable solid-state switches has, in turn, led to the development of very *inexpensive* digital computers. The availability of inexpensive digital computers has led to *increasing* employment demand for individuals who know how to use computers to solve society's problems. The ability to use a digital computer, particularly for engineering applications, requires a knowledge of binary numbers. Digital does *not* mean binary. Every number system has digits, and a computer based on a four-digit number system would also be a digital computer.

We normally study number systems in grammar school. After some confusing exposure to the "new" math, we settle down to using one number system—the decimal number system. Decimal numbers become so familiar to us that we begin to think they are the only numbers possible. Thus we find ourselves, many years past grammar school, beginning all over again with the "new" math.

The Binary Number System

A binary number system uses two decimal numerals, 0 and 1. A binary number is made up of a collection of binary numerals using a radix of 2. Binary digits are also called *bits,* which is a contraction of the words *Binary* and *digITS.*

In an electronic circuit, 1 and 0 might correspond to a switch that is off or on, or to a circuit that has a high voltage or a low voltage output. The assignment of binary 1 to off, for example, or binary 0 to a high voltage is arbitrary. A binary 1 may correspond to a switch that is off, or one that is on. Standards for the computer industry, however, have generally followed the practice that a circuit that is in a high-voltage state is a binary 1, and one that is in a low-voltage state is binary 0.

Conversions Between Decimal and Binary Numbers

The decimal equivalent of a binary number is formed by multiplying each bit in the binary number by the binary radix of 2 raised to the position's power. The result of each multiplication is expressed as a decimal number. The individual decimal numbers are added to obtain the decimal equivalent of the binary number. For instance, the binary number 111110100 can be converted to a decimal equivalent number as shown in Table 2.6.

The reverse process, that of converting a number with a decimal radix to a number with a binary radix, is done by repeated division of the decimal number by the binary radix of 2. The decimal number is repeatedly divided by 2, and the remainder becomes a bit of the equivalent binary number. Any decimal number divided by 2 must have a remainder of 0 if the decimal number is even, or 1 if the decimal number is odd.

TABLE 2.6

Digit	Position	$2^{Position}$	Decimal Equivalent
1	8	256	1 × 256 = 256
1	7	128	1 × 128 = 128
1	6	64	1 × 64 = 64
1	5	32	1 × 32 = 32
1	4	16	1 × 16 = 16
0	3	8	0 × 8 = 0
1	2	4	1 × 4 = 4
0	1	2	0 × 2 = 0
0	0	1	0 × 1 = 0
			500

The remainder of the first division operation yields the *least significant bit* (LSB) of the binary number. The number left after the first division by 2 (the first quotient) is again divided by 2, yielding a second remainder and a second quotient. The second remainder is the next most significant bit of the binary equivalent number. The process of dividing the quotient by 2 and keeping the remainder as a bit of the binary equivalent number is continued until the *quotient* is zero. The final remainder is the *most significant bit* (MSB) of the binary equivalent number.

To demonstrate conversion from a decimal number to an equivalent binary number, we shall take the decimal 500 number from the last example and reconvert it to binary. See Table 2.7.

Hexadecimal Numbers

The *larger* the radix of a number system, the more *compact* is the expression for any number in the system. Because each digit is multiplied by a larger radix raised to the positional power of the digit, the value of the number grows rapidly as digits are added.

A radix 5 number is more compact than a binary number, and a decimal radix number is more compact than a radix 5 number. For example, the number of digits required to express decimal 57 in systems of radix 2, 3, and 5 are shown next:

111001 radix 2 = 57 decimal
2010 radix 3 = 57 decimal
212 radix 5 = 57 decimal

Radix 12 numbers, in turn, are even more compact than decimal numbers. But, a problem arises when we begin to use numbers that have a radix larger than 10: we must invent *new* symbols for the numerals greater than decimal 9. One popular method used to invent new numerals is to borrow other well-known *symbols from the alphabet.*

Alphabetic letters can be adapted for each new numeral, such as A for decimal 10, B for decimal 11, C for decimal 12, and so on until all the new numerals have a symbol. Number systems that use a radix greater than 10 have not proved useful to date, except one, known as *hexadecimal.*

Hexadecimal numbers have a radix of *16, with numerals equivalent to decimal number 0 to decimal number 15.* Hexadecimal numbers are useful for binary work because *one* hexadecimal numeral is equivalent to a 4-bit binary number. Hexadecimal numbers then, may be thought of as a type of binary shorthand. Each hexadecimal numeral can be replaced by its equivalent binary number, or the reverse. (Hexadecimal is also commonly referred to as *hex* by most programmers.)

TABLE 2.7

Dividend ÷ 2		Quotient	Remainder	
$\frac{500}{2}$	=	250 Remainder	0	(LSB)
$\frac{250}{2}$	=	125 Remainder	0	
$\frac{125}{2}$	=	62 Remainder	1	
$\frac{62}{2}$	=	31 Remainder	0	
$\frac{31}{2}$	=	15 Remainder	1	
$\frac{15}{2}$	=	7 Remainder	1	
$\frac{7}{2}$	=	3 Remainder	1	
$\frac{3}{2}$	=	1 Remainder	1	
$\frac{1}{2}$	=	0 Remainder	1	(MSB)
			111110100	

A hexadecimal radix of 16 requires 16 symbols for the 16 numerals used in the system. The decimal number system supplies numeral symbols 0 – 9 for the first 10 hexadecimal digits, and letters A – F are borrowed from the alphabet for the remaining 6 hexadecimal numerals. The hexadecimal numeral symbols and their binary and decimal equivalents are shown in Table 2.8.

TABLE 2.8

Hexadecimal Numeral	Binary Equivalent	Decimal Equivalent
0	0000	0
1	0001	1
2	0010	2
3	0011	3
4	0100	4
5	0101	5
6	0110	6
7	0111	7
8	1000	8
9	1001	9
A	1010	10
B	1011	11
C	1100	12
D	1101	13
E	1110	14
F	1111	15

Conversions Between Hexadecimal, Binary, and Decimal Numbers

Because each hexadecimal *numeral* is equivalent to a 4-bit binary *number,* conversion between hexadecimal and binary numbers involves no calculations. Converting from a hexadecimal number to a binary number means simply replacing each hexadecimal numeral with its 4-bit binary equivalent. Converting from a binary number to a hexadecimal number involves replacing each *group* of 4 binary bits with their equivalent hexadecimal numeral.

Converting from Binary to Hexadecimal

Table 2.8 can be used to convert binary numbers to hex numbers by dividing the binary number into groups of 4 bits—*starting at the LSB and grouping to the left.* Table 2.8 then shows how to replace each 4-bit group with its equivalent hex digit. *Leading zeroes* can be added to the leftmost bit (MSB) as needed to fill out the final (leftmost) group of 4 bits.

As an example, take the binary number for 500 decimal that was just computed above and convert it to hex as follows:

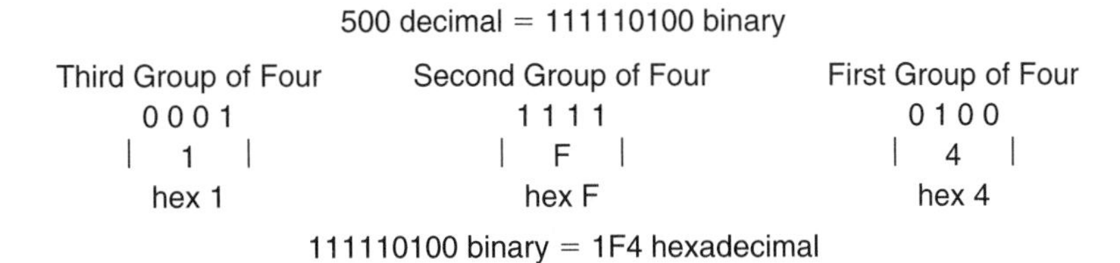

The binary number 111110100 is converted to the hexadecimal number 1F4 by forming groups of 4 binary bits, beginning at the LSB, and grouping to the *left*. Note that a binary 000 was added to the third group to fill it out to 4 bits.

Converting from Hexadecimal to Binary

To convert from a hex number to a binary number requires that a 4-bit binary number from Table 2.8 be inserted in place of each hex numeral. Each numeral of the hex number becomes a 4-bit binary number. For example, hex 3DEC is converted to binary as shown here:

Original hex number	3	D	E	C
Binary equivalent from Table 2.8	| 0011 |	| 1101 |	| 1110 |	| 1100 |

Table 2.8 should be committed to memory. The language of digital computers is filled with hexadecimal numerals and their binary number equivalents.

Converting Hexadecimal Numbers to Decimal Numbers

Conversions between hex numbers and decimal numbers proceed as in previous discussions involving binary numbers. Compact hex numbers expand into large decimal numbers because of the large radix size of a hex number. Conversion from hex to decimal is begun by converting each hex numeral of the hex number into its decimal equivalent found in Table 2.8. Each decimal number is then multiplied by hex radix 16 raised to its positional power. As an example, converting the hex number 5DECA to a decimal number is shown in Table 2.9.

Converting Decimal Numbers to Hexadecimal Numbers

Converting a decimal number into a hexadecimal number involves repeated steps whereby a decimal quotient is divided by hex radix 16. At each step, the decimal remainder is replaced with its equivalent hex numeral. The process

TABLE 2.9

Hex Number		Decimal Equivalent of Hex Numeral		$16^{Position}$		Decimal Equivalent
5×16^4	=	5	×	65,536	=	327,680
$D \times 16^3$	=	13	×	4096	=	53,248
$E \times 16^2$	=	14	×	256	=	3,584
$C \times 16^1$	=	12	×	16	=	192
$A \times 16^0$	=	10	×	1	=	10
						384,714

TABLE 2.10

Dividend ÷ 16		Quotient	Remainder (decimal)	Hex equivalent	$16^{Position}$	
$\frac{384,714}{16}$	=	24,044	Remainder of 10	= A	1	(LSD)
$\frac{24,044}{16}$	=	1,502	Remainder of 12	= C	16	
$\frac{1,502}{16}$	=	93	Remainder of 14	= E	256	
$\frac{93}{16}$	=	5	Remainder of 13	= D	4,096	
$\frac{5}{16}$	=	0	Remainder of 5	= 5	65,536	(MSD)
				5DECA		

begins by dividing the original decimal number by 16 and obtaining a decimal quotient and a decimal remainder. The remainder is converted into a hex numeral using Table 2.8, and the quotient is again divided by 16. Each division step produces a quotient, which is fed to the next division step, and a remainder, which is converted to a hex numeral; the conversion ends when the quotient reaches zero. The remainder of the first division step is the LSD of the hex number; the remainder of the last division step (the one that results in a quotient of zero) is the MSD of the hex number.

For example, converting the decimal number 384,714 back to a hex number is shown in Table 2.10.

Hexadecimal numbers are so convenient to use in place of binary numbers that almost all binary numbers longer than 2 or 3 bits are usually expressed in hex. *This book uses hexadecimal numbers* in place of binary numbers almost exclusively, in keeping with standard practice.

The only other popular nondecimal number system still in use is *octal (radix 8)*. Exercises involving octal number conversions will be left for the end of the chapter.

2.4 Fractional Binary Numbers

Integer (whole) number binary-decimal conversions are readily handled by the conversion techniques of multiply by raised radix or divide by radix discussed in previous examples. Fractional (less than unity) numbers are converted from one number system to another in exactly the same manner, except that the positions of the digits are numbered as negative, not positive, numbers.

Before beginning a discussion of fractional binary-decimal number conversions, however, the concept of the decimal point (binary point) has to be investigated. The point (.) is used to *separate* the positive radix positions of a number from the negative radix positions. To the left of the point are increasingly positive position digits, and to the right of the point are increasingly negative position digits.

Converting Binary Fractions to Decimal

Converting from a binary number that has a fractional part to a decimal number is done by multiplying each bit by the binary radix 2 raised to the appropriate positional power. But, remember that positional powers to the *right* of the binary point are *negative,* not positive, numbers. The individual results of each multiplication step are then added to obtain the final decimal number. As an example, a fractional binary number, 1101.1011, is converted to decimal as shown below.

2^{Position}	8	4	2	1	.	.5	.25	.125	.0625
Position	3	2	1	0	.	−1	−2	−3	−4
Bit	1	1	0	1	.	1	0	1	1

Integer Part		*Fractional Part*	
$1 \times 8 =$	8	$1 \times .5 =$	.5000
$1 \times 4 =$	4	$0 \times .25 =$	.0000
$0 \times 2 =$	0	$1 \times .125 =$	.1250
$1 \times 1 =$	1	$1 \times .0625 =$	.0625
	13	+	$.6875 = 13.6875$

As the fractional example shows, as one moves to the right of the binary point, the power of the radix, 2, is raised to the power -1 to yield $\frac{1}{2}$, to -2 to yield $\frac{1}{4}$, to -3 to yield $\frac{1}{8}$, and so on.

Converting Hexadecimal Fractions to Decimal

Hexadecimal to decimal fractional conversions are done using the hex radix, 16, raised to negative powers. The preceding binary number, 1101.1011, converts to hex number D.B, which is converted to decimal as shown next:

$1101.1011 = \text{D.B}$
$\text{D} = 13$
$.\text{B} = \text{B} \times 16^{-1} = \frac{11}{16} = .6875$
$\text{D.B} = 13 + .6875 = 13.6875$

When converting from hex fractional numbers to decimal numbers, the negative powers of the hex radix 16 begin with $\frac{1}{16}$ followed by $\frac{1}{256}$, $\frac{1}{4096}$, and so on. The last example again demonstrates how binary-decimal conversions are quickly performed when a hex number is used in place of its binary equivalent.

A second example, converting the hex number 3E.4FC to a decimal number, follows next:

Integer Part

3×16		$=48$
$\text{E} \times 1$	$= 14 \times 1$	$=14$
		62

Fractional Part

4×16^{-1}	$= \frac{4}{16}$	$= .2500000000$
$\text{F} \times 16^{-2}$	$= \frac{15}{256}$	$= .0585937500$
$\text{C} \times 16^{-3}$	$= \frac{12}{4096}$	$= .0029296875$
		$.3115234375$

$3\text{E.4FC} = 62 + .3115234375 = 62.3115234375$

Converting Decimal Fractions to Binary and Hexadecimal

Decimal *fractional* numbers are converted to fractional binary or hex equivalents by repeated *multiplication* of the decimal fraction by 2 (binary) or 16 (hex). Every multiplication step will result in a product that has an integer part and a fractional part. The integer part of each product is saved as a binary (hex) numeral. The fractional part of each product is again multiplied by 2 (binary) or 16 (hex) until the fractional part of a product is zero, or it becomes clear that the *process will never end.* The first multiplication yields the MSD of the fractional part of the binary or hex number.

There is one important fact to note when converting from decimal to binary fractional numbers: *exact conversions may not be possible.* Exact binary equivalents of decimal fractions will only be exact if the multiplication product, at *some* stage of the conversion process, ends as an integer product with *no* fractional part. Should repeated multiplication by 2 (or 16) of the decimal fraction result in a *repeating* pattern of decimal remainders, the binary fraction is *infinite in length.*

For instance, the decimal number 21.1, which has a fractional part of decimal .1, will be shown to have *no* exact binary equivalent. The decimal number, 21.1, when converted to binary, is processed as shown next.

Convert integer part of decimal number to binary by repeated division:

```
21
-- = 10        Remainder of 1   (LSB)
 2                            |
                              |
                              |
10
-- = 5       Remainder of 0  |
 2                          | |
                            | |
                            | |
 5
-- = 2     Remainder of 1 | |
 2                       | | |
                         | | |
                         | | |
 2
-- = 1   Remainder of 0 | | |
 2                     | | | |
                       | | | |
                       | | | |
 1
-- = 0 Remainder of 1 | | | |
 2                   | | | | |
                     ---------
      Integer Part = 10101
```

Convert fractional part of decimal number to binary by repeated multiplication:

```
                                  Fractional Part = .0 0011 0011 0
                                              (MSB)   | |||| |||| |
                      .1 × 2 = 0.2 Integer = 0 _______| |||| |||| |
------------------------------------------------------|||| |||| |
                      .2 × 2 = 0.4 Integer = 0 ___________|||| |||| |
First pattern is 0011 .4 × 2 = 0.8 Integer = 0 ____________||| |||| |
                      .8 × 2 = 1.6 Integer = 1 _____________|| |||| |
                      .6 × 2 = 1.2 Integer = 1 ______________| |||| |
-----------------------------------------------------------|||| |
                      .2 × 2 = 0.4 Integer = 0 _________________|||| |
Repeat pattern is 0011 .4 × 2 = 0.8 Integer = 0 __________________||| |
                      .8 × 2 = 1.6 Integer = 1 ___________________|| |
                      .6 × 2 = 1.2 Integer = 1 ____________________| |
---------------------------------------------------------------------|
And so on forever     .2 × 2 = 0.4 Integer = 0 ______________________|
```

Using the binary digits of the preceding decimal-binary fractional example yields the *inexact* equivalent of 21.1 decimal as 10101.000110011 binary.

As can be seen in the example of repeated multiplications of the decimal fraction .1, a repeating pattern of integers is established that will continue for as long as the multiplication continues.

The accuracy of the binary equivalent of a decimal fraction may, of course, be as precise as may be desired. Increased accuracy is obtained by carrying the calculations out to as many binary places as needed. For the prior example of converting .1 decimal to 9 binary places, the conversion from binary back to decimal becomes the following:

$$
\begin{aligned}
.000110011 = \; & 0 \times \tfrac{1}{2} &= .00000000 \\
& 0 \times \tfrac{1}{4} &= .00000000 \\
& 0 \times \tfrac{1}{8} &= .00000000 \\
& 1 \times \tfrac{1}{16} &= .06250000 \\
& 1 \times \tfrac{1}{32} &= .03125000 \\
& 0 \times \tfrac{1}{64} &= .00000000 \\
& 0 \times \tfrac{1}{128} &= .00000000 \\
& 1 \times \tfrac{1}{256} &= .00390625 \\
& 1 \times \tfrac{1}{512} &= \underline{.00195312} \\
& & .09960937
\end{aligned}
$$

The accuracy of the decimal fraction to binary fraction conversion, from .1 decimal to 9-bit binary places, is seen to be 99.61%.

The same fractional decimal, .1, when converted to 3 hex digits is processed by repeated multiplication by 16, as follows:

```
                              (MSD) .199  hex
.1 × 16 = 1.6  Integer = 1_____| | |
.6 × 16 = 9.6  Integer = 9_______| |
.6 × 16 = 9.6  Integer = 9_________|
```

.1 decimal = .199 hex (.0001 1001 1001 binary)

The decimal .1 fraction to hex conversion shows the appearance of the repeating integer (9) pattern, which signals that the process will have no finite end.

To check the accuracy of the conversion of .1 decimal to 3 hex digits, the hex fraction is converted back to decimal as shown next:

$$
\begin{aligned}
\tfrac{1}{16} &= .06250000 \\
\tfrac{9}{256} &= .03515625 \\
\tfrac{9}{4096} &= \underline{.00219726} \\
& .09985351
\end{aligned}
$$

The accuracy of converting .1 decimal to 3 hex places is 99.85%.

Note that hex conversion routines proceed much *faster* than do binary conversion routines, because of the larger size of the hex radix.

2.5 Number System Notation

Normally, in everyday life, numbers do not have to have any *notation*. Notation involves some special written clues as to the radix base of the number system in use. Everyone *assumes* that all numbers used in conversation, or seen in any publication, are decimal. Decimal numbers are so common that a special "decimal" notation does *not* have to be used. This book, however, uses *three* number systems: decimal, binary, and hexadecimal (hex). To help you distinguish among the three number systems, the following notation is used throughout the text:

- Decimal numbers are written with *no* special notation. For instance, decimal 123,456, is written 123,456, with no identifying notation.
- Hexadecimal numbers are written with an *h, after* the number, to denote hexadecimal. For example, hex ABCD will be written ABCDh, or abcdh. Hex numerals may be written using capital letters for the digits greater than 9 or using lowercase, as, for instance, 12ach. In this book, capital letters will normally be used in *explanatory* text and lowercase letters in *programs*.
- Binary numbers will be written with a small *b* after the binary number. For instance, binary 1011 will be written 1011b. The *b* notation for binary numbers is a feature of the A51 assembler.

2.6 Binary Addition and Subtraction

Binary math, when done by a digital computer, involves some ordinary, and extraordinary, concepts. Binary arithmetic calculations are extremely simple because there are only two numerals, 0 and 1. Interesting problems arise, however, when dealing with positive and negative binary numbers. Additional challenges may also appear because of the fact that every binary number has a finite size inside a computer.

Binary Number Addition and Subtraction

There are four possible results when two binary numerals are added together or one numeral is subtracted from the other.

Addition of two binary numerals may have one of the four results shown next:

```
 Carry       Carry       Carry       Carry
   |           |           |           |
   1)1         0)0         0)1         0)0
add 1       add 1       add 0       add 0
  ---         ---         ---         ---
    0           1           1           0
```

As shown when adding 1 to 1, the *possibility* of a carry from one bit position to the next higher order bit position must be included in the addition process. (*Carries* are shown using a right parenthesis to indicate the carry bit.)

Adding one 8-bit number to another 8-bit number, for instance, may produce a number of carries from each bit to the next, as the following example illustrates:

```
     Carry  =  1)1)1)1)1)1)1)1)
       +95  =  0 0 1 0 1 1 1 1 1 b
  add +189  =  0 1 0 1 1 1 1 0 1 b
      ----     -------------------
      +284  =  1)0 0 0 1 1 1 0 0 b = 256) + 28 = +284
```

Note that the *final* carry, from bit position 7 of the sum to bit position 8, must be included as part of the total, or the sum is in error.

The last example began by adding two 8-bit numbers but obtained a result that requires 9 bits to hold the total sum. In general, every time an n-bit number is added to an n-bit number, there is the possibility of an $n+1$ bit sum. But, if the computer is not *designed* to hold an $n+1$ bit number, then errors can occur.

Binary subtraction, involving 2 bits, also has four possible outcomes, as shown next:

```
Borrow      Borrow      Borrow      Borrow
  |           |           |           |
 0}1         1}0         0}1         0}0
sub 1       sub 1       sub 0       sub 0
   --          --          --          --
    0           1           1           0
```

A borrow, from the next higher order bit position to the lower order bit position, when 1 is subtracted from 0, is a possibility that must be considered during the subtraction process. (*Borrows* are shown using a right curly bracket.)

Subtraction of one 8-bit number from another 8-bit number also may generate borrows from higher order bits. For example, by reversing the previous addition problem, subtracting binary equivalent 95 from binary equivalent 189 proceeds as shown next:

```
   Borrow  =  0}1}0}1}1}1}1}0}
     +189  =    1 0 1 1 1 1 0 1 b
  sub +95  =    0 1 0 1 1 1 1 1 b
     ----       -----------------
      +94  =  0}0 1 0 1 1 1 1 0 b = 0} + 94 = 94
```

Subtracting a smaller number from a larger number presents no problems, because the result is less than either of the two original numbers. However, reversing the numbers in the previous subtraction problem so that +189 is subtracted from +95 poses some difficulties when the subtraction is done in binary, as shown next:

```
   Borrow = 1}0}1}0}0}0}0}0}
      +95 =   0 1 0 1 1 1 1 1 b
 sub +189 =   1 0 1 1 1 1 0 1 b
      - 94 = 1}1 0 1 0 0 0 1 0 b = 1} + 162
```

The result of a borrow plus 162 as the answer can be converted to −94 by using the borrow. If the borrow into the 8th binary place is understood to be -2^8, or −256, then −256 + 162 is −94. As was the case for 8-bit addition, however, subtracting two 8-bit numbers can result in a 9-bit answer.

To subtract +189 from +95, manually in the decimal system, we *actually* subtract +95 from +189 and then affix a negative *sign* to the answer to arrive at −94. By inspection, we know that +189 is larger than +95, and we subtract the smaller number from the larger, adjusting the sign of the answer as needed. The same process (subtracting the smaller from the larger number) in a computer binary circuit works just as well as manual decimal subtraction, but a *negative sign* for the answer is needed.

The possibility of generating negative and positive binary numbers, as well as the formation of $n+1$ bit numbers from n-bit numbers, leads to the need for a system of *signed* binary numbers of *finite* size.

Signed Binary Numbers

Arithmetic defines both positive and negative numbers, as well as the operations of addition, subtraction, multiplication, and division. Addition, multiplication, and division operations with positive numbers involve no awareness of signs; all such positive number operations have positive results. Subtraction of a large positive number from a smaller positive number, however, introduces the need for numbers that are less than zero, the so-called *negative* numbers. The introduction of signed numbers requires that the result of any arithmetic operation be adjusted to reflect the sign.

Not only must the signs of results be adjusted, but some symbol must be invented to indicate the sign of numbers. Pencil-and-paper arithmetic carries the sign of a number as a preceding positive (+) or negative (−) symbol. Usually, if a number is positive, *no* sign is attached. It is often assumed that a number is positive unless preceded by a negative sign.

Positive and *negative* sign symbols should *not* be confused with the math operations of *addition* and *subtraction.* It is unfortunate that we tend to use the same symbols for addition (+) and subtraction (−) as we do for positive (+) and negative (−). Fortunately, computer programs can spell out arithmetic opera-

TABLE 2.11

Decimal		Unsigned Binary
000	=	00000000b
080	=	01010000b
123	=	01111011b
128	=	10000000b
255	=	11111111b

tions to be done, such as ADD or SUB, and so signs are never confused with math *operations.* To prevent confusion in this book, the words *add* and *subtract* (or *sub*) will be used to differentiate between a mathematical *operation* and a magnitude *sign.*

Note that binary numbers do not *have* to be signed. The computer *programmer* decides, when the program is *written,* if the binary numbers involved should be signed. The programmer may decide, for instance, that the numbers used in a program are to be unsigned 8-bit integers. Each program number is then an 8-bit positive binary number that ranges from 00h to FFh. Some examples of 8-bit binary, positive, unsigned integer numbers, and their decimal equivalents, are shown in Table 2.11.

All the bits of an unsigned binary number are used as part of the *magnitude* for the number, and the number is *assumed* to be positive.

Numbers in a program may be unsigned positive integers (such as those just listed), signed integers, signed or unsigned fractions, packed BCD, unpacked BCD, or ASCII. The choice of what type of numbers to use, and how to code the numbers, is *entirely up to the programmer.* The programmer can also decide how *large* the numbers are to be in the program. Binary numbers extend away from both sides of the binary point. Large numbers, and exact fractions, require more bits to represent them than do small numbers and rounded fractions. The *programmer, and the problem at hand,* determine to how many significant binary places program numbers will be extended.

Sign-Magnitude Binary Numbers

There are two ways to express signed numbers: *sign-magnitude* signed numbers and *two's complement* signed numbers. Both sign-magnitude and two's complement numbers use the most significant bit of the number as a *sign* bit. The sign bit is a 0 for a positive number and a 1 for a negative number.

Sign-magnitude numbers are the type of numbers taught in grammar school. To make a sign-magnitude number, simply place the *sign* of the number in front of the *magnitude* of the number. For example, sign-magnitude numbers -123 and $+123$ both use the *magnitude* 123 and affix the proper sign. Some exam-

TABLE 2.12

Sign-Magnitude Decimal		Sign-Magnitude Binary
+000	=	00000000b
+096	=	01100000b
+127	=	01111111b
−000	=	10000000b
−001	=	10000001b
−087	=	11010111b
−127	=	11111111b

ples of 8-bit binary sign-magnitude numbers, and their decimal equivalents, are listed in Table 2.12.

To make an 8-bit sign-magnitude number, 7 bits are used for the magnitude of the number, and the MSB is the *sign.* The magnitude of any such 8-bit number is thus limited to 127. Note that sign-magnitude binary numbers may define both a positive *and* negative 0. Sign-magnitude numbers are the type of numbers we deal with in everyday decimal number systems. Decimal systems use subtraction of sign-magnitude numbers, and adjust the sign of the answer to that of the largest signed number.

Sign-magnitude binary numbers, although perfectly feasible for use in computer programs, are *not* employed to any great extent. One difficulty with using sign-magnitude numbers in a program is excluding the sign bit from arithmetic operations. To use an 8-bit sign-magnitude number, for instance, the sign bit must be removed, bit 7 made a binary 0, the operation performed, the sign of the result calculated, and then bit 7 of the result re-inserted.

Most computer programs that use negative numbers use *complementary negative* numbers. Complementary numbers automatically adjust the sign of the result for addition and subtraction operations. Complementary numbers can be used for arithmetic in all number systems, including decimal, but are not usually taught in grammar school.

Complementary Numbers

Complementary numbers enable subtraction to be done using *addition.* Complementary numbers can do subtraction by addition, because subtraction is done when a complementary negative number is *formed.*

Complementary number theory became popular because early computers could only perform one mathematical operation, that of addition. Complementary numbers and complementary arithmetic became highly developed to serve the arithmetic needs of the first computers. Complementary math, although it appears awkward at first inspection, is actually much simpler than conventional "grammar school" math.

Complementary numbers are also *finite* numbers, that is, the *programmer must decide how large the largest number in the program needs to be* and then use complementary numbers based on the size of the largest number. If it turns out that the programmer erred, and larger numbers are needed, then *all* of the numbers in the program must be re-sized.

Ten's Complement Numbers

We begin our study of complementary numbers using the familiar decimal radix 10 number system.

First, a decision as to the *largest possible* number needed for our discussion must be made. *Assume* that the largest number needed for our purposes is positive 49, and the smallest number needed is negative 50. Each number in the system will use the most significant decimal digit of the number to indicate the sign of the number. All numbers that are 49 or *less,* down to 00, are *positive.* All numbers that are 50 or *greater,* up to 99, are *negative.*

A partial listing of some complementary decimal numbers in our system is shown in Table 2.13.

Positive numbers, in our system, are formed by simply writing the numbers from 00 to 49. The smallest positive number is 00, and the largest is 49, as expected. Positive numbers are said to be in *true* form.

Negative numbers, in our system, are formed by *subtracting* the *magnitude* of the negative number from 100. The number formed by subtracting the magnitude of the negative number from 100 is called the *ten's complement* of the negative number. Unexpectedly, our largest (in magnitude) negative number is 50, and the smallest is 99. Ten's complement negative numbers are said to be in *complementary* form.

The ten's complement form of the example negative numbers is formed as follows:

TABLE 2.13

Positive Numbers	Negative Numbers
+00 = 00	−01 = 99
+01 = 01	−05 = 95
+10 = 10	−10 = 90
+20 = 20	−20 = 80
+40 = 40	−30 = 70
+49 = 49	−50 = 50

−01 = 100 subtract 01 = 99
−05 = 100 subtract 05 = 95
−10 = 100 subtract 10 = 90
−20 = 100 subtract 20 = 80
−30 = 100 subtract 30 = 70
−50 = 100 subtract 50 = 50

Negative and positive numbers carry *no* sign attached. The most significant digit of each number determines its sign.

Negative numbers are formed by *pre-subtracting the magnitude* of the negative number from the system radix raised to a power equal to the number of digits in the negative number. For instance, there are 2 digits in negative 01 to negative 50 in our ten's complement example, so all negative magnitudes are subtracted from 10 raised to the 2nd power, or 100.

To see the advantages that complementary numbers enjoy over "standard" sign-magnitude numbers, consider the following *conventional* subtraction of number *B* from number *A:*

1. The sign of *B* is changed.
2. The signs of both *A* and *B* are compared.
3. If the signs are the *same,* the numbers are added, and the sign of the result is the same as the signs of *A* and *B.*
4. If the signs of A and B are *different,* then the smaller *magnitude* is subtracted from the larger, and the sign of the difference is the sign of the larger magnitude number.

Conventional addition follows the same process, except the first step, changing the sign of *B,* is omitted.

Complementary subtraction proceeds in a more direct manner, because no consideration is given to the relative sizes of either *A* or *B.* Complementary subtraction proceeds as follows:

1. *B* is complemented (or is already in complementary form).
2. B is added to A.

Complementary addition involves simply adding *B* to *A.* Complementary math is simpler than conventional math because complementary operations of addition and subtraction automatically *adjust* the sign of the result.

The *downside* to complementary numbers is a condition called *overflow* (discussed later). Complementary number systems always involve the concept of a *finite* number limit. Pencil-and-paper math, however, always assumes that numbers can be made arbitrarily large, or arbitrarily small, simply by writing more digits.

Subtraction by Addition of Ten's Complement Numbers

Several examples of adding ten's complement numbers together using complement numbers that vary from 00 to 99 demonstrate how subtraction can be done by addition.

EXAMPLE 1 Subtract Positive 12 from Positive 32

Subtracting +12 from +32 is the same as adding 32 to the ten's complement of 12.

```
   Standard Math          Complementary Math
        +32                       32
subtract +12  =  (100 − 12)  =  add 88  (complementary form)
        ----                        --
         20                       1)20
                                  ↑
                               discard
```

The result of the complementary subtraction has a carry of 1, and result of +20.

Adding positive 32 to the ten's complement of 12 is the same as subtracting 12 from 32, as shown here:

```
32 + (100 − 12) = 100 + 32 − 12 = 100 + (32 − 12) = 1)20
                                                    ↑
                                                 discard
```

The result is 32 subtract 12, the desired operation. Because the largest number in our system is 49, the carry of 1 into the hundreds place in the answer is, conveniently, thrown away. Subtraction by addition "works" for complementary numbers because the negative form of a number has been "pre-subtracted" when the negative number is formed. ◆

EXAMPLE 2 Subtract Positive 32 from Positive 12

The reverse of Example 1 will yield a negative number, in ten's complement form. The subtraction proceeds, for standard and ten's complement numbers, as shown next.

```
   Standard Math          Complementary Math
        +12                       12
subtract +32  =  (100 − 32)  =  add 68 (complementary form)
        ----                        --
        −20                       0)80 = (100 − 20) = negative 20
```

There is no carry from the operation, therefore the result is the ten's complement negative number for 20 (100 − 20 = 80). Note that the size of the most significant digit, 8, correctly shows that the result is *negative.* ◆

EXAMPLE 3 Add Positive Numbers and Overflow

Complementary number mathematics will work correctly as long as the sum of any addition does not *exceed* the largest finite number allowed when the system is set up.

Examples 1 and 2, above, will work correctly as long as no result is smaller than −50 (50) or larger than +49 (49). Should any operation result in a number out of the allowed *bounds,* an *overflow* is said to have occurred. An overflow condition is a serious *error* on the part of the programmer: A number has been generated that is larger than planned for by the *programmer.*

Usually, when an overflow occurs, the programmer must rewrite the program or, at least, resize all of the numbers in the program. Overflows are the result of using signed numbers in a program. Unsigned numbers cannot overflow, because there is no finite limit to their size, except the total memory capacity of the computer.

To see an overflow occur in our example ten's complement system, we shall add +30 to +21, and get +51, which exceeds the largest positive number, 49, *allowed* in the system. The overflow occurs as shown next:

Complementary Math

```
      +30
add   +21
      ---
     0)51
```

The resulting number is 51, with no carry. A 51 is supposed to be a negative 49 in our ten's complement system, therefore the answer is in *error.* We have added two legal-size positive numbers and gotten an illegal negative result, because our largest legal positive number is 49.

Overflow can also occur if two *negative* numbers are added together and a positive result is reached. For example, adding a negative 30 to a negative 21 will result in a positive number, and an overflow condition, as shown next:

Complementary Math

```
    −30 = (100 − 30) =  70 (complementary form)
add −21 = (100 − 21) =  79 (complementary form)
                       ---
                      1)49
```

The result has a carry and a sum of 49, or positive 49. An overflow error has occurred because the sum of two legal-size negative numbers has resulted in an illegal positive number.

Overflows can occur only when two positive or two negative numbers (in ten's complement form) are added, yielding numbers larger than those allowed by the system. Additions involving one negative ten's complement number and one positive number will always result in a correct answer, because the result must be smaller than either of the parts.

To solve the overflow problem of the preceding examples, a *larger* signed number system that uses three digits, based on numbers from 000 to 999, could be used. The new system would then have positive numbers from 000 to 499, and negative numbers from 999 (−001) to 500 (−500). Negative numbers in this larger system are formed by subtracting positive magnitudes from 10 raised to the 3rd power, or 1,000.

Our preceding overflow examples would then be legal, as shown next:

Complementary Math Addition

```
     +030
add  +021
     ----
     0)051
```

The answer, 051, is within the allowable range of positive numbers that range from 000 to 499.

Complementary Math Subtraction

```
     −30 = (1000 − 30) = 970
add  −21 = (1000 − 21) = 979
     ---                 ---
                      1)949 = −51
```

The answer, 949, is −51 in ten's complement form for a system with three digits. No overflow has occurred, because the result fits into a legal-size answer. ◆

Two's Complement Numbers

Negative *two's complement* numbers are formed by subtracting the binary *magnitude* of the negative numbers from 2 raised to a power equal to the number of bits used in the system. For example, if the *programmer* decides that the largest binary number in a program is to be positive 01111111, or negative 10000000, then an 8-bit two's complement number system has been decided on. The two's complement of any number in an 8-bit system is found by subtracting the number from 2 raised to the 8th power, or 100000000, a 9-bit number.

The two's complement of a binary number may also be found by inverting every bit in the number (also called the *one's complement*), and adding 1 to the least significant bit of the inverted number. For example, the two's complement of an 8-bit number equal to 80 decimal is formed using both methods, as shown next:

80 decimal = 50h = 01010000b

Invert and add

```
              01010000b
complement    ||||||||
              10101111b
add                  1b
              _________
              10110000b = B0h = −80 decimal
```

Subtract from 2^8

```
              1 00000000
subtract        01010000
              __________
              0 10110000 = B0h = −80 decimal
```

Several numbers in an 8-bit two's complement number system are shown in Table 2.14, in true and two's complement form.

Using 8 bits to represent all numbers in a program results in numbers that vary from as small as −128 (10000000b) to as large as +127 (01111111b). Note that, for an 8-bit number system, there are 128 positive numbers (0 to +127) and 128 negative numbers (−1 to −128). All 256 possible binary numbers, from 00h to FFh, are used to represent positive and negative numbers.

All negative numbers, in two's complement form, *begin* with a binary *1* in the *most significant bit* position of the number. All positive numbers *begin* with a binary *0* in the *most significant bit* position. The *most significant bit* of a *signed* binary number is named the *sign bit*. However, if the number is *not designed* to be a signed number (by the programmer), then the most significant bit of the number is a magnitude bit, not a sign bit.

The programmer chooses to use signed numbers, or not. If signed numbers are chosen then the programmer must also choose a range of signed binary

TABLE 2.14

	Decimal Value	Binary Value	Hex Value
Range of positive numbers	+127	0111 1111	7F
	+080	0101 0000	50
	+050	0011 0010	32
	+001	0000 0001	01
	+000	0000 0000	00
Range of negative numbers	−001	1111 1111	FF
	−050	1100 1101	CE
	−080	1010 0000	B0
	−127	1000 0001	81
	−128	1000 0000	80

numbers as large as may be needed by the program. The programmer chooses as follows:

- A *group* of bytes is chosen by the programmer to represent the largest number to be needed by the program.
- Bits 0 to 6 of the most significant *byte* (MSBY), and all 8 bits of any other lower order bytes, express the *magnitude* of the number.
- Signed numbers use a 1 in bit position 7 of the *MSBY* as a negative sign and a 0 as a positive sign.
- All negative numbers are in two's complement form.

When doing signed arithmetic, the programmer must *know in advance* how large the largest number is to be; that is, how many bytes are needed for the largest *possible* number the program can generate.

Two's Complement Mathematics

Addition

Signed numbers may be added two ways: the addition of like (same sign) signed numbers, and the addition of unlike signed numbers. An example of adding two 8-bit signed positive numbers that does not produce an overflow condition is as follows:

```
     +045 = 00101101b
add  +075 = 01001011b
     +120   01111000b = 120
```

If two 8-bit positive numbers are added, there is always the *possibility* that the sum will exceed +127 and an overflow will occur. An overflow condition is demonstrated in the example below when 50 is added to 100:

```
     +100 = 01100100b
add  +050 = 00110010b
     +150   10010110b = −106 (two's complement form)
```

There is an overflow because the sum of the positive numbers (150) exceeds the largest positive number (127) allowed in an 8-bit system.

If *unlike* sign 8-bit numbers are added, then it is not possible that the result can be larger than −128 or +127, and the sign of the result will be correct. For example:

```
     −001 = 11111111b                  −128 = 10000000b
add  +027 = 00011011b             add  +127 = 01111111b
     +026   00011010b = +026           −001   11111111b = −001
```

The result of adding two negative numbers together for a sum that does *not* exceed the negative limit is shown in this example:

```
    −030 = (100h − 1Eh) = E2h =   11100010b
add −050 = (100h − 32h) = CEh =   11001110b
    −080 = (100h − 50h) = B0h = 1)10110000b = −080
```

The carry is discarded, and the result is correct.

On adding two negative numbers whose sum *does* exceed −128, we have the following overflow condition:

```
    −070 = (100h − 46h) = BAh =   10111010b
add −070 = (100h − 46h) = BAh =   10111010b
    −140 Exceeds allowed range  1)01110100b = Carry and +116
```

An overflow error has occurred by adding two negative numbers and receiving a positive result.

By changing our numbers from 8-bit to 9-bit, or decimal values from 000 to 511, the overflow conditions generated by our previous examples may be avoided. In a 9-bit system, positive numbers are those that start at 000000000b (000 decimal) and extend up to 011111111b (255 decimal). Negative numbers, in a 9-bit system, extend from 111111111b (−001 decimal) to 100000000b (−256 decimal).

Most programmers do not size numbers to the nearest bit needed, however, but to the nearest byte.

Multiple-Byte Addition

Number size is increased by using more bytes to express the number. Multiple-byte addition involves adding carries from low-order sums to higher-order sums.

Using multi-byte numbers in *unsigned* addition means that carries between bytes are propagated from the highest-order bit of a low-order byte to the lowest-order bit of the next-higher-order byte. Carries are propagated from byte to byte by the simple technique of adding the carry bit to the next higher byte.

For example, a pair of 2-byte unsigned numbers is added as shown next:

```
       High-order byte         Low-order byte
Carry  1)1)0)1)1)1)1)1)1) ←    1)1)1)1)0)0)0)
         1 1 0 1 0 1 1 1       0 1 1 1 1 0 1 0 b
add      0 1 0 0 1 0 0 1       1 0 0 1 1 1 0 1 b
       1)0 0 1 0 0 0 0 1       0 0 0 1 0 1 1 1 b
```

As can be seen in the example, the carry out from the most significant bit of the low-order byte sum is added to the lowest-order bit of the next higher byte.

Should there have been a carry out of the high byte of the example, it could be added to a higher-order byte, and so on. The process of adding the carry to the next-higher-order byte is continued until the last (highest-order) byte in the number is reached.

Signed numbers appear to behave as unsigned numbers until the highest-order byte is reached. Signed numbers use bit 7 of the *highest*-order byte in the number as the sign. If the sign bit is negative, then the *entire* number, from the lowest-order byte upward, is in two's complement form. For example, the smallest and largest positive and negative numbers of a 2-byte (1-word) signed number are expressed as follows:

$$
\begin{aligned}
+32767 &= 01111111\ 11111111b = 7FFFh \\
+00000 &= 00000000\ 00000000b = 0000h \\
-00001 = 2^{16} - 00000000\ 00000001b &= 11111111\ 11111111b = FFFFh \\
-32768 = 2^{16} - 10000000\ 00000000b &= 10000000\ 00000000b = 8000h
\end{aligned}
$$

Note that the *low*-order bytes of the numbers 00000 and −32768 are *exactly alike,* as are the low-order bytes for +32767 and −00001. The difference between positive and negative numbers appears at the sign bit of the high-order byte for each number.

For multi-byte signed number arithmetic then, the lower bytes are treated as unsigned numbers. All checks for overflow are done only for the *highest*-order byte, which contains the sign bit.

Subtraction

Binary subtraction *can* be done, as noted in the section on two's complement addition, by taking the two's complement of the number (the *subtrahend*) to be subtracted. The subtrahend is *added* to the number (the *minuend*) from which the subtrahend is to be subtracted. Subtraction using addition is an obsolete concept, however. Most modern computers can carry out subtraction operations just as they carry out addition operations.

As was the case for addition, the programmer may choose to use signed or unsigned numbers, according to the needs of the program. An overflow condition, if signed numbers are in use, indicates that an error has occurred when two numbers of *unlike* signs are *subtracted.* Overflows cannot occur when unsigned numbers are subtracted, because all unsigned numbers are, essentially, positive.

Unsigned Subtraction

Unsigned subtraction begins with a positive minuend and subtrahend, but the possibility of a negative result is always present.

The result of an unsigned subtraction operation will be in *true* (positive) form, with no borrow, if the subtrahend is *smaller* than the minuend. The result of unsigned subtraction will be in two's complement form, with a borrow, if the subtrahend is *larger* than the minuend. A negative result in two's complement form is *not* a signed number, as all bits of the number are used for magnitude, and there is *no* sign bit.

For example, using 8-bit unsigned numbers, 100 is subtracted from 15 as shown next:

```
         015 =    00001111b
subtract 100 =    01100100b
        -085 = 1}10101011b = Borrow, ABh = Borrow, 171
```

The borrow signifies that the 8-bit magnitude result of the subtraction is in two's complement form, with a sign bit of 1. (The 2^8 two's complement of 85 is 171.)

The same result, with no borrow, could have been reached by subtracting 1 from 172, as demonstrated in the next example:

```
         172 =    10101100b
subtract 001 =    00000001b
         171 = 0}10101011b = No Borrow, ABh = 171
```

No borrow signifies the result is in true, positive form. Note the first bit of the unsigned result is a 1, yet the number is *unsigned positive.*

Signed Subtraction

As was the case for signed addition, there are two possible combinations of signed numbers when subtracting. One may subtract numbers of like or unlike signs.

When numbers of *like* sign are subtracted, it is impossible for the result to exceed the positive or negative magnitude limits of the signed number system in use. The result must be smaller than either the minuend or subtrahend. For example, subtraction involving two 8-bit signed positive numbers is shown next:

```
         +100 =    01100100b
subtract +126 =    01111110b
         -026 = 1}11100110b = E6h = 2^8 - (+026)
                |
             discard
```

The result, −26 in two's complement form, is smaller than the minuend, +100, or the subtrahend, +126. Note the sign bit of the result is a 1, signifying the result is a signed negative number in the system.

The next example, using two 8-bit signed negative numbers, shows a correct result when numbers of like signs are subtracted:

$$
\begin{array}{rrcr}
 & -061 = 2^8 - 00111101\text{b} & = & 11000011\text{b} \\
\text{subtract} & \underline{-116} = 2^8 - 01110100\text{b} & = & \underline{10001100\text{b}} \\
 & +55 & & 0\}00110111\text{b} = 37\text{h} = +55
\end{array}
$$

The answer, after subtracting −116 from −61, is +55, which is numerically smaller than the minuend or subtrahend. Note that the sign bit of the result is a 0, signifying that the result is a signed positive number in the system.

An overflow becomes possible when subtracting numbers of opposite sign because the situation becomes one of adding numbers of like signs. This can be demonstrated in the following example that uses 8-bit signed numbers:

$$
\begin{array}{rrl}
 & -099 = & 10011101\text{b} \\
\text{subtract} & \underline{+100} = & \underline{01100100\text{b}} \\
 & -199 & 1\}00111001\text{b} = 39\text{h} = +057 \\
 & & | \\
 & & \text{discard}
\end{array}
$$

The result, −199, exceeds the size of the maximum negative number (−128) allowed in an 8-bit signed number system. An overflow has occurred, because two legal-size numbers, that should have generated a negative result, have generated an illegal positive result.

An example of an overflow resulting from subtracting a negative 8-bit signed number from a positive 8-bit number is shown next:

$$
\begin{array}{rrl}
 & +087 = & 01010111\text{b} \\
\text{subtract} & \underline{-052} = & \underline{11001100\text{b}} \\
 & +139 & 1\}10001011\text{b} = 8\text{Bh} = -117 \\
 & & | \\
 & & \text{discard}
\end{array}
$$

The result, +139, exceeds the maximum positive number (+127) allowed in an 8-bit signed number system. An overflow has occurred because two legal-size numbers, that should have generated a positive result, have combined into an illegal negative result.

Again it must be emphasized: When an overflow occurs in a program that uses signed numbers, an *error* has been made in the *estimation* of the largest number needed to successfully operate the program. Theoretically, the program could stop and re-size every signed number. Re-sizing every number in a program, as the program operates, is rarely feasible.

Recovering from an Overflow

The remedy for an overflow error is to make the range of signed numbers larger. Increasing the range of signed numbers means increasing the number of bits

TABLE 2.15

Number	8-Bit Form	16-Bit Form
+52	00110100	00000000 00110100
+120	01111000	00000000 01111000
+870	Too big	00000011 01100110
−52	11001100	11111111 11001100
−120	10001000	11111111 10001000
−870	Too big	11111100 10011010

used in the number system. For instance, examples of the same signed numbers in an 8-bit and a 16-bit system are listed in Table 2.15.

Note that numbers, such as 52 or 120, may be expanded from 8-bit size to 16-bit size by *extending* their sign bits. Sign extension is done from a low-order byte by copying the sign bit of the low-order byte into every bit of succeeding higher-order bytes. Thus, as shown in the last example, −52 in 8-bit form (11001100) becomes −52 in 16-bit form (11111111 11001100) by copying the negative sign bit (1) into the high-order byte.

Expanding the size of signed numbers is easily done before the computer program begins to run. Once in operation, however, it is very difficult for the program to correct the effects of an overflow. If an overflow occurs, the programmer has made a *mistake* and has made no provisions for a number as large as the one that generated the overflow. Some error acknowledgment procedure, or user notification, should be included in the program if an overflow is a possibility. Most pocket calculators, for instance, will show some kind of error message in the display should the user try to generate a number larger than the calculator can hold.

2.7 Binary Multiplication and Division

Binary multiplication and division proceed as in other number systems. Multiplication can be carried out as a repeated series of shift-and-add sequences. Division can be carried out as a sequence of trial subtractions.

Overflow conditions are generally not a possibility for multiplication or division operations, unless *division by 0 is attempted.* Overflows occur when two legal-sized signed numbers are added or subtracted, generating an illegal result that is of the *wrong* sign. Multiplication and division operations handle sign and magnitude separately, and the sign of the result can be made to be correct.

Binary Multiplication

The multiplication table for binary numbers involves four entries, as shown next:

Binary Multiplication

$0 \times 0 = 0$
$0 \times 1 = 0$
$1 \times 0 = 0$
$1 \times 1 = 1$

Binary numbers can be multiplied together by a repetitious shift-and-add process. If the number is signed, the sign bits can be separated from the magnitude bits, and the multiplication carried out. The sign of the result may be found by XORing the sign bits.

An example of multiplying two 8-bit unsigned binary numbers is shown as follows:

```
   130 = 10000010b
 × 240 = 11110000b
 -----------------
         00000000
        00000000
       00000000
      00000000
     10000010
    10000010
   10000010
  10000010
0111100111100000 = 79E0h = 31,200
```

Every multiplication operation must make allowances for a result that is up to twice as large as each component part. Note that multiplying two 8-bit numbers together, as in the example, yields a result that can be as large as 16 bits in length. Multiplying the two largest possible 8-bit unsigned numbers (255 by 255) yields an answer of 65,025, or FE01h.

Multiplication (and division), in primitive computers, is done using *software* programs to perform the shifting and adding of each multiplication step. Modern processors, such as in the 8051, do multiplication and division in *hardware.*

Binary Division

Binary division is a repeated process of *trial* subtractions. If the divisor will not divide into the partial dividend by 1, then the quotient bit is 0, another bit is *added* to the partial dividend, and the trial subtraction re-tried.

An example of dividing an unsigned 16-bit number by an unsigned 8-bit number is shown next:

```
                          0000000011101010 = EAh = 234 (Quotient)
32780                    _________________
----- = 10001100 | 1000000000001100
 140                 10001100|||||||
                    ---------|||||||
                    011101000||||||
                     10001100||||||
                     ---------||||||
                     010111000|||||
                      10001100|||||
                      ---------|||||
                      010110001|||
                       10001100|||
                       ---------|||
                       0010010110|
                         10001100|
                         ---------|
                          00010100 = 14h = 20 (Remainder)
```

Every division operation results in a quotient and a remainder. Note that the quotient can be as large as the dividend (when the divisor is 1) and that the remainder can contain as may bits as the divisor. Division by 0 is undefined, and many computers contain circuits that detect and act on division by 0.

2.8 Binary Codes

A *code* is any arrangement by which one set of items, say numbers, is uniquely assigned to members of another set of items. Human beings are assigned code numbers in many ways. We all have our social security codes, telephone codes, home address codes, and driver's license codes, for instance. The word *code* usually brings to mind spies, secrets, and intrigue, but there are many more common uses for codes, particularly in the binary world.

Character Codes

Each letter, number, space, and punctuation mark, as well as each drawing seen in this book, has been *coded* as a set of binary numbers and stored on a magnetic disk. Things such as letters, numbers, and punctuation marks are all human-invented symbols called *characters.* Characters let us communicate with each other using standard symbols.

Clearly, a binary digital computer cannot internally store a character in its written form, say the letter *a*. The computer must store some binary number made up of 1s and 0s that is the *code* for the letter *a*. In order to store all possible characters, a unique code number for each letter, number, punctuation mark, and any other symbol humans might use must be established. Keyboards, for instance, are often the means by which humans communicate with computers using keys that have been imprinted with common human characters. Each key placed on a keyboard has a unique binary code assigned to it; therefore the keys are said to be *encoded*.

In order to avoid confusion in the computer industry as to how each character is to be encoded, standards have been agreed on that define which binary number is assigned to which character. The character code used in the United States, and most other countries, is the *ASCII* character code. Appendix F lists the standard binary ASCII code for characters.

The ASCII Character Code

The acronym *ASCII* stands for the American Standard Code for Information Interchange. Every industry attempts to set standards for its products. There are standards for tires, paper, thread, print fonts, milk, shirts, fishhooks, pedigreed dogs and cats, beer, doorknobs, gasoline, and just about anything that is produced and consumed in great quantities. About the only thing that has not been standardized is human beings.

Various industries band together in voluntary associations to set standards for their products. A United States national standards body exists, whose name is the American National Standard Institute or ANSI. ANSI, in collaboration with the electronic communications and computer industry, has established the ASCII standard. Most standards are *voluntary*, but only the largest and financially strongest manufacturer can choose to ignore industry standards.

Standards basically exist in order to expand markets. Consumers will not tolerate mass consumption products that are "one-of-a-kind." Most of us, for instance, would not buy a car that had to use a certain gasoline that was made only by the car manufacturer.

The original ASCII character standard assigned a 7-bit binary number (00h–7Fh) to code 127 characters. The original 7-bit 127-character ASCII code has been expanded to 8 bits (80h–FFh) to code international and graphics characters. Most of the ASCII characters are the familiar symbols seen on keyboards, such as letters, numbers, and punctuation marks. Some of the ASCII characters are called *control* characters, or *nonprintable* characters. Control character symbols are not seen on key caps, computer monitor screens, or the printed page, thus they are *nonprintable*.

Two of the most common ASCII control characters are the *carriage return* (CR) and *line feed* (LF) characters. Each line of visible text characters stored in the disk file for this book ends in invisible control characters CR and LF. Every time the Enter key is struck on the computer keyboard used to write this book, a CR, LF character sequence is generated to the word processor program. The word processor program then knows that it is time to start a new line on the computer screen. When the word processor program stores visible book text characters to the disk file, many control characters, such as tabs and the CR, LF control characters are stored also. Word processors may also *add* other ASCII control characters (or "happy faces" as they sometimes appear on the screen) in order to add features such as underlines, superscripts, boldface, and the like. A *pure* ASCII file will contain nothing but printable ASCII characters and CR, LF control characters. Most word processor files are *not* pure ASCII but may be saved on a disk as pure ASCII as a user option.

The ASCII code was originally designed to handle a typewriter-style keyboard. As computer keyboards added new keys, such as function keys and cursor control keys, new key codes have been defined. The new codes are often called *extended* codes. Extended codes use an ASCII 00h byte *followed* by a second byte to represent the new computer keyboard keys. The extended codes have not yet been standardized by ANSI.

Table 2.16 lists the ASCII codes for some common keyboard characters. ASCII is a 7-bit code for characters from 0000000b (NUL) to 1111111b (DEL).

Note that the decimal numerals 0 – 9 differ from their true form only by the number *30h*. The decimal numeral *8*, for instance, is ASCII *38h*, and decimal numeral *3* is coded as ASCII *33h*. ASCII decimal numeral codes can easily be

TABLE 2.16

ASCII Code (hex)	Keyboard Character
0A	LF
0D	CR
20	(SPACE)
30	0
31	1
32	2
33	3
38	8
39	9
41	A
42	B
61	a
62	b
63	c
77	w

converted to the binary code for the decimal numeral by subtracting 30h from the ASCII code.

Numeric Codes

Binary Coded Decimal (BCD) Code

Decimal numerals play such an important part in computer programs that several *codes for decimal numerals* have evolved. The ASCII code for decimal numerals, shown in part in Table 2.16, adds a 30h to the binary equivalent number for each decimal numeral. Another popular code for decimal numerals is named the *Binary Coded Decimal* or BCD code. Table 2.17 lists the BCD code for the 10 decimal numerals, 0 through 9.

In essence, the BCD code is the radix 2 binary numbers for decimal numerals 0 to 9. In the BCD code shown in Table 2.17, 4 binary bits (1 nibble) are used to express each decimal numeral. The BCD code stops at decimal 9, or 1001b. Decimal 17, for instance, requires two BCD numbers of 0001b (1) and 0111b (7). The BCD code "wastes" the remaining 4-bit binary numbers greater than 1001, so that *none* of the six nondecimal hex numerals A–F are used in a BCD scheme.

A BCD number is a binary number in which each group of 4 bits codes a decimal numeral from 0 to 9. Two BCD numerals require two 4-bit codes (a byte), three BCD numerals require three 4-bit codes, and so on. The inefficiency of the BCD code can be seen by noting that the largest BCD byte-sized number is 99 decimal, whereas the largest byte-sized hex number is FFh, or 255 decimal.

BCD numbers, although not as bit-efficient as pure binary or hex numbers, have a place in computer programs that must deal with decimal-oriented humans.

TABLE 2.17

BCD Code (binary)	Decimal Numeral
0000	0
0001	1
0010	2
0011	3
0100	4
0101	5
0110	6
0111	7
1000	8
1001	9

Packed and Unpacked BCD Numbers

ASCII numbers and BCD numbers are closely related. The BCD code for decimal 8, for instance, is 1000b; the ASCII code for decimal 8 is 0011 1000b (38h). Adding 30h (0011 0000b) to any *unpacked* BCD number yields the equivalent ASCII character for the decimal number. Unpacked BCD numbers are those numbers that occupy an *entire* byte. The 4 *least* significant bits of the byte hold a binary number ranging from 0000b to 1001b, and the 4 *most* significant bits are set to 0000b. For example, an unpacked BCD 5 is 05h or 0000 0101b. When 30h is added to 05h, the ASCII character for decimal 5, 35h, is generated.

Packed BCD numbers are those BCD numbers that use *both nibbles* of the byte to hold a BCD numeral. Packed BCD numbers range from 00h to 99h, with no nibble greater than binary 1001. To convert from packed BCD to unpacked BCD requires that each nibble of the packed number be put into a single byte that begins with a 0000b nibble. For example, the packed BCD number 87h becomes the unpacked bytes 08h and 07h.

Conversions from packed BCD numbers to unpacked BCD and ASCII numbers are often useful for programs that deal primarily with decimal numbers.

Other Binary Codes

Several specialized binary codes have been formulated for use with position sensors known as *encoders*. Encoders serve to convert rotary motion in degrees, for example, into binary codes. One degree of rotation might be encoded as a single binary number, so that 360 binary numbers would encode a complete revolution of a shaft. Many encoders use optical pickups to read a circular pattern of tracks. Each track is made up of black-and-white patches that represent binary 0 and 1. There is one track for each significant bit of the encoder binary number.

A popular encoder code is known as the *Gray* code. Table 2.18 shows a Gray code for the first 16 hex numerals.

A careful inspection of Table 2.18 reveals that, as the count progresses from 0 hex to F hex, only one bit at a time changes between *adjacent* code numbers. For instance, when the count goes from 8 to 9, the code changes from 1100b to 1101b. Only a single bit, the LSB, changes from 1 to 0 as the count changes.

Gray codes permit slight misalignment of encoder readout pickups and make changes between a 1 and a 0 on each encoder disk track at least 2 bits long. For instance, assume that the LSB track pickup is 1 bit ahead of the MSB track pickup, and the encoder goes from 7 to 8. The Gray code correctly changes from 0100 to 1100, as only the MSB changes. If the encoder used a natural binary code, the numbers would incorrectly change from 0110 to 1001.

Other standard industrial binary codes include Universal Product Codes (UPC) bar codes, excess-three codes, and Hamming error-correcting codes.

TABLE 2.18

Digit	Gray Code
0	0000
1	0001
2	0011
3	0010
4	0110
5	0111
6	0101
7	0100
8	1100
9	1101
A	1111
B	1110
C	1010
D	1011
E	1001
F	1000
	0000

Error Correction and Detection Codes

Many coding schemes exist for detecting and correcting errors in binary numbers, particularly numbers that are subject to distortion when they are transmitted over great distances by communication systems. Discussion of the most common error detection and correction code system, Hamming codes, is beyond the scope of this book. Hamming codes, in essence, using *redundant* bits, enable the receiver to detect and correct errors in a binary message.

Parity

One very simple error-detection code is the *parity* code concept. Parity involves adding a single bit to a string of coded bits. The bit that is added, called the *parity bit,* makes the total *number* of 1 bits in the string, *including the parity bit, odd* or *even.* If the parity bit makes the total number of bits in the string, *including* the parity bit, odd, then the code is called an *odd parity* code. If the parity bit added to the string makes the total number of bits in the string, *including* the parity bit, even, then the code is called an *even parity* code. The parity bit is referred to in each case as an *odd* parity bit or an *even* parity bit. Parity is used extensively in binary data transmission schemes and memory systems in computers.

Parity bits are usually added at the end of a string of bits to be parity coded during transmission. The value of the parity bit is calculated based on the preceding bits and added at the end. ASCII parity codes, for instance, make the parity bit the most significant bit of the character byte, bit 7. ASCII transmissions send each ASCII byte low-bit first, so that the last bit transmitted is the parity bit.

Parity bits enable a receiver to determine if one bit in the bit string is in error. For instance, if the ASCII character 9 is transmitted, using even parity, then the ASCII character is encoded by the sender as:

```
ASCII 9, even parity = 0011 1001b = 39h
                       |
                   Parity bit
```

The ASCII code for 9 is 0111001b, or 39h using 7-bit ASCII code. There are four 1 bits in 39h, so there are an even number of bits in the character. A 0 parity bit is added at bit position 7 to keep the number of 1 bits an even number.

Assume that an error occurred in bit position 0 of a character 9 transmitted in ASCII, and the character was received as:

Received character, even parity = 0011 1000b = 38h

The received character is ASCII 8, but there are only three 1 bits in the character. The receiver knows that the parity is supposed to be even, counts the number of 1s, and detects that an error has been made. The receiver can request that the character be retransmitted.

Parity will *not* always detect errors involving more than 1 bit. For instance, had 2 bits of the ASCII 9 been changed during transmission to yield an ASCII 3, the received character would be:

Received character, even parity = 0011 0011b = 33h

The receiver would count four 1 bits, determine that the number of 1 bits are an even number, and be unaware that an error had taken place.

ASCII character codes may be coded even, odd, or *no parity.* No parity ASCII means that bit 7 of the character is always a binary 0. The ASCII code shown in Appendix F is for no parity ASCII.

2.9 Summary

The binary number system is a positional number system that consists of two digits, 0 and 1. The radix of the binary system is 2. Binary numbers are expressed as:

$$b_n \times 2^n + b_{n-1} \times 2^{n-1} \ldots b_1 \times 2^1 + b_0 \times 2^0 . b_{-1} \times 2^{-1} + b_{-2} \times 2^{-2} + \ldots b_{-m} \times 2^{-m}$$

TABLE 2.19

Binary Code	Use
ASCII	Codes decimal, alphabetic, and control characters
BCD	Codes the decimal numerals 0 – 9 in binary
Gray	Used in position sensors
Parity	Detects errors in binary data

Binary numbers may be treated as pure numbers, or they may be used to encode other quantities. Common binary codes are listed in Table 2.19.

Binary math may be done using signed or unsigned numbers. Unsigned numbers are assumed to be positive numbers. Signed numbers consist of positive numbers in true form and negative numbers in two's complement form. The two's complement of a binary number is found by subtracting the number from 2 raised to a power equal to the number of bits in the number.

Signed numbers use the most significant bit of the most significant byte of the number as the sign bit. Positive numbers are signified by a leading 0 bit, and negative numbers by a 1 bit.

Forming two's complement negative numbers by subtraction from 2 raised to a finite power results in signed numbers that are limited to programmer-defined magnitudes. Any operation involving signed numbers that exceeds the programmer-defined maximum number size results in an error condition called an overflow. Overflow errors are usually serious programming blunders.

Most contemporary microprocessors perform the arithmetic operations of addition, subtraction, multiplication, and division using signed or unsigned binary numbers.

2.10 Problems

1. Express the following decimal numbers as separate numerals multiplied by 10 raised to the positional power of the numeral.
   a. 123 **b.** 124,678 **c.** 3,204 **d.** 4,096 **e.** 1,896,573 **f.** 1,024
2. Convert each number of Problem 1 to a number based on a radix of 6. Check your answer by converting the radix 6 number back to decimal.
3. Convert each number of Problem 1 to a number based on a radix of 8 (the octal number system). Check your answer by converting the radix 8 number back to decimal.
4. Convert the following decimal numbers to binary. Check your answers by converting the binary answer back to decimal.
   a. 127 **b.** 89 **c.** 356 **d.** 16 **e.** 3,289 **f.** 487,941 **g.** 255

5. Convert the following binary numbers to decimal. Check your answer by converting the decimal answer back to binary.
   a. 101101011 **b.** 111111011 **c.** 10101 **d.** 10101100 **e.** 111001
6. Convert the decimal numbers of Problem 4 to hexadecimal. Check your answer by converting the hex answer back to decimal.
7. Convert the binary numbers of Problem 5 to hexadecimal, and then to decimal. Check your answers by converting the decimal answers back to hexadecimal, and then to binary.
8. Convert the following binary fractions to decimal. Check your answers by converting the decimal fractions back to binary.
   a. .10110 **b.** .11111 **c.** .00101 **d.** .01011 **e.** .10001 **f.** .1101
9. Convert the following decimal fractions to binary. Limit the binary fraction to 5 binary places. Check your answers by converting the binary fractions back to decimal.
   a. .123 **b.** .4 **c.** .999 **d.** .125 **e.** .034 **f.** .078
10. Convert the binary fractions of Problem 8 to hex, and the hex fractions to decimal. Check your answers by converting the decimal fractions back to hex, and then to binary.
11. Convert the decimal fractions of Problem 9 to hexadecimal fractions. Limit the hex fractions to 3 places. Check your answers by converting the hex fractions back to decimal.
12. Write the first 10 and last 10 positive and negative numbers of an 8-bit signed number system.
13. Add together the following sets of unsigned binary numbers.

   a. 10110110110
10000011101 **b.** 1100000101
0011111011 **c.** 101010100011
100001111001

   d. 111111111111
000000000001 **e.** 0111111111
1000000000 **f.** 10100011110
10000000011

14. Add together the numbers of Problem 13 as sets of signed numbers. Indicate which sums result in an overflow error.
15. Subtract the numbers of Problem 13 as unsigned numbers. Subtract the top number from the bottom number, and the reverse.
16. Subtract the numbers of Problem 13 as signed numbers. Subtract the top number from the bottom number, and the reverse. Indicate the results that generate an overflow error.
17. Sign extend the negative number 10111000 to 2 bytes in length.
18. Sign extend the positive number 011111011 to 3 bytes in length.
19. Multiply the numbers of Problem 13 together.
20. Divide each number of Problem 13 by 1011 binary.

21. Convert every letter of this sentence to no parity ASCII.
22. Decode this message in ASCII: 20h 48h 65h 6Ch 6Ch 6Fh 20h.
23. Convert 99d to ASCII, then to unpacked BCD and packed BCD.
24. Convert packed BCD 54H to ASCII.
25. Take all of the no parity ASCII codes for lowercase letters *a* through *d* and convert them to even parity ASCII codes.

3

The 8051 Architecture

Chapter Outline

CHAPTER OBJECTIVES

On successful completion of this chapter you will be able to:

- Describe the hardware features of the 8051 microcontroller.
- List the internal registers of the 8051 microcontroller and their functions.
- Draw the machine cycle for the 8051 microcontroller.
- State the physical differences between the Port 0, 1, 2, and 3 I/O pins.
- Describe the various operating modes of the timer/counters and associated control registers.
- Describe the various operating modes of the UART, and associated control registers.
- List the types of interrupts, the interrupt program addresses, and the interrupt control registers.

3.0 Introduction

The first task faced when learning to use a new computer is to become familiar with the capability of the machine. The features of the computer are best learned by studying the internal hardware design, also called the architecture of the device, to determine the type, number, and size of the registers and other circuitry.

The hardware is manipulated by an accompanying set of program instructions, or software, which is usually studied next. Once familiar with the hardware and software, the system designer can then apply the microcontroller to the problems at hand.

A natural question during this process is "What do I do with all this stuff?" Similar to attempting to write a poem in a foreign language before you have a vocabulary and rules of grammar, writing meaningful programs is not possible until you have become acquainted with both the hardware and the software of a computer.

This chapter provides a broad overview of the architecture of the 8051. In subsequent chapters, we will cover in greater detail the interaction between the hardware and the software.

3.1 8051 Microcontroller Hardware

The 8051 microcontroller generic part number actually includes a whole family of microcontrollers that have numbers ranging from 8031 to 8751 and are available in N-Channel Metal Oxide Silicon (NMOS) and Complementary Metal Oxide

Silicon (CMOS) construction in a variety of package types. An enhanced version of the 8051, the 8052, also exists with its own family of variations and even includes one member that can be programmed in BASIC. An inspection of Appendix E shows that there are dozens of other variations on the "core" 8051 architecture. This galaxy of parts, the result of desires by the manufacturers to leave no market niche unfilled, would require many chapters to cover. In this chapter, we will study a "generic" 8051, housed in a 40-pin DIP, and direct the investigation of a particular type to the data books. The block diagram of the 8051 in Figure 3.1a shows all of the features unique to microcontrollers:

Internal ROM and RAM

I/O ports with programmable pins

Timers and counters

Serial data communication

The figure also shows the usual CPU components: program counter, ALU, working registers, and clock circuits.[1]

The 8051 architecture consists of these specific features:

- Eight-bit CPU with registers A (the accumulator) and B
- Sixteen-bit program counter (PC) and data pointer (DPTR)
- Eight-bit program status word (PSW)
- Eight-bit stack pointer (SP)
- Internal ROM or EPROM (8751) of 0 (8031) to 4K (8051)
- Internal RAM of 128 bytes:
    - Four register banks, each containing eight registers
    - Sixteen bytes, which may be addressed at the bit level
    - Eighty bytes of general-purpose data memory
- Thirty-two input/output pins arranged as four 8-bit ports: P0 – P3
- Two 16-bit timer/counters: T0 and T1
- Full duplex serial data receiver/transmitter: SBUF
- Control registers: TCON, TMOD, SCON, PCON, IP, and IE
- Two external and three internal interrupt sources
- Oscillator and clock circuits

[1] Knowledge of the details of circuit operation that cannot be affected by any instruction or external data, although intellectually stimulating, tends to confuse the student new to the 8051. For this reason, this text concentrates on the essential features of the 8051; the more advanced student may wish to refer to manufacturers' data books for additional information.

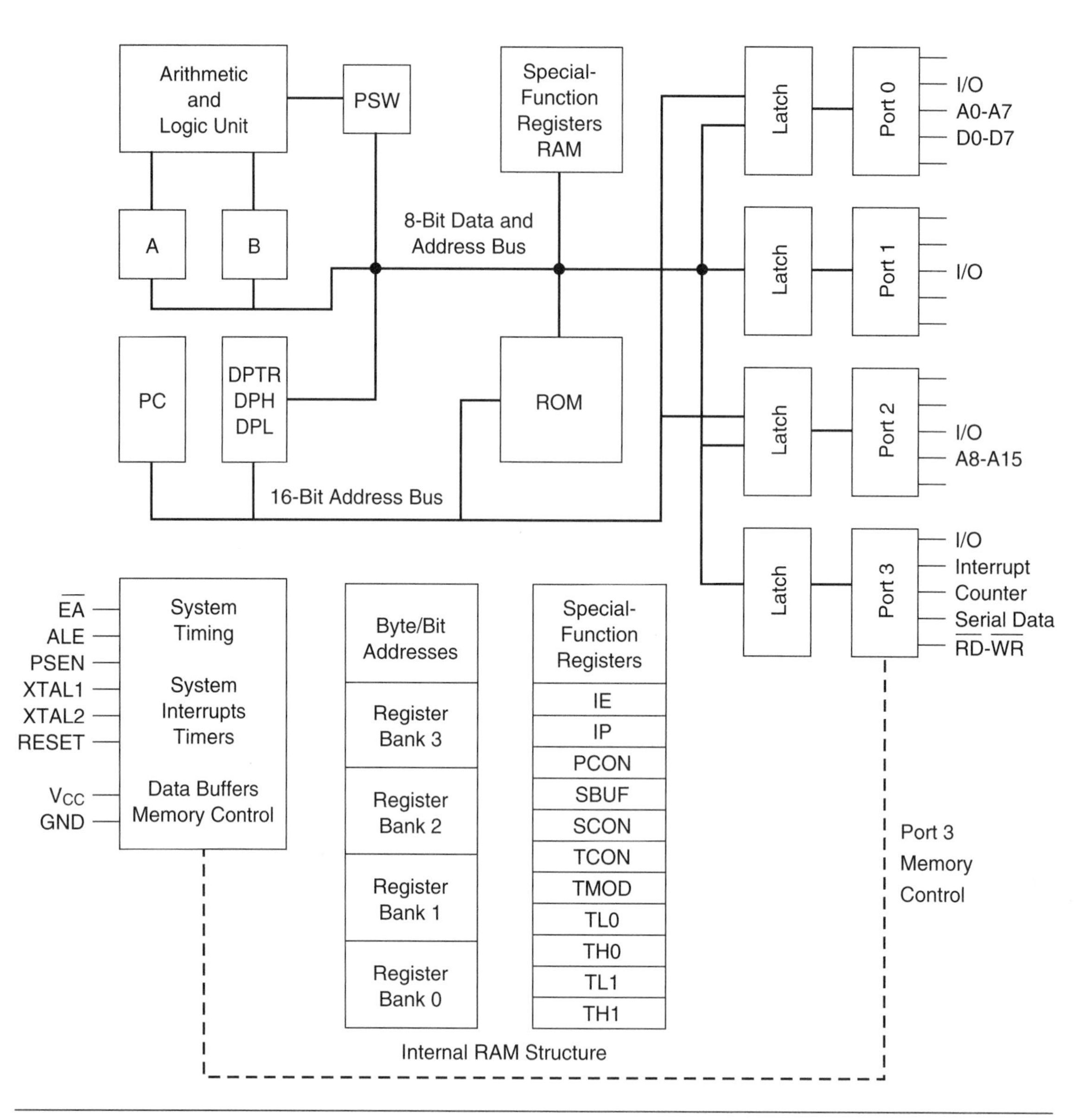

FIGURE 3.1A ◆ 8051 Block Diagram

The programming model of the 8051 in Figure 3.1b shows the 8051 as a collection of 8- and 16-bit registers and 8-bit memory locations. These registers and memory locations can be made to operate using the software instructions that are incorporated as part of the design. The program instructions have to do with the control of the registers and digital data paths that are physically con-

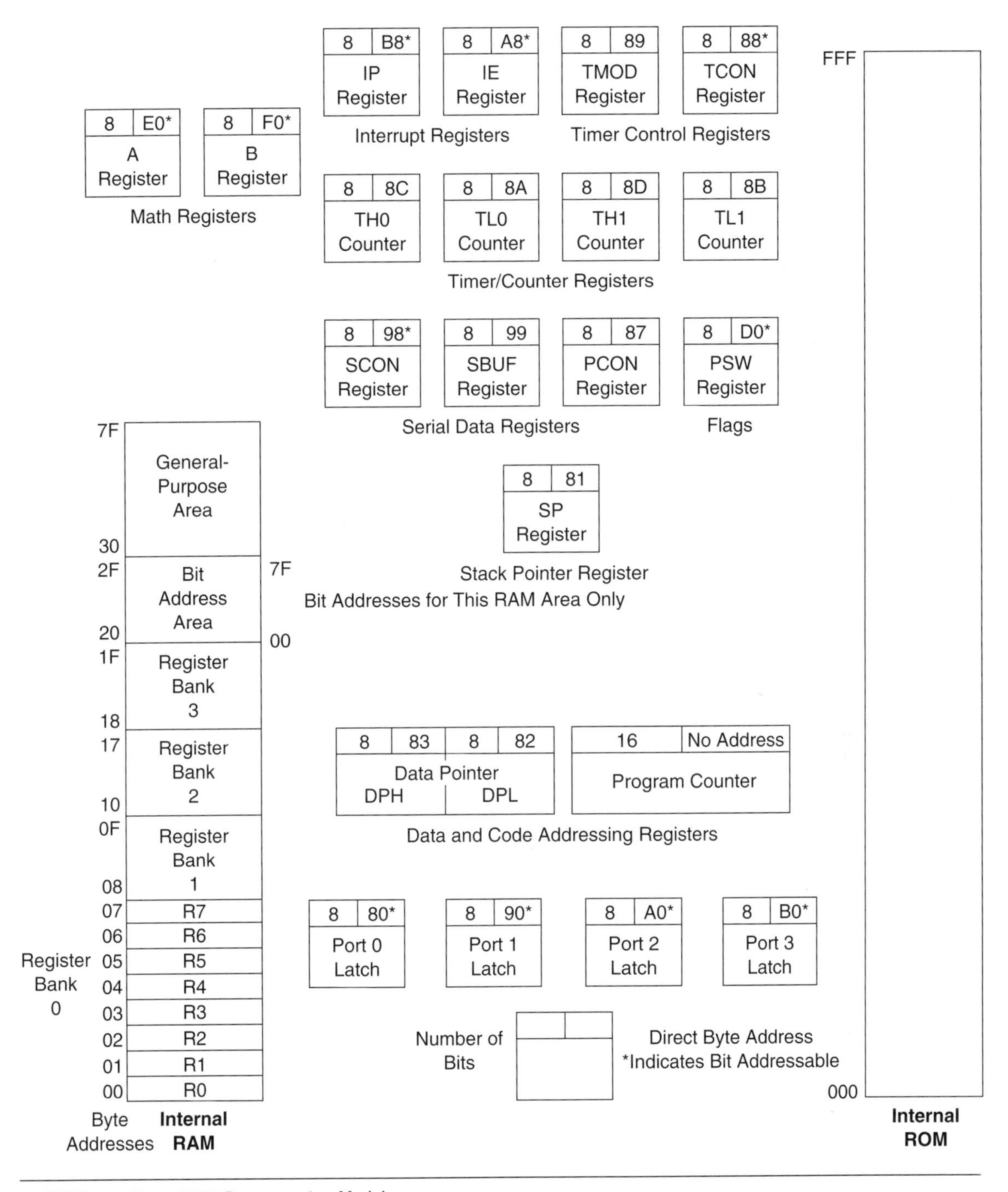

FIGURE 3.1B ◆ 8051 Programming Model

tained inside the 8051, as well as memory locations that are physically located outside the 8051.

The model is complicated by the number of special-purpose registers that must be present to make a microcomputer a microcontroller. A cursory inspection of the model is recommended for the first-time viewer; return to the model as needed while progressing through the remainder of the text.

Most of the registers have a specific function; those that do occupy an individual block with a symbolic name, such as A or TH0 or PC. Others, which are generally indistinguishable from each other, are grouped in a larger block, such as internal ROM or RAM memory.

Each register, with the exception of the program counter, has an internal 1-byte address assigned to it. Some registers (marked with an asterisk* in Figure 3.1b) are both byte and bit addressable. That is, the entire byte of data at such register addresses may be read or altered, or individual bits may be read or altered. Software instructions are generally able to specify a register by its address, its symbolic name, or both.

A pinout of the 8051 packaged in a 40-pin DIP is shown in Figure 3.2 with the full and abbreviated names of the signals for each pin. It is important to note that many of the pins are used for more than one function (the alternate functions are shown in parentheses in Figure 3.2). Not all of the possible 8051 features may be used *at the same time.*

Programming instructions or physical pin connections determine the use of any multifunction pins. For example, port 3 bit 0 (abbreviated P3.0) may be used as a general-purpose I/O pin, or as an input (RXD) to SBUF, the serial data receiver register. The system designer decides which of these two functions is to be used and designs the hardware and software affecting that pin accordingly.

The 8051 Oscillator and Clock

The heart of the 8051 is the circuitry that generates the clock pulses by which all internal operations are synchronized. Pins XTAL1 and XTAL2 are provided for connecting a resonant network to form an oscillator. Typically, a quartz crystal and capacitors are employed, as shown in Figure 3.3. The crystal frequency is the basic internal clock frequency of the microcontroller. The manufacturers make available 8051 designs that can run at specified maximum and *minimum* frequencies, typically 1 megahertz to 16 megahertz. Minimum frequencies imply that some internal memories are dynamic and must always operate above a minimum frequency or data will be lost.

Serial data communication needs often dictate the frequency of the oscillator because of the requirement that internal counters must divide the basic clock rate to yield standard communication bit per second (baud) rates. If the basic clock frequency is not divisible without a remainder, then the resulting communication frequency is not standard.

Function	Pin	Pin	Function
Port 1 Bit 0	1 P1.0	Vcc 40	+ 5V
Port 1 Bit 1	2 P1.1	(AD0)P0.0 39	Port 0 Bit 0 (Address/Data 0)
Port 1 Bit 2	3 P1.2	(AD1)P0.1 38	Port 0 Bit 1 (Address/Data 1)
Port 1 Bit 3	4 P1.3	(AD2)P0.2 37	Port 0 Bit 2 (Address/Data 2)
Port 1 Bit 4	5 P1.4	(AD3)P0.3 36	Port 0 Bit 3 (Address/Data 3)
Port 1 Bit 5	6 P1.5	(AD4)P0.4 35	Port 0 Bit 4 (Address/Data 4)
Port 1 Bit 6	7 P1.6	(AD5)P0.5 34	Port 0 Bit 5 (Address/Data 5)
Port 1 Bit 7	8 P1.7	(AD6)P0.6 33	Port 0 Bit 6 (Address/Data 6)
Reset Input	9 RST	(AD7)P0.7 32	Port 0 Bit 7 (Address/Data 7)
Port 3 Bit 0 (Receive Data)	10 P3.0(RXD)	(Vpp)/EA 31	External Enable (EPROM Programming Voltage)
Port 3 Bit 1 (XMIT Data)	11 P3.1(TXD)	(PROG)ALE 30	Address Latch Enable (EPROM Program Pulse)
Port 3 Bit 2 (Interrupt 0)	12 P3.2($\overline{\text{INT0}}$)	$\overline{\text{PSEN}}$ 29	Program Store Enable
Port 3 Bit 3 (Interrupt 1)	13 P3.3($\overline{\text{INT1}}$)	(A15)P2.7 28	Port 2 Bit 7 (Address 15)
Port 3 Bit 4 (Timer 0 Input)	14 P3.4(T0)	(A14)P2.6 27	Port 2 Bit 6 (Address 14)
Port 3 Bit 5 (Timer 1 Input)	15 P3.5(T1)	(A13)P2.5 26	Port 2 Bit 5 (Address 13)
Port 3 Bit 6 (Write Strobe)	16 P3.6($\overline{\text{WR}}$)	(A12)P2.4 25	Port 2 Bit 4 (Address 12)
Port 3 Bit 7 (Read Strobe)	17 P3.7($\overline{\text{RD}}$)	(A11)P2.3 24	Port 2 Bit 3 (Address 11)
Crystal Input 2	18 XTAL2	(A10)P2.2 23	Port 2 Bit 2 (Address 10)
Crystal Input 1	19 XTAL1	(A9)P2.1 22	Port 2 Bit 1 (Address 9)
Ground	20 Vss	(A8)P2.0 21	Port 2 Bit 0 (Address 8)

Note: Alternate functions are shown below the port name (in parentheses). Pin numbers and pin names are shown inside the DIP package.

FIGURE 3.2 ◆ 8051 DIP Pin Assignments

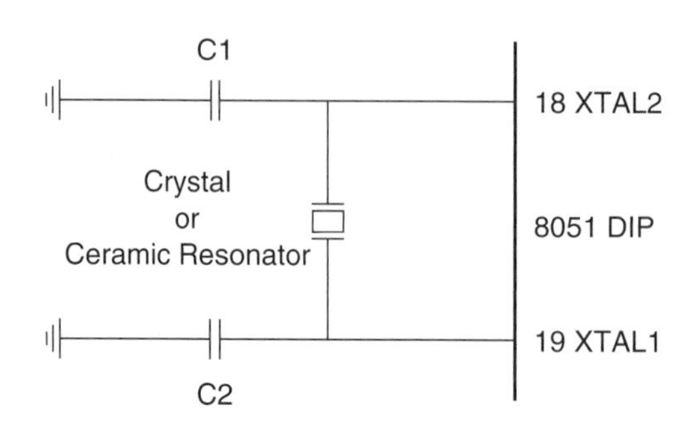

Crystal or Ceramic Resonator Oscillator Circuit

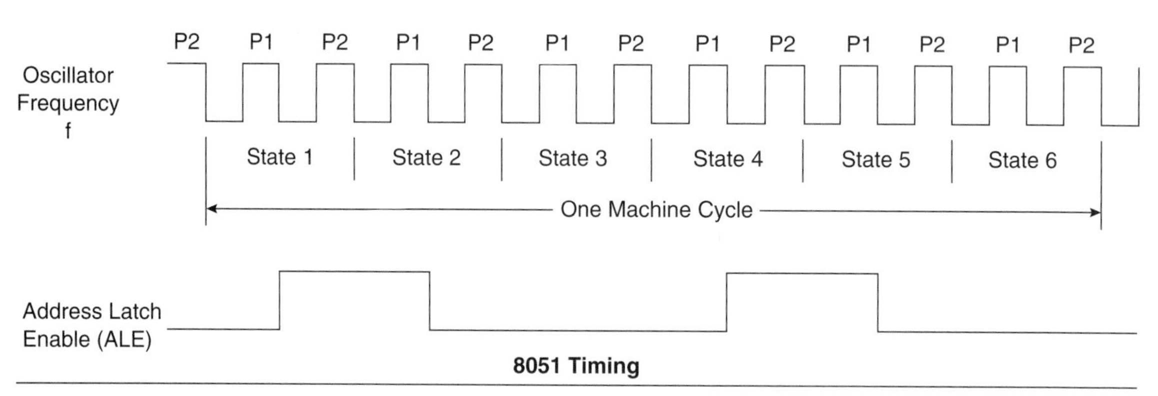

FIGURE 3.3 ◆ Oscillator Circuit and Timing

Ceramic resonators may be used as a low-cost alternative to crystal resonators. However, decreases in frequency stability and accuracy make the ceramic resonator a poor choice if high-speed serial data communication with other systems, or critical timing, is to be done.

The oscillator formed by the crystal, capacitors, and an on-chip inverter generates a pulse train at the frequency of the crystal, as shown in Figure 3.3.

The clock frequency, *f*, establishes the smallest interval of time within the microcontroller, called the pulse, *P*, time. The smallest interval of time to accomplish any simple instruction, or part of a complex instruction, however, is the machine cycle. The machine cycle is itself made up of six states. A state is the basic time interval for discrete operations of the microcontroller such as fetching an opcode byte, decoding an opcode, executing an opcode, or writing a data byte. Two oscillator pulses define each state.

Program instructions may require one, two, or four machine cycles to be executed, depending on the type of instruction. Instructions are fetched and executed by the microcontroller automatically, beginning with the instruction located at ROM memory address 0000h at the time the microcontroller is first reset.

To calculate the time any particular instruction will take to be executed, find the number of cycles, *C*, from the list in Appendix A. The time to execute that instruction is then found by multiplying *C* by 12 and dividing the product by the crystal frequency:

$$T_{inst} = \frac{C \times 12d}{\text{crystal frequency}}$$

For example, if the crystal frequency is 16 megahertz, then the time to execute an ADD A, R1 one-cycle instruction is .75 microseconds. A 12 megahertz crystal yields the convenient time of 1 microsecond per cycle. An 11.0592 megahertz crystal, although seemingly an odd value, yields a cycle frequency of 921.6 kilohertz, which can be divided evenly by the standard communication baud rates of 19200, 9600, 4800, 2400, 1200, and 300 hertz.

Note, in Figure 3.3, there are two ALE pulses per machine cycle. The ALE pulse, which is primarily used as a timing pulse for external memory access, indicates when every instruction byte is fetched. Two bytes of a single instruction may thus be fetched, and executed, in one machine cycle. Single-byte instructions are not executed in a half cycle, however. Single-byte instructions "throw-away" the second byte (which is the first byte of the next instruction). The next instruction is then fetched in the following cycle.

Program Counter and Data Pointer

The 8051 contains two 16-bit registers: the program counter (PC) and the data pointer (DPTR). Each is used to hold the address of a byte in memory.

Program instruction bytes are fetched from locations in memory that are addressed by the PC. Program ROM may be on the chip at addresses 0000h to 0FFFh, external to the chip for addresses that exceed 0FFFh, or totally external for all addresses from 0000h to FFFFh. The PC is automatically incremented after every instruction byte is fetched and may also be altered by certain instructions. The PC is the only register that does not have an internal address.

The DPTR register is made up of two 8-bit registers, named DPH and DPL, which are used to furnish memory addresses for internal and external code access and external data access. The DPTR is under the control of program instructions and can be specified by its 16-bit name, DPTR, or by each individual byte name, DPH and DPL. DPTR does not have a single internal address; DPH and DPL are each assigned an address.

A and B CPU Registers

The 8051 contains 34 general-purpose, or working, registers. Two of these, registers A and B, hold results of many instructions, particularly math and logical operations, of the 8051 central processing unit (CPU). The other 32 are arranged as part of internal RAM in four banks, B0 – B3, of eight registers and comprise the mathematical core.

The A (accumulator) register is the most versatile of the two CPU registers and is used for many operations, including addition, subtraction, integer multiplication and division, and Boolean bit manipulations. The A register is also used for all data transfers between the 8051 and any external memory. The B register is used with the A register for multiplication and division operations and has no other function other than as a location where data may be stored.

Flags and the Program Status Word (PSW)

Flags are 1-bit registers provided to store the results of certain program instructions. Other instructions can test the condition of the flags and make decisions based on the flag states. In order that the flags may be conveniently addressed, they are grouped inside the program status word (PSW) and the power control (PCON) registers.

The 8051 has four math flags that respond automatically to the outcomes of math operations and three general-purpose user flags that can be set to 1 or cleared to 0 by the programmer as desired. The math flags include Carry (C), Auxiliary Carry (AC), Overflow (OV), and Parity (P). User flags are named F0, GF0, and GF1; they are general-purpose flags that may be used by the programmer to record some event in the program. Note that all of the flags can be set and cleared by the programmer at will. The math flags, however, are also affected by math operations.

The program status word is shown in Figure 3.4. The PSW contains the math flags, user program flag F0, and the register select bits that identify which of the four general-purpose register banks is currently in use by the program. The remaining two user flags, GF0 and GF1, are stored in PCON, which is shown in Figure 3.13.

Detailed descriptions of the math flag operations will be discussed in chapters that cover the opcodes that affect the flags. The user flags can be set or cleared using data move instructions covered in Chapter 5.

Internal Memory

A functioning computer must have memory for program code bytes, commonly in ROM, and RAM memory for variable data that can be altered as the program runs. The 8051 has internal RAM and ROM memory for these functions. Additional memory can be added externally using suitable circuits.

Unlike microcontrollers with Von Neumann architectures, which can use a *single* memory address for either program code or data, *but not for both,* the 8051 has a Harvard architecture, which uses the *same address,* in *different* memo-

7	6	5	4	3	2	1	0
CY	AC	F0	RS1	RS0	OV	—	P

The Program Status Word (PSW) Special Function Register

Bit	Symbol	Function		
7	CY	Carry flag; used in arithmetic, jump, rotate, and Boolean instructions		
6	AC	Auxiliary Carry flag; used for BCD arithmetic		
5	F0	User flag 0		
4	RS1	Register bank select bit 1		
3	RS0	Register bank select bit 0		
		RS1	RS0	
		0	0	Select register bank 0
		0	1	Select register bank 1
		1	0	Select register bank 2
		1	1	Select register bank 3
2	OV	Overflow flag; used in arithmetic instructions		
1	—	Reserved for future use		
0	P	Parity flag; shows parity of register A: 1 = Odd Parity		

Bit addressable as PSW.0 to PSW.7

FIGURE 3.4 ◆ PSW Program Status Word Register

ries, for code and data. Internal circuitry accesses the correct memory based on the nature of the operation in progress.

Internal RAM

The 128-byte internal RAM, which is shown generally in Figure 3.1 and in detail in Figure 3.5, is organized into three distinct areas:

1. Thirty-two bytes from address 00h to 1Fh that make up 32 working registers organized as four banks of eight registers each. The four register banks are numbered 0 to 3 and are made up of eight registers named R0 to R7. Each register can be addressed by name (when its bank is selected) or by its RAM address. Thus R0 of bank 3 is R0 (if bank 3 is currently selected) or address 18h (whether bank 3 is selected or not). Bits RS0 and RS1 in the PSW determine which bank of registers is currently in use at any time when the program is running. Register banks not selected can be used as general-purpose RAM. Bank 0 is selected on reset.
2. A *bit*-addressable area of 16 bytes occupies RAM *byte* addresses 20h to 2Fh, forming a total of 128 addressable bits. An addressable bit may be

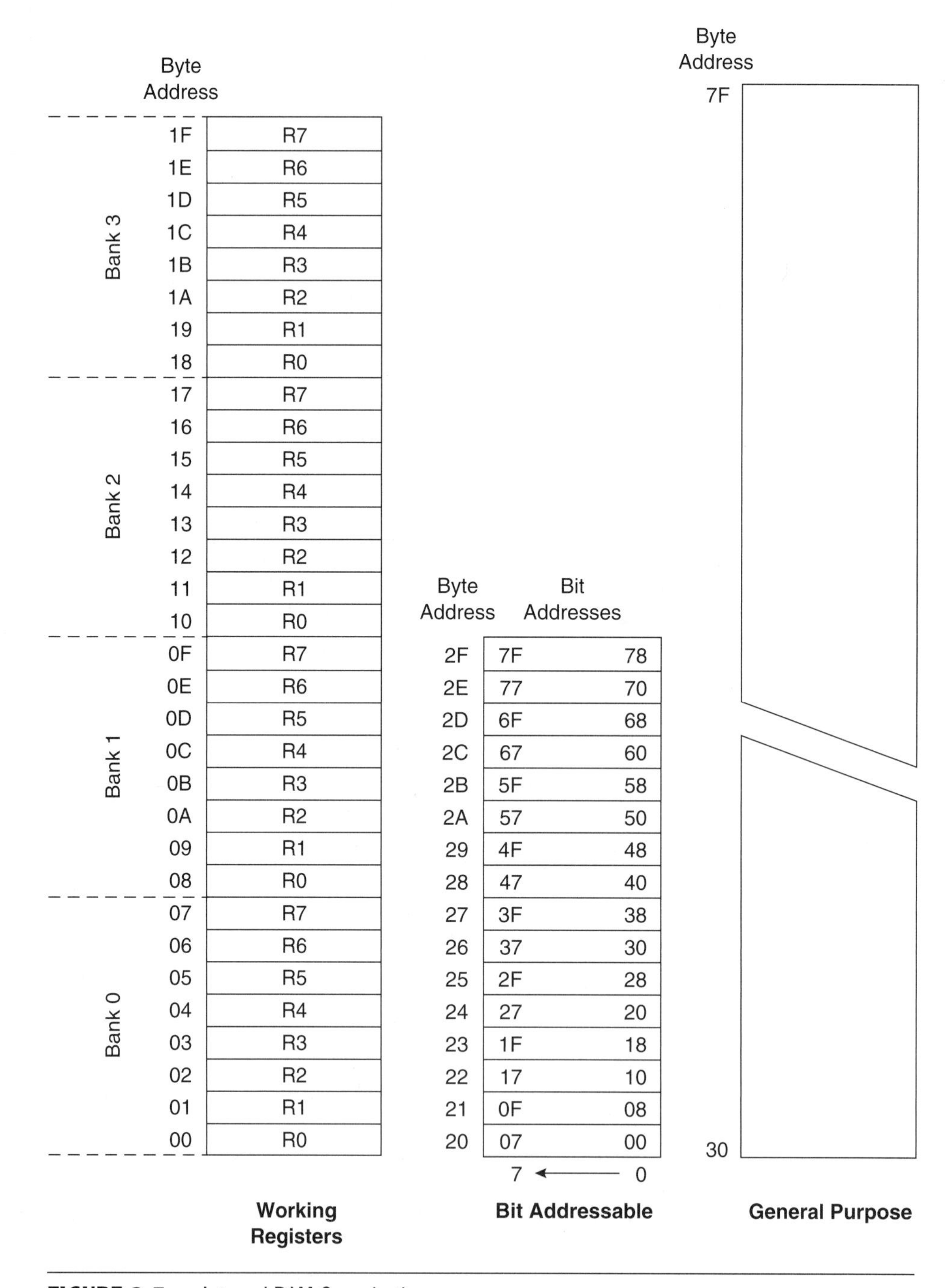

FIGURE 3.5 ◆ Internal RAM Organization

specified by its *bit* address of 00h to 7Fh, or 8 bits may form any *byte* address from 20h to 2Fh. Thus, for example, bit address 4Fh is also bit 7 of byte address 29h. Addressable bits are useful when the program need only remember a binary event (switch on, light off, etc.). Internal RAM is in short supply as it is, so why use a byte when a bit will do?

3. A general-purpose RAM area above the bit area, from 30h to 7Fh, addressable as bytes.

The Stack and the Stack Pointer

The stack refers to an area of internal RAM that is used in conjunction with certain opcodes to store and retrieve data quickly. The 8-bit Stack Pointer (SP) register is used by the 8051 to hold an internal RAM address that is called the *top of the stack*. The address held in the SP register is the location in internal RAM where the last byte of data was stored by a stack operation.

When data is to be placed on the stack, the SP increments *before* storing data on the stack so that the stack grows *up* as data is stored. As data is retrieved from the stack, the byte is read from the stack, and then the SP decrements to point to the next available byte of stored data.

Operation of the stack and the SP is shown in Figure 3.6. The SP is set to 07h when the 8051 is reset and can be changed to any internal RAM address by the programmer, using a data move command from Chapter 5.

The stack is limited in height to the size of the internal RAM. The stack has the potential (if the programmer is not careful to limit its growth) to overwrite valuable data in the register banks, bit-addressable RAM, and scratch-pad RAM areas. The programmer is responsible for making sure the stack does not grow beyond predefined bounds!

The stack is normally placed high in internal RAM, by an appropriate choice of the number placed in the SP register, to avoid conflict with the register, bit, and scratch-pad internal RAM areas.

Special Function Registers

The 8051 operations that do not use the internal 128-byte RAM addresses from 00h to 7Fh are done by a group of specific internal registers, each called a Special-Function register (SFR), which may be addressed much like internal RAM, using addresses from 80h to FFh.

Some SFRs (marked with an asterisk* in Figure 3.1b) are also bit addressable, as is the case for the bit area of RAM. This feature allows the programmer to change only what needs to be altered, leaving the remaining bits in that SFR unchanged.

Not all of the addresses from 80h to FFh are used for SFRs, and attempting to use an address that is not defined, or *empty*, results in unpredictable results.

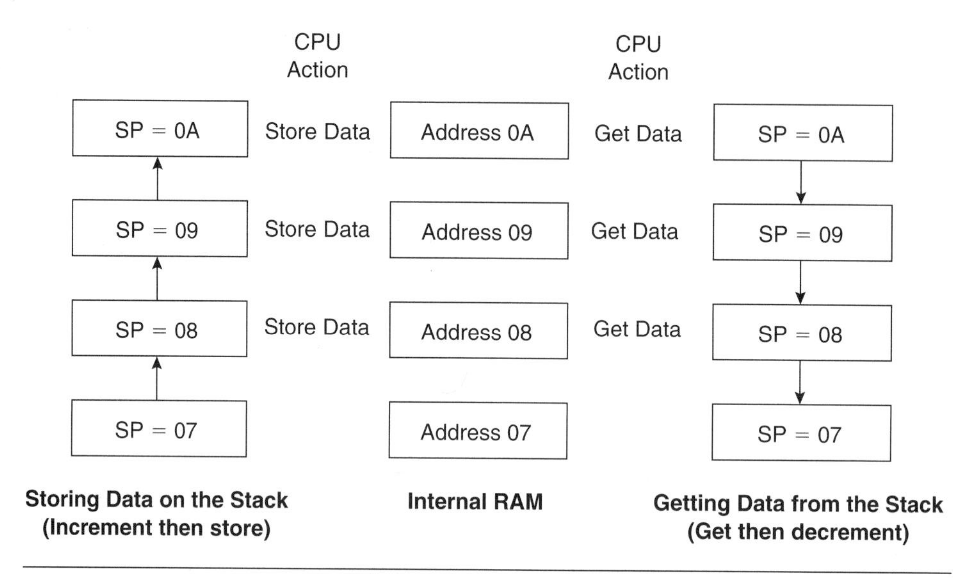

FIGURE 3.6 ◆ Stack Operation

In Figure 3.1b, the SFR addresses are shown in the upper right corner of each block. The SFR names and equivalent internal RAM addresses are given in Table 3.1. Note that the PC is not part of the SFR and has no internal RAM address. See also Appendix F.

SFRs are named in certain opcodes by their functional names, such as A or TH0, and are referenced by other opcodes by their addresses, such as 0E0h or 8Ch. Note that *any* address used in the program *must* start with a number; thus address E0h for the A SFR begins with 0. Failure to use this number convention will result in an assembler error when the program is assembled.

Internal ROM

The 8051 is organized so that data memory and program code memory can be in two entirely different physical memory entities. *Each* has the same address ranges.

The structure of the internal RAM has been discussed previously. A corresponding block of internal program code, contained in an internal ROM, occupies code address space 0000h to 0FFFh. The PC is ordinarily used to address program code bytes from addresses 0000h to FFFFh. Program addresses higher than 0FFFh, which exceed the internal ROM capacity, will cause the 8051 to automatically fetch code bytes from external program memory. Code bytes can

TABLE 3.1

Name	Function	Internal RAM Address (HEX)
A	Accumulator	0E0
B	Arithmetic	0F0
DPH	Addressing external memory	83
DPL	Addressing external memory	82
IE	Interrupt enable control	0A8
IP	Interrupt priority	0B8
P0	Input/output port latch	80
P1	Input/output port latch	90
P2	Input/output port latch	A0
P3	Input/output port latch	0B0
PCON	Power control	87
PSW	Program status word	0D0
SCON	Serial port control	98
SBUF	Serial port data buffer	99
SP	Stack pointer	81
TMOD	Timer/counter mode control	89
TCON	Timer/counter control	88
TL0	Timer 0 low byte	8A
TH0	Timer 0 high byte	8C
TL1	Timer 1 low byte	8B
TH1	Timer 1 high byte	8D

also be fetched exclusively from an external memory, addresses 0000h to FFFFh, by connecting the external access pin (EA pin 31 on the DIP) to ground. The PC does not care where the code is; the circuit designer decides whether the code is found totally in internal ROM, totally in external ROM, or in a combination of internal and external ROM.

3.2 Input/Output Pins, Ports, and Circuits

One major feature of a microcontroller is the versatility built into the input/output (I/O) circuits that connect the 8051 to the outside world. As noted in Chapter 1, microprocessor designs must add additional chips to interface with external circuitry; this ability is built into the microcontroller.

To be commercially viable, the 8051 had to incorporate as many functions as were technically and economically feasible. The main constraint that limits numerous functions is the number of pins available to the 8051 circuit designers. The DIP has 40 pins, and the success of the design in the marketplace was determined by the flexibility built into the use of these pins.

For this reason, 24 of the pins may each be used for one of two entirely different functions, yielding a total pin configuration of 64. The function a pin performs at any given instant depends, first, on what is physically connected to it and, then, on what software commands are used to "program" the pin. Both of these factors are under the complete control of the 8051 programmer and circuit designer.

Given this pin flexibility, the 8051 may be applied simply as a single component with I/O only, or it may be expanded to include additional memory, parallel ports, and serial data communication by using the alternate pin assignments. The key to programming an alternate pin function is the port pin circuitry shown in Figure 3.7.

Each port has a D-type output latch for each pin. The SFR for each port is made up of these eight latches, which can be addressed at the SFR address for that port. For instance, the eight latches for port 0 are addressed at location 80h; port 0 pin 3 is bit 2 of the P0 SFR. The port latches should not be confused with the port pins; the data on the latches does *not* have to be the same as that on the pins.

The two data paths are shown in Figure 3.7 by the circuits that read the latch or pin data using two entirely separate buffers. The upper buffer is enabled when latch data is read, and the lower buffer, when the pin state is read. The status of each latch may be read from a latch buffer, while an input buffer is connected directly to each pin so that the pin status may be read independently of the latch state.

Different opcodes access the latch or pin states as appropriate. Port operations are determined by the manner in which the 8051 is connected to external circuitry.

Programmable port pins have completely different alternate functions. The configuration of the control circuitry between the output latch and the port pin determines the nature of any particular port pin function. An inspection of Figure 3.7 reveals that only port 1 cannot have alternate functions; ports 0, 2, and 3 can be programmed.

The ports are not capable of driving loads that require currents in the tens of milliamperes (mA). As previously mentioned, the 8051 has many family members, and many are fabricated in varying technologies. An example range of logic-level currents, voltages, and total device power requirements is given in Table 3.2.

These figures tell us that driving more than two LSTTL inputs degrades the noise immunity of the ports and that careful attention must be paid to buffering the ports when they must drive currents in excess of those listed. Again, we must refer to the manufacturers' data books when designing a "real" application.

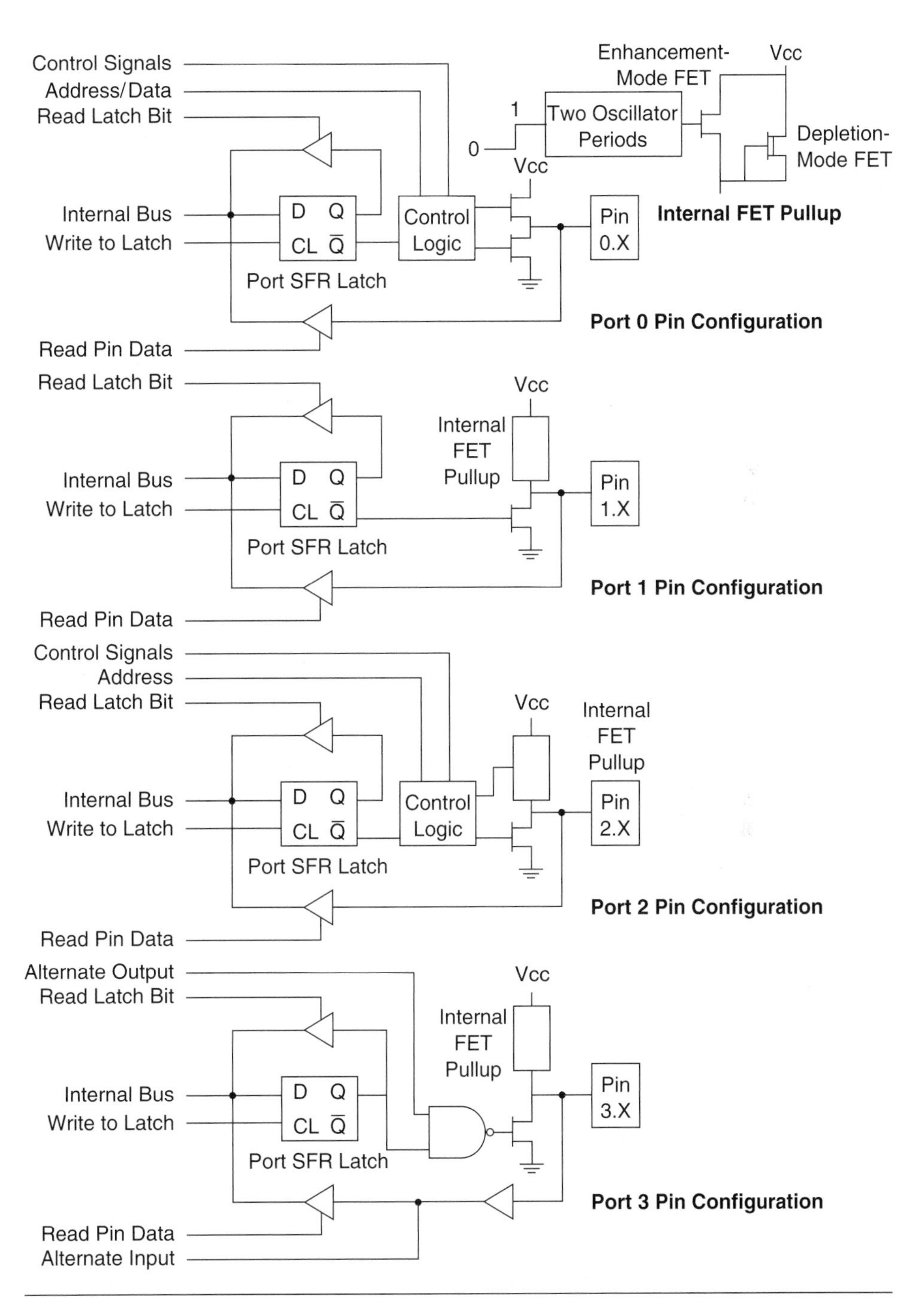

FIGURE 3.7 ◆ Port Pin Circuits

TABLE 3.2

Parameter	V_{oh}	I_{oh}	V_{ol}	I_{ol}	V_{il}	I_{il}	V_{ih}	I_{ih}	P_t
CMOS	2.4 V	−60 μA	.45 V	1.6 mA	.9 V	|10 μA|	1.9 V	|10 μA|	50 mW
NMOS	2.4 V	−80 μA	.45 V	1.6 mA	.8 V	−800 μA	2.0 V	10 μA	800 mW

Port 0

Port 0 pins may serve as inputs, outputs, or, when used together, as a bidirectional low-order address and data bus for external memory. For example, when a pin is to be used as an input, a 1 *must be* written to the corresponding port 0 latch by the program, thus turning both of the output transistors off, which in turn causes the pin to "float" in a high-impedance state, and the pin is essentially connected to the input buffer.

When used as an output, the pin latches that are programmed to a 0 will turn on the lower FET, grounding the pin. All latches that are programmed to a 1 still float; thus, external pullup resistors will be needed to supply a logic high when using port 0 as an output.

When port 0 is used as an address bus to external memory, internal control signals switch the address lines to the gates of the Field Effect Transistors (FETs). A logic 1 on an address bit will turn the upper FET on and the lower FET off to provide a logic high at the pin. When the address bit is a zero, the lower FET is on and the upper FET off to provide a logic low at the pin. After the address has been formed and latched into external circuits by the Address Latch Enable (ALE) pulse, the bus is turned around to become a data bus. Port 0 now reads data from the external memory and must be configured as an input, so a logic 1 is automatically written by internal control logic to all port 0 latches.

Port 1

Port 1 pins have no dual functions. Therefore, the output latch is connected directly to the gate of the lower FET, which has an FET circuit labeled Internal FET Pullup as an active pullup load.

Used as an input, a 1 is written to the latch, turning the lower FET off; the pin and the input to the pin buffer are pulled high by the FET load. An external circuit can overcome the high-impedance pullup and drive the pin low to input a 0 or leave the input high for a 1.

If used as an output, the latches containing a 1 can drive the input of an external circuit high through the pullup. If a 0 is written to the latch, the lower

FET is on, the pullup is off, and the pin can drive the input of the external circuit low.

To aid in speeding up switching times when the pin is used as an output, the internal FET pullup has another FET in parallel with it. The second FET is turned on for two oscillator time periods during a low-to-high transition on the pin, as shown in Figure 3.7. This arrangement provides a low impedance path to the positive voltage supply to help reduce rise times in charging any parasitic capacitances in the external circuitry.

Port 2

Port 2 may be used as an input/output port similar in operation to port 1. The alternate use of port 2 is to supply a high-order address byte in conjunction with the port 0 low-order byte to address external memory.

Port 2 pins are momentarily changed by the address control signals when supplying the high byte of a 16-bit address. Port 2 latches remain stable when external memory is addressed, as they do not have to be turned around (set to 1) for data input as is the case for port 0.

Port 3

Port 3 is an input/output port similar to port 1. The input and output functions can be programmed under the control of the P3 latches or under the control of various other special function registers. The port 3 alternate uses are shown in Table 3.3.

Unlike ports 0 and 2, which can have external addressing functions and change all eight port bits when in alternate use, each pin of port 3 may be individually programmed to be used either as I/O or as one of the alternate functions.

TABLE 3.3

Pin	Alternate Use	SFR
P3.0 – RXD	Serial data input	SBUF
P3.1 – TXD	Serial data output	SBUF
P3.2 – $\overline{\text{INT0}}$	External interrupt 0	TCON.1
P3.3 – $\overline{\text{INT1}}$	External interrupt 1	TCON.3
P3.4 – T0	External timer 0 input	TMOD
P3.5 – T1	External timer 1 input	TMOD
P3.6 – $\overline{\text{WR}}$	External memory write pulse	—
P3.7 – $\overline{\text{RD}}$	External memory read pulse	—

3.3 External Memory

The system designer is not limited by the amount of internal RAM and ROM available on chip. Two separate external memory spaces are made available by the 16-bit PC and DPTR and by different control pins for enabling external ROM and RAM chips. Internal control circuitry accesses the correct physical memory, depending on the machine cycle state and the opcode being executed.

There are several reasons for adding external memory, particularly program memory, when applying the 8051 in a system. When the project is in the prototype stage, the expense—in time and money—of having a masked internal ROM made for each program "try" is prohibitive. To alleviate this problem, the manufacturers make available an EPROM version, the 8751, which has 4K of on-chip EPROM that may be programmed and erased as needed as the program is developed. The resulting circuit board layout will be identical to one that uses a factory-programmed 8051. The only drawbacks to the 8751 are the specialized EPROM programmers that must be used to program the nonstandard 40-pin part, and the limit of "only" 4096 bytes of program code.

The 8751 solution works well if the program will fit into 4K. Unfortunately, many times, particularly if the program is written in a high-level language, the program size exceeds 4K, and an external program memory is needed. Again, the manufacturers provide a version for the job, the ROMless 8031. The $\overline{EA}$ pin is grounded when using the 8031, and all program code is contained in an external EPROM that may be as large as 64K and that can be programmed using standard EPROM programmers.

External RAM, which is accessed by the DPTR, may also be needed when 128 bytes of internal data storage is not sufficient. External RAM, up to 64K, may also be added to any chip in the 8051 family.

Connecting External Memory

Figure 3.8 shows the connections between an 8031 and an external memory configuration consisting of 16K of EPROM and 8K of static RAM. The 8051 accesses external RAM whenever certain program instructions are executed. External ROM is accessed whenever the $\overline{EA}$ (external access) pin is connected to ground or when the PC contains an address higher than the last address in the internal 4K ROM (0FFFh). 8051 designs can thus use internal and external ROM automatically; the 8031, having no internal ROM, must have $\overline{EA}$ grounded.

Figure 3.9 shows the timing associated with an external memory access cycle. During any memory access cycle, port 0 is time multiplexed. That is, it first provides the lower byte of the 16-bit memory address, then acts as a bidirectional data bus to write or read a byte of memory data. Port 2 provides the high byte of the memory address during the entire memory read/write cycle.

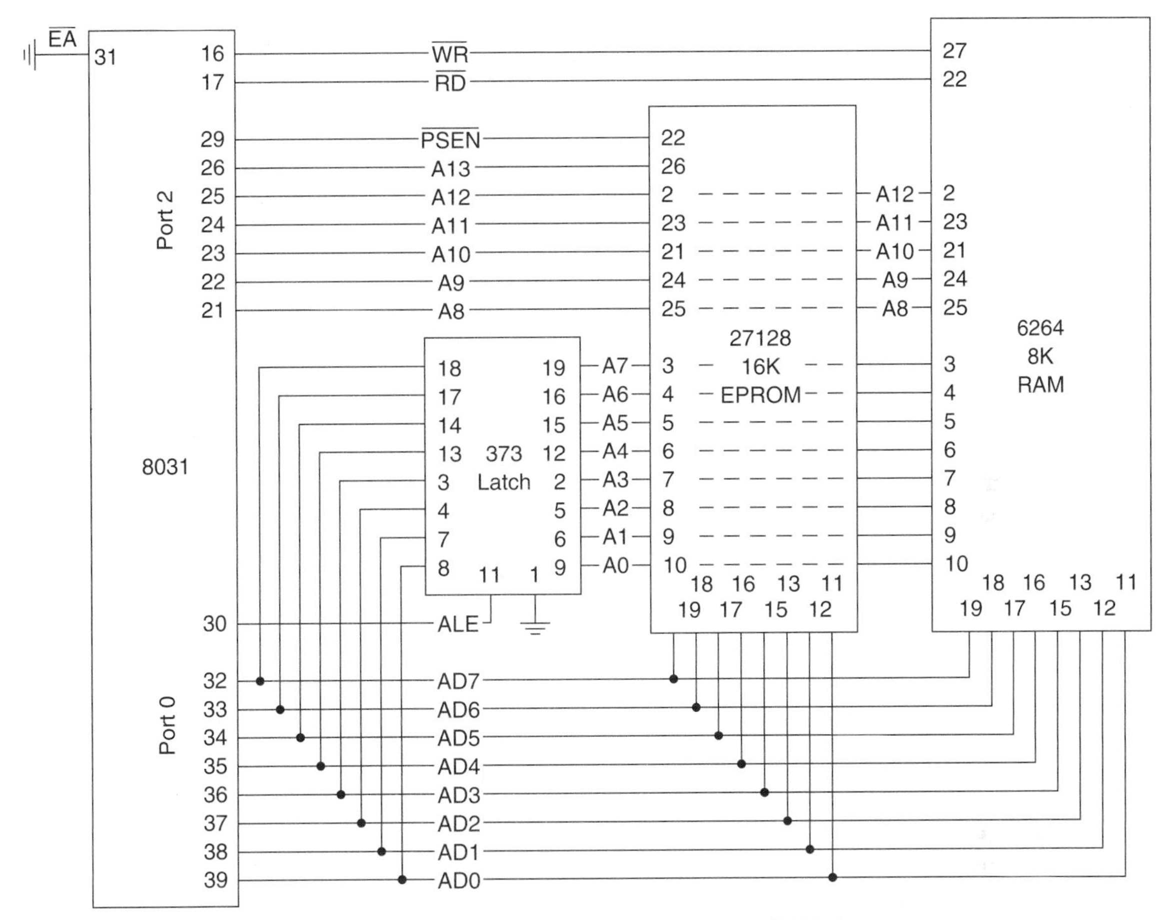

FIGURE 3.8 ◆ External Memory Connections

The lower address byte from port 0 must be latched into an external register to save the byte. Address byte save is accomplished by the ALE clock pulse that provides the correct timing for the '373 type data latch. The port 0 pins then become free to serve as a data bus.

If the memory access is for a byte of program code in the ROM, the $\overline{\text{PSEN}}$ (program store enable) pin will go low to enable the ROM to place a byte of program code on the data bus. If the access is for a RAM byte, the $\overline{\text{WR}}$ (write) or $\overline{\text{RD}}$ (read) pins will go low, enabling data to flow between the RAM and the data bus.

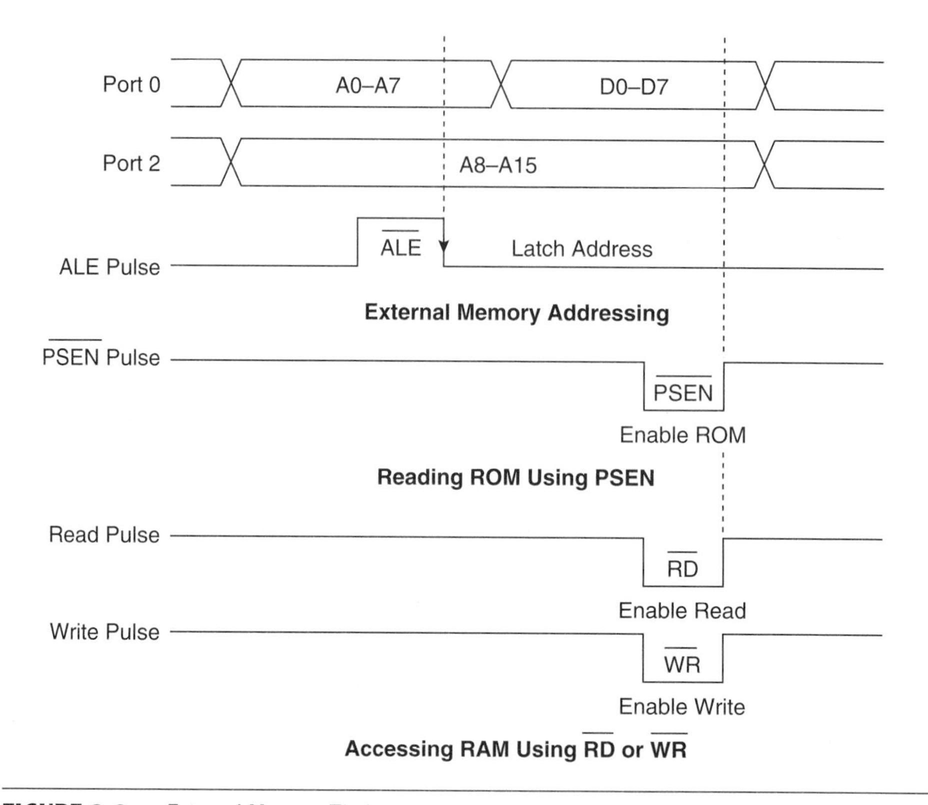

FIGURE 3.9 ◆ External Memory Timing

The ROM may be expanded to 64K by using a 27512 type EPROM and connecting the remaining port 2 upper address lines A14 – A15 to the chip.

SRAM capacity may be expanded to 64K by using a 62864-type chip.

Note that the $\overline{\text{WR}}$ and $\overline{\text{RD}}$ signals are alternate uses for port 3 pins 16 and 17. Also, port 0 is used for the lower address byte and data; port 2 is used for upper address bits. The use of external memory consumes many of the port pins, leaving only port 1 and parts of port 3 for general I/O.

3.4 Counters and Timers

Many microcontroller applications require the counting of external events, such as the frequency of a pulse train, or the generation of precise internal time delays between computer actions. Both of these tasks can be accomplished using software techniques, but software loops for counting or timing keep the processor occupied so that other, perhaps more important, functions are not done. To relieve the processor of this burden, two 16-bit *up* counters, named T0 and T1, are provided for the general use of the programmer. Each counter may be programmed to count internal clock pulses, acting as a timer, or programmed to count external pulses as a counter.

The counters are divided into two 8-bit registers called the timer low (TL0, TL1) and high (TH0, TH1) bytes. All counter action is controlled by bit states in the timer mode control register (TMOD), the timer/counter control register (TCON), and certain program instructions.

TMOD is dedicated solely to the two timers and can be considered to be two duplicate 4-bit registers, each of which controls the action of one of the timers. TCON has control bits and flags for the timers in the upper nibble, and control bits and flags for the external interrupts in the lower nibble. Figure 3.10 shows the bit assignments for TMOD and TCON.

Timer Counter Interrupts

The counters have been included on the chip to relieve the processor of timing and counting chores. When the program wishes to count a certain number of internal pulses or external events, a number is placed in one of the counters. The number represents the maximum count *less* the desired count, plus 1. The counter increments from the initial number to the maximum and then rolls over to 0 on the final pulse and also sets a timer flag. The flag condition may be tested by an instruction to tell the program that the count has been accomplished, or the flag may be used to interrupt the program.

Timing

If a counter is programmed to be a timer, it will count the internal clock frequency of the 8051 oscillator divided by 12d. As an example, if the crystal frequency is 6.0 megahertz, then the timer clock will have a frequency of 500 kilohertz.

The resultant timer clock is gated to the timer by means of the circuit shown in Figure 3.11. In order for oscillator clock pulses to reach the timer, the $C/\overline{T}$ bit in the TMOD register must be set to 0 (timer operation). Bit TRX in the TCON register must be set to 1 (timer run), and the *gate* bit in the TMOD register must be 0, or external pin $\overline{INTX}$ must be a 1. In other words, the counter is configured as a timer, then the timer pulses are gated to the counter by the run bit *and* the gate bit *or* the external input bits $\overline{INTX}$.

Timer Modes of Operation

The timers may operate in any one of four modes that are determined by the mode bits, M1 and M0, in the TMOD register. Figure 3.12 shows the four timer modes.

7	6	5	4	3	2	1	0
TF1	TR1	TF0	TR0	IE1	IT1	IE0	IT0

The Timer Control (TCON) Special Function Register

Bit	Symbol	Function
7	TF1	Timer 1 Overflow flag. Set when timer rolls from all 1s to 0. Cleared when processor vectors to execute interrupt service routine located at program address 001Bh.
6	TR1	Timer 1 run control bit. Set to 1 by program to enable timer to count; cleared to 0 by program to halt timer. Does not reset timer.
5	TF0	Timer 0 Overflow flag. Set when timer rolls from all 1s to 0. Cleared when processor vectors to execute interrupt service routine located at program address 000Bh.
4	TR0	Timer 0 run control bit. Set to 1 by program to enable timer to count; cleared to 0 by program to halt timer. Does not reset timer.
3	IE1	External interrupt 1 Edge flag. Set to 1 when a high-to-low edge signal is received on port 3 pin 3.3 ($\overline{INT1}$). Cleared when processor vectors to interrupt service routine located at program address 0013h. Not related to timer operations.
2	IT1	External interrupt 1 signal type control bit. Set to 1 by program to enable external interrupt 1 to be triggered by a falling edge signal. Set to 0 by program to enable a low-level signal on external interrupt 1 to generate an interrupt.
1	IE0	External interrupt 0 Edge flag. Set to 1 when a high-to-low edge signal is received on port 3 pin 3.2 ($\overline{INT0}$). Cleared when processor vectors to interrupt service routine located at program address 0003h. Not related to timer operations.
0	IT0	External interrupt 0 signal type control bit. Set to 1 by program to enable external interrupt 0 to be triggered by a falling edge signal. Set to 0 by program to enable a low-level signal on external interrupt 0 to generate an interrupt.

Bit addressable as TCON.0 to TCON.7

7	6	5	4	3	2	1	0
Gate	$C/\overline{T}$	M1	M0	Gate	$C/\overline{T}$	M1	M0
[Timer 1			]	[Timer 0			]

The Timer Mode Control (TMOD) Special Function Register

Bit	Symbol	Function
7/3	Gate	OR gate enable bit which controls RUN/STOP of timer 1/0. Set to 1 by program to enable timer to run if bit TR1/0 in TCON is set and signal on external interrupt $\overline{INT1/0}$ pin is high. Cleared to 0 by program to enable timer to run if bit TR1/0 in TCON is set.
6/2	$C/\overline{T}$	Set to 1 by program to make timer 1/0 act as a counter by counting pulses from external input pins 3.5 (T1) or 3.4 (T0). Cleared to 0 by program to make timer act as a timer by counting internal frequency.
5/1	M1	Timer/counter operating mode select bit 1. Set/cleared by program to select mode.
4/0	M0	Timer/counter operating mode select bit 0. Set/cleared by program to select mode.

M1	M0	Mode
0	0	0
0	1	1
1	0	2
1	1	3

TMOD is not bit addressable

FIGURE 3.10 ◆ TCON and TMOD Function Registers

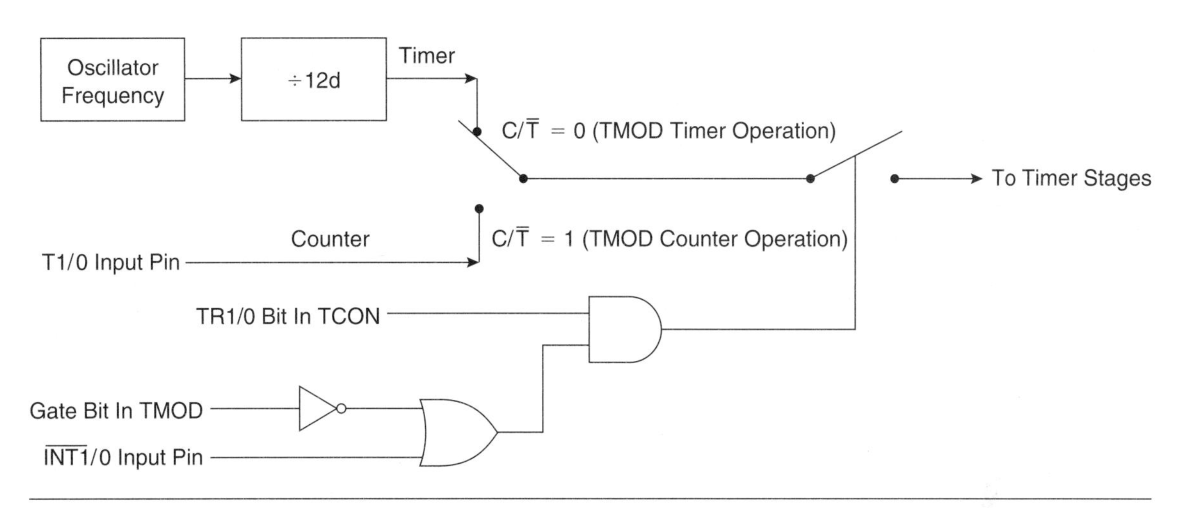

FIGURE 3.11 ◆ Timer/Counter Control Logic

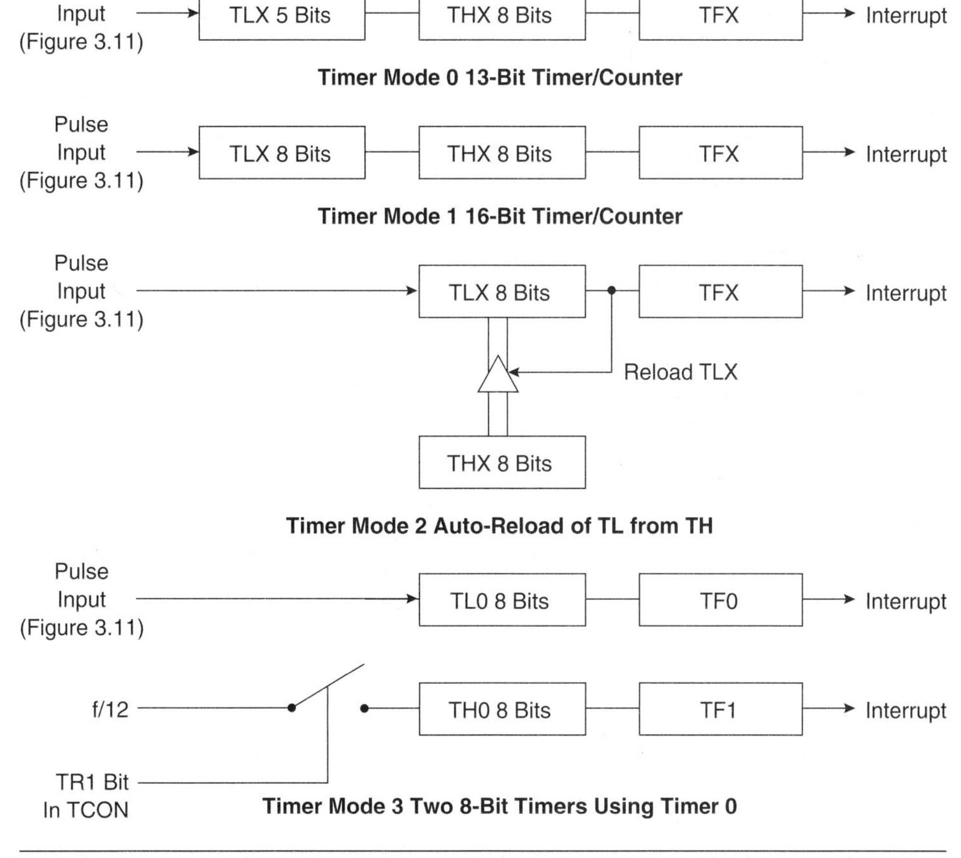

FIGURE 3.12 ◆ Timer 1 and Timer 0 Operation Modes

Timer Mode 0

Setting timer X mode bits to 00b in the TMOD register results in using the THX register as an 8-bit counter and TLX as a 5-bit counter; the pulse input is divided by 32d in TL so that TH counts the original oscillator frequency reduced by a total 384d. As an example, the 6 megahertz oscillator frequency would result in a final frequency to TH of 15625 hertz. The Timer flag is set whenever THX goes from FFh to 00h, or in .0164 seconds for a 6 megahertz crystal if THX starts at 00h.

Timer Mode 1

Mode 1 is similar to mode 0 except TLX is configured as a full 8-bit counter when the mode bits are set to 01b in TMOD. The Timer flag would be set in .1311 seconds using a 6 megahertz crystal.

Timer Mode 2

Setting the mode bits to 10b in TMOD configures the timer to use only the TLX counter as an 8-bit counter. THX is used to hold a value that is loaded into TLX every time TLX overflows from FFh to 00h. The Timer flag is also set when TLX overflows.

This mode exhibits an auto-reload feature: TLX will count up from the number in THX, overflow, and be initialized again with the contents of THX. For example, placing 9Ch in THX will result in a delay of exactly .0002 seconds before the Overflow flag is set if a 6 megahertz crystal is used.

Timer Mode 3

Timers 0 and 1 may be programmed to be in mode 0, 1, or 2 independently of a similar mode for the other timer. This is not true for mode 3; the timers do not operate independently if mode 3 is chosen for timer 0. Placing timer 1 in mode 3 causes it to stop counting; the control bit TR1 and the timer 1 flag TF1 are then used by timer 0.

Timer 0 in mode 3 becomes two completely separate 8-bit counters. TL0 is controlled by the gate arrangement of Figure 3.11 and sets timer flag TF0 whenever it overflows from FFh to 00h. TH0 receives the timer clock (the oscillator divided by 12) under the control of TR1 only and sets the TF1 flag when it overflows.

Timer 1 may still be used in modes 0, 1, and 2, while timer 0 is in mode 3 with one important exception: *No interrupts* will be generated by timer 1 while timer 0 is using the TF1 overflow flag. Switching timer 1 to mode 3 will stop it (and hold whatever count is in timer 1). Timer 1 can be used for baud rate generation for the serial port, or any other mode 0, 1, or 2 function that does not depend on an interrupt (or any other use of the TF1 flag) for proper operation.

Counting

The only difference between counting and timing is the source of the clock pulses to the counters. When used as a timer, the clock pulses are sourced from the oscillator through the divide-by-12d circuit. When used as a counter, pin T0 (P3.4) supplies pulses to counter 0, and pin T1 (P3.5) to counter 1. The $C/\overline{T}$ bit in TMOD must be set to 1 to enable pulses from the TX pin to reach the control circuit shown in Figure 3.11.

The input pulse on TX is sampled during P2 of state 5 every machine cycle. A change on the input from high to low between samples will increment the counter. Each high and low state of the input pulse must thus be held constant for at least one machine cycle to ensure reliable counting. Since this takes 24 pulses, the maximum input frequency that can be accurately counted is the oscillator frequency divided by 24. For our 6 megahertz crystal, the calculation yields a maximum external frequency of 250 kilohertz.

3.5 Serial Data Input /Output

Computers must be able to communicate with other computers in modern multiprocessor distributed systems. One cost-effective way to communicate is to send and receive data bits serially. The 8051 has a serial data communication circuit that uses register SBUF to hold data. Register SCON controls data communication, register PCON controls data rates, and pins RXD (P3.0) and TXD (P3.1) connect to the serial data network.

SBUF is physically two registers. One is write only and is used to hold data to be transmitted *out* of the 8051 via TXD. The other is read only and holds received data *from* external sources via RXD. Both mutually exclusive registers use address 99h.

There are four programmable modes for serial data communication that are chosen by setting the SMX bits in SCON. Baud rates are determined by the mode chosen. Figure 3.13 shows the bit assignments for SCON and PCON.

Serial Data Interrupts

Serial data communication is a relatively slow process, occupying many milliseconds per data byte to accomplish. In order not to tie up valuable processor time, Serial Data flags are included in SCON to aid in efficient data transmission and reception. Notice that data transmission is under the complete control of the program, but reception of data is unpredictable and at random times that are beyond the control of the program.

The serial data flags in SCON, TI and RI, are set whenever a data byte is transmitted (TI) or received (RI). These flags are ORed together to produce an interrupt to the program. The program must read these flags to determine which caused the interrupt and then clear the flag. This is unlike the timer flags that are cleared automatically; it is the responsibility of the programmer to write routines that handle the serial data flags.

7	6	5	4	3	2	1	0
SM0	SM1	SM2	REN	TB8	RB8	TI	RI

The Serial Port Control (SCON) Special Function Register

Bit	Symbol	Function
7	SM0	Serial port mode bit 0. Set/cleared by program to select mode.
6	SM1	Serial port mode bit 1. Set/cleared by program to select mode.

SM0	SM1	Mode	Description
0	0	0	Shift register; baud = $f/12$
0	1	1	8-bit UART; baud = variable
1	0	2	9-bit UART; baud = $f/32$ or $f/64$
1	1	3	9-bit UART; baud = variable

Bit	Symbol	Function
5	SM2	Multiprocessor communications bit. Set/cleared by program to enable multiprocessor communications in modes 2 and 3. When set to 1 an interrupt is generated if bit 9 of the received data is a 1; no interrupt is generated if bit 9 is a 0. If set to 1 for mode 1, no interrupt will be generated unless a valid stop bit is received. Clear to 0 if mode 0 is in use.
4	REN	Receive enable bit. Set to 1 to enable reception; cleared to 0 to disable reception.
3	TB8	Transmitted bit 8. Set/cleared by program in modes 2 and 3.
2	RB8	Received bit 8. Bit 8 of received data in modes 2 and 3; stop bit in mode 1. Not used in mode 0.
1	TI	Transmit Interrupt flag. Set to one at the end of bit 7 time in mode 0, and at the beginning of the stop bit for other modes. Must be cleared by the program.
0	RI	Receive Interrupt flag. Set to one at the end of bit 7 time in mode 0, and halfway through the stop bit for other modes. Must be cleared by the program.

Bit addressable as SCON.0 to SCON.7

7	6	5	4	3	2	1	0
SMOD	—	—	—	GF1	GF0	PD	IDL

The Power Mode Control (PCON) Special Function Register

Bit	Symbol	Function
7	SMOD	Serial baud rate modify bit. Set to 1 by program to double baud rate using timer 1 for modes 1, 2, and 3. Cleared to 0 by program to use timer 1 baud rate.
6-4	—	Not implemented.
3	GF1	General purpose user flag bit 1. Set/cleared by program.
2	GF0	General purpose user flag bit 0. Set/cleared by program.
1	PD	Power down bit. Set to 1 by program to enter power down configuration for CHMOS processors.
0	IDL	Idle mode bit. Set to 1 by program to enter idle mode configuration for CHMOS processors.

PCON is not bit addressable.

FIGURE 3.13 ◆ SCON and PCON Function Registers

Data Transmission

Transmission of serial data bits begins *anytime* data is written to SBUF. TI is set to a 1 when the data has been transmitted and signifies that SBUF is empty (for transmission purposes) and that another data byte can be sent. If the program fails to wait for the TI flag and overwrites SBUF while a previous data byte is in the process of being transmitted, the results will be unpredictable (a polite term for "garbage out").

Data Reception

Reception of serial data will begin *if* the receive enable bit (REN) in SCON is set to 1 for all modes. In addition, for mode 0 *only*, RI must be cleared to 0. Receiver Interrupt flag RI is set after data has been received in all modes. Setting REN is the only direct program control that limits the reception of unexpected data; the requirement that RI also be 0 for mode 0 prevents the reception of new data until the program has dealt with the old data and reset RI.

Reception can *begin* in modes 1, 2, and 3 if RI is set when the serial stream of bits begins. RI must have been reset by the program before the *last* bit is received or the *incoming* data will be lost. Incoming data is not transferred to SBUF until the last data bit has been received so that the previous transmission can be read from SBUF while new data is being received.

Serial Data Transmission Modes

The 8051 designers have included four modes of serial data transmission that enable data communication to be done in a variety of ways and a multitude of baud rates. Modes are selected by the programmer by setting the mode bits SM0 and SM1 in SCON. Baud rates are fixed for mode 0 and variable, using timer 1 and the serial baud rate modify bit (SMOD) in PCON, for modes 1, 2, and 3.

Serial Data Mode 0—Shift Register Mode

Setting bits SM0 and SM1 in SCON to 00b configures SBUF to receive or transmit eight data bits using pin RXD for *both* functions. Pin TXD is connected to the internal shift frequency pulse source to supply shift pulses to external circuits. The shift frequency, or baud rate, is fixed at 1/12 of the oscillator frequency, the same rate used by the timers when in the timer configuration. The TXD shift clock is a square wave that is low for machine cycle states S3 – S4 – S5 and high for S6 – S1 – S2. Figure 3.14 shows the timing for mode 0 shift register data transmission.

When transmitting, data is shifted *out* of RXD; the data changes on the *falling* edge of S6P2, or one clock pulse after the *rising* edge of the output TXD

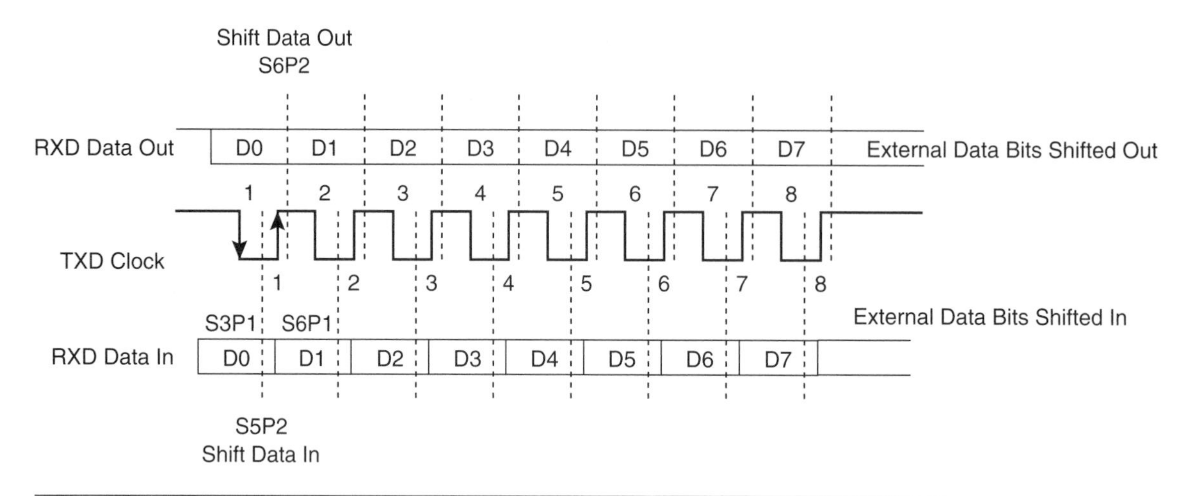

FIGURE 3.14 ◆ Shift Register Mode 0 Timing

shift clock. The system designer must design the external circuitry that receives this transmitted data to receive the data reliably based on this timing.

Received data comes *in* on pin RXD and should be synchronized with the shift clock produced at TXD. Data is sampled on the *falling* edge of S5P2 and shifted in to SBUF on the *rising* edge of the shift clock.

Mode 0 is *not* intended for data communication between computers, but as a high-speed serial data-collection method using discrete logic to achieve high data rates. The baud rate used in mode 0 will be much higher than standard for any reasonable oscillator frequency; for a 6 megahertz crystal, the shift rate will be 500 kilohertz.

Serial Data Mode 1—Standard UART

When SM0 and SM1 are set to 01b, SBUF becomes a 10-bit full-duplex receiver/ transmitter that may receive and transmit data at the same time. Pin RXD receives all data, and pin TXD transmits all data. Figure 3.15 shows the format of a data word.

Transmitted data is sent as a start bit, eight data bits (least significant bit, LSB, first), and a stop bit. Interrupt flag TI is set once all ten bits have been sent. Each bit interval is the inverse of the baud rate frequency, and each bit is maintained high or low over that interval.

Received data is obtained in the same order; reception is triggered by the falling edge of the start bit and continues if the stop bit is true (0 level) halfway through the start bit interval. This is an anti-noise measure; if the reception circuit is triggered by noise on the transmission line, the check for a low after half a bit interval should limit false data reception.

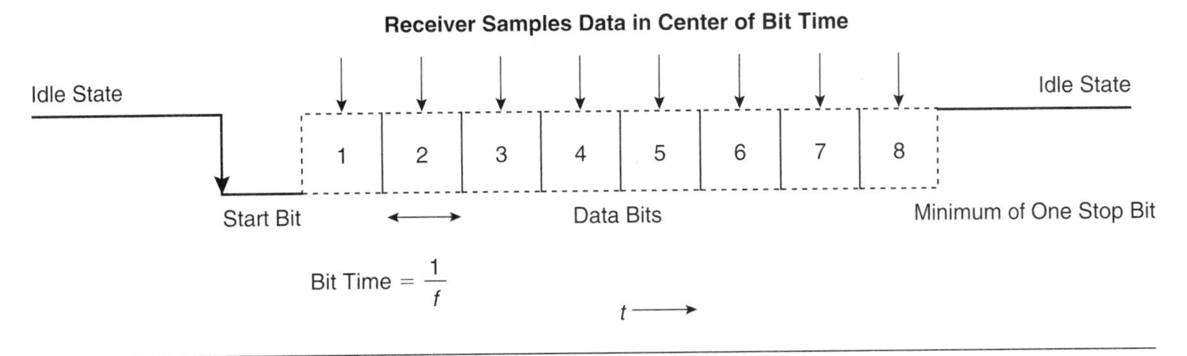

FIGURE 3.15 ◆ Standard UART Data Word

Data bits are shifted into the receiver at the programmed baud rate, and the data word will be loaded to SBUF *if* the following conditions are true: RI *must* be 0, *and* mode bit SM2 is 0 *or* the stop bit is 1 (the normal state of stop bits). RI set to 0 implies that the program has read the previous data byte and is ready to receive the next; a normal stop bit will then complete the transfer of data to SBUF regardless of the state of SM2. SM2 set to 0 enables the reception of a byte with any stop-bit state, a condition that is of limited use in this mode, but very useful in modes 2 and 3. SM2 set to 1 forces reception of only "good" stop bits, an anti-noise safeguard.

Of the original ten bits, the start bit is discarded, the eight data bits go to SBUF, and the stop bit is saved in bit RB8 of SCON. RI is set to 1, indicating a new data byte has been received.

If RI is found to be set at the end of the reception, indicating that the previously received data byte has not been read by the program, or if the other conditions listed are not true, the *new* data will not be loaded and will be *lost.*

Mode 1 Baud Rates

Timer 1 is used to generate the baud rate for mode 1 by using the Overflow flag of the timer to determine the baud frequency. Typically, timer 1 is used in timer mode 2 as an autoload 8-bit timer that generates the baud frequency:

$$f_{\text{baud}} = \frac{2^{\text{SMOD}}}{32\text{d}} \times \frac{\text{oscillator frequency}}{12\text{d} \times [256\text{d} - (\text{TH1})]}$$

SMOD is the control bit in PCON and can be 0 or 1, which raises the 2 in the equation to a value of 1 or 2.

If timer 1 is not run in timer mode 2, then the baud rate is

$$f_{\text{baud}} = \frac{2^{\text{SMOD}}}{32\text{d}} \times (\text{timer 1 overflow frequency})$$

and timer 1 can be run using the internal clock or as a counter that receives clock pulses from any external source via pin T1.

The oscillator frequency is chosen to help generate both standard and nonstandard baud rates. If standard baud rates are desired, then an 11.0592 megahertz crystal could be selected. To get a standard rate of 9600 hertz then, the setting of TH1 may be found as follows:

$$\text{TH1} = 256\text{d} - \left(\frac{2^0}{32\text{d}} \times \frac{11.0592 \times 10^6}{12 \times 9600\text{d}}\right) = 253.000\text{d} = 0\text{FDh}$$

if SMOD is cleared to 0. Note that the frequency that is generated by the timer is 16 (SMOD = 1) or 32 (SMOD = 0) times the actual serial data communication rate. The UART must be fed a clock frequency that is much higher than the serial baud rate in order to be able to sample close to the center of each received bit. Clearly, a UART clock rate equal to the baud rate would not be "fine" enough to slice each serial bit into pieces.

Serial Data Mode 2—Multiprocessor Mode

Mode 2 is similar to mode 1 except 11 bits are transmitted: a start bit, nine data bits, and a stop bit, as shown in Figure 3.16. The ninth data bit is copied from bit TB8 in SCON during transmit and stored in bit RB8 of SCON when data is received. Both the start and stop bits are discarded.

The baud rate is programmed as follows:

$$f_{\text{baud2}} = \frac{2^{\text{SMOD}}}{64\text{d}} \times \text{oscillator frequency}$$

Here, as in the case for mode 0, the baud rate is much higher than standard communication rates. This high data rate is needed in many multiprocessor ap-

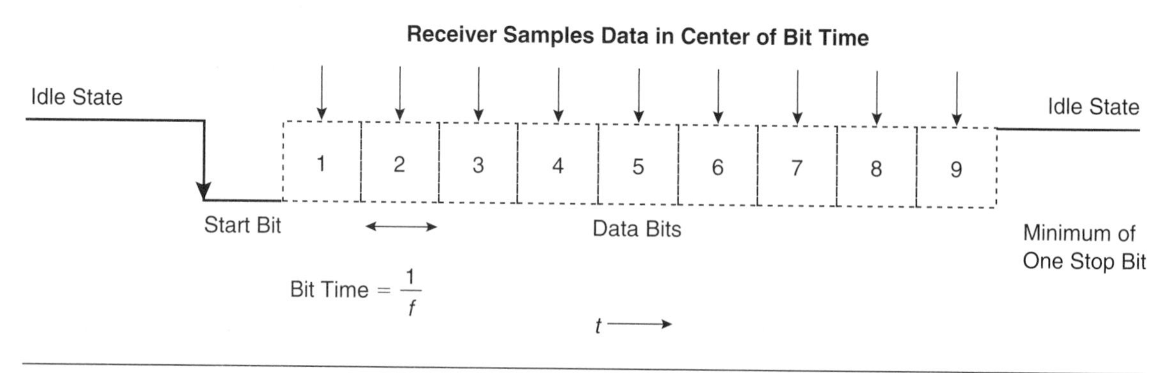

FIGURE 3.16 ◆ Multiprocessor Data Word

plications. Data can be collected quickly from an extensive network of communicating microcontrollers if high baud rates are employed.

The conditions for setting RI for mode 2 are similar to mode 1: RI must be 0 before the last bit is received, *and* SM2 must be 0 *or* the ninth data bit must be a 1. Setting RI based on the state of SM2 in the receiving 8051 and the state of bit 9 in the transmitted message makes multiprocessing possible by enabling some receivers to be interrupted by certain messages, while other receivers ignore those messages. Only those 8051s that have SM2 set to 0 will be interrupted by received data that has the ninth data bit set to 0; those with SM2 set to 1 will not be interrupted by messages with data bit 9 at 0. *All* receivers will be interrupted by data words that have the ninth data bit set to 1; the state of SM2 will not block reception of such messages.

This scheme allows the transmitting computer to "talk" to selected receiving computers without interrupting other receiving computers. Receiving computers can be commanded by the "talker" to "listen" or "deafen" by transmitting coded byte(s) with the ninth data bit set to 1. The 1 in data bit 9 interrupts all receivers, instructing those that are programmed to respond to the coded byte(s) to program the state of SM2 in their respective SCON registers. Selected listeners then respond to the bit 9 set to 0 messages, while all other receivers ignore these messages. The talker can change the mix of listeners by transmitting bit 9 set to 1 messages that instruct new listeners to set SM2 to 0, while others are instructed to set SM2 to 1.

Serial Data Mode 3

Mode 3 is identical to mode 2 except that the baud rate is determined exactly as in mode 1, using timer 1 to generate communication frequencies.

3.6 Interrupts

A computer program has only two ways to determine the conditions that exist in internal and external circuits. One method uses software instructions that jump to subroutines on the states of flags and port pins. The second method responds to hardware signals, called interrupts, that force the program to call a subroutine. Software techniques use up processor time that could be devoted to other tasks; interrupts take processor time only when action by the program is needed. Most applications of microcontrollers involve responding to events quickly enough to control the environment that generates the events (generically termed *real-time programming*). Interrupts are often the only way in which real-time programming can be done successfully.

Interrupts may be generated by internal chip operations or provided by external sources. Any interrupt can cause the 8051 to perform a hardware call to an interrupt-handling subroutine that is located at a predetermined (by the 8051 designers) absolute address in program memory.

Five interrupts are provided in the 8051. Three of these are generated automatically by internal operations: Timer flag 0, Timer flag 1, and the serial port interrupt (RI *or* TI). Two interrupts are triggered by external signals provided by circuitry that is connected to pins $\overline{\text{INT0}}$ and $\overline{\text{INT1}}$ (port pins P3.2 and P3.3).

All interrupt functions are under the control of the program. The programmer is able to alter control bits in the Interrupt Enable register (IE), the Interrupt Priority register (IP), and the Timer Control register (TCON). The program can block all or any combination of the interrupts from acting on the program by suitably setting or clearing bits in these registers. The IE and IP registers are shown in Figure 3.17.

After the interrupt has been handled by the interrupt subroutine, which is placed by the programmer at the interrupt location in program memory, the interrupted program must resume operation at the instruction where the interrupt took place. Program resumption is done by storing the interrupted PC address on the stack in RAM before changing the PC to the interrupt address in ROM. The PC address will be restored from the stack after an RETI instruction is executed at the end of the interrupt subroutine.

Timer Flag Interrupt

When a timer/counter overflows, the corresponding Timer flag, TF0 or TF1, is set to 1. The flag is cleared to 0 when the resulting interrupt generates a program call to the appropriate timer subroutine in memory.

Serial Port Interrupt

If a data byte is received, an interrupt bit, RI, is set to 1 in the SCON register. When a data byte has been transmitted an interrupt bit, TI, is set in SCON. These are ORed together to provide a single interrupt to the processor: the serial port interrupt. These bits are *not* cleared when the interrupt-generated program call is made by the processor. The program that handles serial data communication *must* reset RI or TI to 0 to enable the next data communication operation.

External Interrupts

Pins $\overline{\text{INT0}}$ and $\overline{\text{INT1}}$ are used by external circuitry. Inputs on these pins can set the Interrupt flags IE0 and IE1 in the TCON register to 1 by two different methods. The IEX flags may be set when the $\overline{\text{INTX}}$ pin signal reaches a low *level*, or the flags may be set when a high-to-low *transition* takes place on the $\overline{\text{INTX}}$ pin. Bits IT0 and IT1 in TCON program the $\overline{\text{INTX}}$ pins for low-level interrupt when set to 0 and program the $\overline{\text{INTX}}$ pins for transition interrupt when set to 1.

Flags IEX will be reset when a transition-generated interrupt is accepted by the processor and the interrupt subroutine is accessed. It is the responsibility of

7	6	5	4	3	2	1	0
EA	—	ET2	ES	ET1	EX1	ET0	EX0

The Interrupt Enable (IE) Special Function Register

Bit	Symbol	Function
7	EA	Enable interrupts bit. Cleared to 0 by program to disable all interrupts; set to 1 to permit individual interrupts to be enabled by their enable bits.
6	—	Not implemented.
5	ET2	Reserved for future use.
4	ES	Enable serial port interrupt. Set to 1 by program to enable serial port interrupt; cleared to 0 to disable serial port interrupt.
3	ET1	Enable timer 1 overflow interrupt. Set to 1 by program to enable timer 1 overflow interrupt; cleared to 0 to disable timer 1 overflow interrupt.
2	EX1	Enable external interrupt 1. Set to 1 by program to enable $\overline{INT1}$ interrupt; cleared to 0 to disable $\overline{INT1}$ interrupt.
1	ET0	Enable timer 0 overflow interrupt. Set to 1 by program to enable timer 0 overflow interrupt; cleared to 0 to disable timer 0 overflow interrupt.
0	EX0	Enable external interrupt 0. Set to 1 by program to enable INT0 interrupt; cleared to 0 to disable $\overline{INT0}$ interrupt.
		Bit addressable as IE.0 to IE.7

7	6	5	4	3	2	1	0
—	—	PT2	PS	PT1	PX1	PT0	PX0

The Interrupt Priority (IP) Special Function Register

Bit	Symbol	Function
7	—	Not implemented.
6	—	Not implemented.
5	PT2	Reserved for future use.
4	PS	Priority of serial port interrupt. Set/cleared by program.
3	PT1	Priority of timer 1 overflow interrupt. Set/cleared by program.
2	PX1	Priority of external interrupt 1. Set/cleared by program.
1	PT0	Priority of timer 0 overflow interrupt. Set/cleared by program.
0	PX0	Priority of external interrupt 0. Set/cleared by program.
		Note: Priority may be 1 (highest) or 0 (lowest)
		Bit addressable as IP.0 to IP.7

FIGURE 3.17 ◆ IE and IP Function Registers

the system designer and programmer to reset *any* level-generated external interrupts when they are serviced by the program. The external circuit *must* remove the low level before an RETI is executed. Failure to remove the low will result in an immediate interrupt after RETI, from the same source.

Reset

A reset can be considered to be the ultimate interrupt because the program may not block the action of the voltage on the RST pin. This type of interrupt is often called *nonmaskable,* because no combination of bits in any register can stop, or mask, the reset action. Unlike other interrupts, the PC is not stored for later program resumption; a reset is an absolute command to jump to program address 0000h and commence running from there.

Whenever a high level is applied to the RST pin, the 8051 enters a reset condition. After the RST pin is brought low, the internal registers will have the values shown in Table 3.4.

Internal RAM contents may change during reset; also, the states of the internal RAM bytes when power is first applied to the 8051 are random. Register bank 0 is selected on reset as all bits in PSW are 0.

TABLE 3.4

Register	Value (hex)
PC	0000
DPTR	0000
A	00
B	00
SP	07
PSW	00
P0 – 3	FF
IP	XXX00000b
IE	0XX00000b
TCON	00
TMOD	00
TH0	00
TL0	00
TH1	00
TL1	00
SCON	00
SBUF	XX
PCON	0XXXXXXXb

Interrupt Control

The program must be able, at critical times, to inhibit the action of some or all of the interrupts so that crucial operations can be finished. The IE register holds the programmable bits that can enable or disable all the interrupts as a group, or if the group is enabled, each individual interrupt source can be enabled or disabled.

Often, it is desirable to be able to set priorities among competing interrupts that may conceivably occur simultaneously. The IP register bits may be set by the program to assign priorities among the various interrupt sources so that more important interrupts can be serviced first should two or more interrupts occur at the same time.

Interrupt Enable/Disable

Bits in the IE register are set to 1 if the corresponding interrupt source is to be enabled and set to 0 to disable the interrupt source. Bit EA is a master, or *global,* bit that can enable or disable all of the interrupts.

Interrupt Priority

Register IP bits determine if any interrupt is to have a high or low priority. Bits set to 1 give the accompanying interrupt a high priority; a 0 assigns a low priority. Interrupts with a high priority can interrupt another interrupt with a lower priority; the lower priority interrupt continues after the higher is finished.

If two interrupts with the same priority occur at the same time, then they have the following ranking:

1. IE0
2. TF0
3. IE1
4. TF1
5. Serial = RI OR TI

For example, the serial interrupt could be given the highest priority by setting the PS bit in IP to 1, and all others to 0.

Interrupt Destinations

Each interrupt source causes the program to do a hardware call to one of the dedicated addresses in program memory. It is the responsibility of the programmer to place a routine at the address that will service the interrupt.

TABLE 3.5

Interrupt	Address (hex)
IE0	0003
TF0	000B
IE1	0013
TF1	001B
SERIAL	0023

The interrupt saves the PC of the program, which is running at the time the interrupt is serviced on the stack in internal RAM. A call is then done to the appropriate memory location. These locations are shown in Table 3.5.

A RETI instruction at the end of the routine restores the PC to its place in the interrupted program and resets the interrupt logic so that another interrupt can be serviced. Interrupts that occur but are ignored because of any blocking condition (IE bit not set or a higher priority interrupt already in process) *must* persist until they are serviced, or they will be *lost*. This requirement applies primarily to the level-activated $\overline{\text{INTX}}$ interrupts.

Software-Generated Interrupts

When *any* Interrupt flag is set to 1 *by any means*, an interrupt is generated unless blocked. This means that the program itself can cause interrupts of any kind to be generated simply by setting the desired Interrupt flag to 1 using a program instruction.

3.7 Summary

The internal hardware configuration of the 8051 registers and control circuits has been examined at the functional block diagram level. The 8051 may be considered to be a collection of RAM, ROM, and addressable registers that have some unique functions. See Table 3.6.

3.8 Questions

Find the following using the information provided in Chapter 3.

1. Size of the internal RAM.
2. Internal ROM size in the 8031.
3. Execution time of a single cycle instruction for a 6 megahertz crystal.

TABLE 3.6

Special-Function Registers

Register	Bit	Primary Function	Bit Addressable
A	8	Math, data manipulation	Y
B	8	Math	Y
PC	16	Addressing program bytes	N
DPTR	16	Addressing code and external data	N
SP	8	Addressing internal RAM stack data	N
PSW	8	Processor status	Y
P0 – P3	8	Store I/O port data	Y
TH0/TL0	8/8	Timer/counter 0	N
TH1/TL1	8/8	Timer/counter 1	N
TCON	8	Timer/counter control	Y
TMOD	8	Timer/counter control	N
SBUF	8	Serial port data	N
SCON	8	Serial port control	Y
PCON	8	Serial port control, user flags	N
IE	8	Interrupt enable control	Y
IP	8	Interrupt priority control	Y

Data and Program Memory

Internal	Bytes	Function
RAM	128	R0 – R7 registers, data storage, stack
ROM	4K	Program storage

External	Bytes	Function
RAM	64K	Data storage
ROM	64K	Program storage

External Connection Pins

		Function
Port pins	36	I/O, external memory, interrupts
Oscillator	2	Clock
Power	2	

4. The 16-bit data addressing registers and their functions.
5. Registers that can do division.
6. The flags that are stored in the PSW.
7. Which register holds the serial data interrupt bits TI and RI.
8. Address of the stack when the 8051 is reset.

9. Number of register banks and their addresses.
10. Ports used for external memory access.
11. The bits that determine timer modes and the register that holds these bits.
12. Address of a subroutine that handles a timer 1 interrupt.
13. Why a low-address byte latch for external memory is needed.
14. How an I/O pin can be both an input and output.
15. Which port has no alternate functions.
16. The maximum pulse rate that can be counted on pin T1 if the oscillator frequency is 6 megahertz.
17. Which bits in which registers must be set to give the serial data interrupt the highest priority.
18. The baud rate for the serial port in mode 0 for a 6 megahertz crystal.
19. The largest possible time delay for a timer in mode 1 if a 6 megahertz crystal is used.
20. The setting of TH1, in timer mode 2, to generate a baud rate of 1200 if the serial port is in mode 1 and an 11.059 megahertz crystal is in use. Find the setting for both values of SMOD.
21. The address of the PCON special-function register.
22. The time it will take a timer in mode 1 to overflow if initially set to 03AEh with a 6 megahertz crystal.
23. Which bits in which registers must be set to 1 to have timer 0 count input pulses on pin T0 in timer mode 0.
24. The register containing GF0 and GF1.
25. The signal that reads external ROM.
26. When used in multiprocessing, which bit in which register is used by a transmitting 8051 to signal receiving 8051s that an interrupt should be generated.
27. The two conditions under which program opcodes are fetched from external, rather than internal, memory.
28. Which bits in which register(s) must be set to make $\overline{\text{INT0}}$ level activated, and $\overline{\text{INT1}}$ edge triggered.
29. The address of the interrupt program for the $\overline{\text{INT0}}$ level-generated interrupt.
30. The bit address of bit 4 of RAM byte 2Ah.

4

Basic Assembly Language Programming Concepts

Chapter Outline

CHAPTER OBJECTIVES

On successful completion of this chapter you will be able to:

- Describe digital computer system organization and operation.
- Explain the function of the CPU and memory.
- Explain the difference between code and data memory.
- Describe the structure of a CPU instruction.
- Explain why assembly language programming is used.
- Write assembly language programs for the PAL CPU.
- Describe the assembly language programming process.
- Use flow chart elements.
- Understand the 8051 assembly language syntax.

4.0 The Forest and the Trees

The next four chapters of this book focus on programming, in *assembly language,* the 8051 family of microcontrollers.

Programming is both a *science* and an *art* and is similar in concept to writing fiction. The science of writing involves rules of grammar, punctuation, spelling, and the structure of a story; the art of writing is how the words are arranged by the author. To write fiction you must first have a set of words to use in writing your story. You must learn how to spell the words correctly and how to grammatically arrange the words. Then you learn to write acceptable fiction by studying common literary techniques, by reading many works of fiction by successful authors, and by much *practice* writing.

You learn to program in the same way you learn to write fiction. First, you must have a set of words, grammar, and punctuation to use in writing your program. Then you study common programming techniques, analyze example programs, and write *many practice* programs. As with fiction, the chances are that the more you practice writing programs, the better you will become. In one sense, programming is a set of rules that are followed in order to get a computer to accomplish some objective. Rules of programming are similar to rules of grammar and punctuation for an author. The art of programming is in how you *arrange* the words.

Programming any computer implies that some method can be found that will enable us, the humans, to get it, a lump of semiconductor, to do what we "tell" it to do. How to tell a lump of sand what to do is the purpose of this book. But, instead of getting bogged down in the root, trunk, branches, limbs, twigs, and leaf details of programming the 8051, let us step back from the tree and view the forest.

Before we get into the exact details of programming a particular computer, such as the 8051, we need to develop the concept of a *generic* computer. Generalization will carry us only so far, however. Computers are all alike, in that they all can produce the same results, and all different in the *exact* way they produce those results. Eventually we must pick a real example of the general model, such as the 8051, in order to do actual programming.

4.1 A Generic Computer

Question: What is a computer? Answer: It depends on whom you ask. Everyone's concept of a computer is not the same. Technically oriented persons tend to forget that, to the "man or woman in the street," a computer is mysterious and unknowable.

In its most basic sense, a digital computer is any machine that can be made to perform a series of binary operations. Common computer *operations* include storing and retrieving binary numbers, addition, subtraction, and Boolean logic. Computer operations may be done in any *sequence.* The arrangement of the *sequence of operations* that the computer is to perform is called a computer *program.*

Computer operations are carried out in a special binary circuit named the *central processing unit* or CPU. Binary numbers, which are needed by the CPU for the operations, are stored in other circuits called the *memory.*

Hardware Concepts: The Central Processing Unit

A computer CPU is made of semiconductor material (usually silicon) that is arranged to form millions of transistors. The transistors are connected together (as an integrated circuit) as gates that perform the Boolean logic operations of AND, OR, and NOT on voltages applied to them. Flip-flops are also made from the transistors and arranged into counters, latches, and registers. Signals applied to the CPU logic circuits are applied in synchronism with internal pulses, called a *clock.* The clock frequency is determined by an oscillator that normally uses a synthetic mineral crystal as a reference.

A CPU is an enormous collection of combinational and sequential logic circuits similar to those studied in logic design courses—it is both the heart and brains of a computer. The scale of the number of CPU circuits is staggering, involving hundreds of thousands of hours of human design work. The CPU provides the *timing pulses* that regulate the rest of the computer and the logical operation circuitry that carries out the program. The CPU is "told what to do," or programmed, by a *sequence of binary numbers.* The *total sequence* of the binary numbers is called, collectively, the *computer program*. The program, however, is not generally stored inside the CPU but is located in the computer system memory. The CPU is designed to automatically *get* the program sequence from memory. The programmer is responsible for the CPU's getting the *right* sequence in memory.

Hardware Concepts: Memory

Creating a complete computer system involves connecting the CPU to many other types of integrated circuits, most notably circuits that are solely designed to *store* binary data in electrical form. Binary bits may be stored as charges in a capacitor, in switched transistors, or in metal fuse patterns. Circuits that are meant to store binary data in electrical form are known as *solid-state* memories and are made up of millions of storage devices, or *cells*. Each cell holds a bit of a binary number and can interchange that bit with the CPU under the *control of the CPU timing circuits*. Typically, bits are transferred between the CPU and memory as positive-logic voltages.

ROM

Memory circuits may *permanently* store binary numbers and are known as *read only memory* or ROM. ROM memory cell contents may not be changed (written to) by the CPU, but they may be used (read) by the CPU. ROM is also called *non-volatile* because it does not depend on electrical power to store the numbers.

RAM

Memory circuits that store numbers that may be both read *and written to* by the CPU are given the name of random access memory or RAM.[1] RAM is also called *volatile* memory, because its contents are lost when power is removed. RAM types include static RAM (SRAM) and dynamic RAM (DRAM).

Software Concepts: Code and Data Memory

The function of memory is to store binary numbers for use by the CPU. The numbers in ROM and RAM are of two types: *program code* numbers and *program data* numbers.

Code Numbers

The bits at a memory address might be used by the programmer to instruct the CPU as to what *program operation* it is to perform. Another name for a program

[1]ROM may be accessed (read) in a random manner also, but the names ROM and RAM are common industrial usage. There are also hybrid electronic memory circuits that may be written to by the CPU but retain bits when power is removed. The hybrid memories are generically named *electrically erasable and programmable read only memories* or EEPROMs.

operation is an *instruction.* If a bit pattern in memory is intended to be used as a CPU instruction, it is generally made up of two parts. The first part of the instruction is the operational code, or *opcode.* The opcode is the "verb" part of the instruction and contains the binary code for a CPU *action.* The second part of an instruction is the *operand(s)* code. The operand code is the "noun" part of the instruction and describes *what* is to be acted on by the CPU. For example: "Go home" has "Go" as the opcode and "home" as the operand. A *complete* CPU instruction contains an opcode and, generally, operands. The complete sequence of program instructions *placed in memory by the programmer* are numbers known, collectively, as *code* and are stored in the area of memory known as *code memory.*

The CPU Fetches Code Numbers

The process of getting an instruction from code memory into the CPU is called *fetching* the instruction. Once an instruction is fetched by the CPU it is then *executed,* that is, the CPU does whatever the instruction tells it to do. The operation of fetching and executing instructions from code memory is *a completely automatic* CPU action, much like the involuntary actions of our own bodies. We have many automatic functions in our bodies, such as digestion, heartbeats, and hearing. We do not have to program our bodies' automatic actions; they are "transparent" to our thought processes. In a similar manner, the CPU will carry out the operation of fetching and executing program code with no direction from the programmer. All other CPU actions, other than fetching and executing instructions, are controlled by the *sequence of instructions* stored in code memory. The original code number fetched from code memory on start-up is the first opcode of the computer program. The CPU executes the operation that is indicated by the first instruction number and then fetches the next instruction. The process of fetching and executing program code continues until the computer is turned off or is otherwise halted.

We, the programmers, are responsible for making sure that the sequence of program instructions work toward some goal. The CPU will execute (run) the program *exactly* as we write it. We hear, via the media, that a "computer error" has caused some disruption of normal life. Computers rarely make errors. Computer programmers, however, often make errors.

Data Numbers

Some addresses in memory may be used by the programmer to be accessed by the CPU under program control. Addresses in memory that are meant to be read by the CPU, or written to by the CPU, are called *data* addresses. The contents of a data address is a data number.

Data numbers may be *variable,* meaning they may be changed as the CPU executes the program, or they may be *constant* and not change during program execution. In general, data is stored in RAM, because RAM can be altered by the CPU as the program runs. Fixed data, however, may also be stored permanently in ROM as a read-only source of data for the program.

If code and data occupy the same memory area (von Neumann) architecture, the CPU has *no* way of discerning between code and data numbers. If, by some mistake on the programmer's part, or by a rare mishap of nature, the CPU should *fetch data* when it was expecting code, it will decode and attempt to execute *whatever* operation is coded by the data number. Such accidents involving mistaking data for code are usually *attributable to the programmer.* Occasionally nature, in the form of electrical noise, will cause an erroneous fetching of data instead of code. Nature is very *rarely* to blame for program mistakes, particularly in well-designed personal computers. Your instructor will not be noticeably moved to pity if you claim, "The *computer* messed up *my* program."

Harvard architectures, which physically separate the code and data, as is done in the 8051, solve the data/code mix-up problem.

Computer Concepts: A Computer Model

A fundamental model of a computer is made up of a CPU and a ROM and RAM memory, as shown in Figure 4.1. Inspection of Figure 4.1 shows that the CPU obtains numbers stored in memory by *addressing* memory locations that contain numbers.

Memory Organization

Memory is organized as a group of *locations* where binary numbers may be *stored.* Each memory location is assigned a unique binary number, called the *address* of the location. The CPU uses addresses to find code and data numbers in memory. Memory addresses are assigned by the computer system designer who connects the CPU and memory together. Memory address numbers may begin at zero and go up to as large a number as the CPU can physically generate. Figure 4.1 illustrates a memory made up of over 64k addresses, from address 0000h up to address FFFFh.

Each individual storage element in a memory is a bit. A bit is one digit of a binary number and may be a 0 or a 1. (Binary numbers, and binary math, are discussed in Chapter 2.) Bits stored at each memory location can be organized into binary numbers of any length. The names given to binary numbers of various lengths are shown in Table 4.1.

Any group of bits, typically a byte, that can all be accessed by the CPU *at one time* are the *contents* of an address in memory.

Memory can be thought of as a huge college dorm building with hundreds of thousands of rooms. Each room (group of people) has a unique address in the building. Any group of people, in any room, can be accessed at the address of the room in which they are placed. Another analogy for computer memory is post office boxes. Each P.O. box has an address number. The contents of each box is mail. We access our mail by addressing our P.O. box.

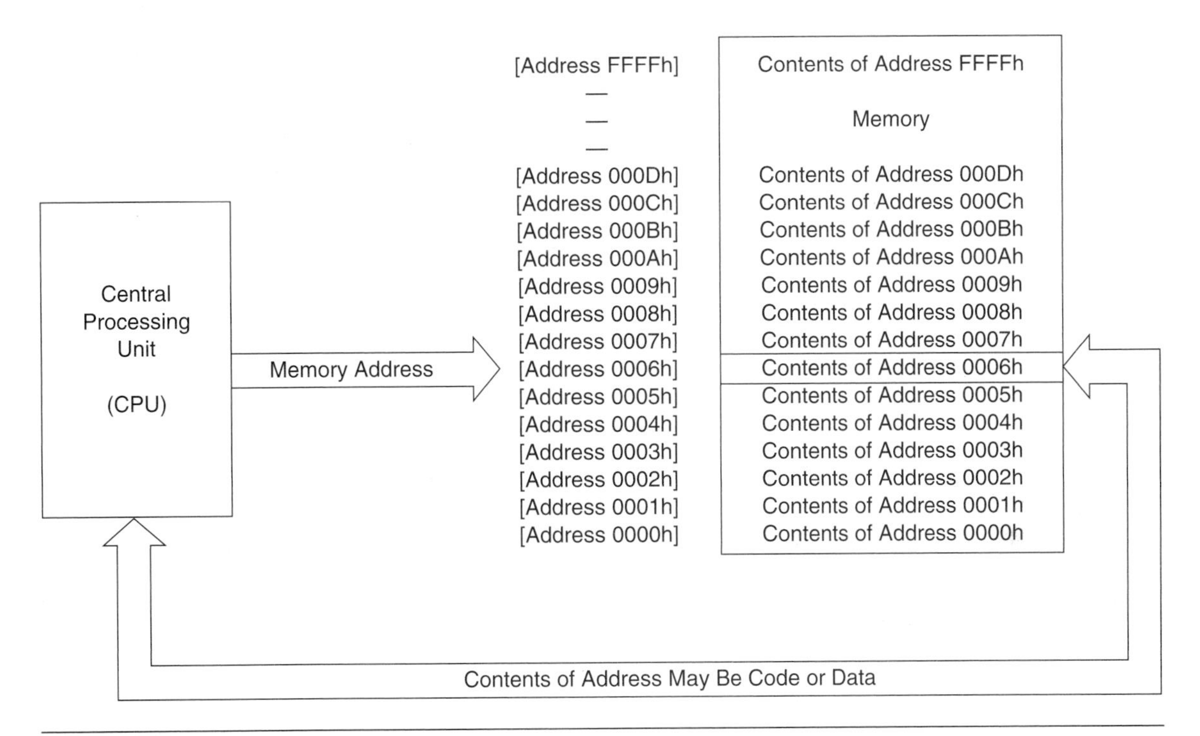

FIGURE 4.1 ◆ A Fundamental Computer Model

The CPU is physically connected to the memory by a number of electrical connections, called *address lines* and *data lines.* The CPU places signals on the address lines so that a unique address in memory can be accessed by the CPU. Information stored at that address is then accessed (fetched, read, or written) using the data lines. The name *data* lines does *not* mean that only data numbers may be accessed by the CPU. The contents of *any* memory address (code or data) are accessed using the data lines.

TABLE 4.1

Name	Length, in bits
Bit	1
Nibble	4
Byte	8
Word[2]	16

[2] Strictly defined, a word's length is the maximum number of bits a specific computer can access in *data* memory *during the execution of a single instruction.* The names used in Table 4.1 apply to the 8051 and are used in this book.

If the memory is RAM, then the CPU may obtain data bits from an address (read the data at the address) or place new data bits in an address (write data to the address).

If the memory is ROM, then the CPU may only read data bits or fetch code bits at an address, because of the read-only nature of ROM.

Keep in mind that reading or writing data at a memory address means that the CPU has been instructed *by the program* to access a certain memory address. Fetching a code number from an address in memory is an *automatic* CPU function.

The concept of code memory and data memory is central to programming a computer. In one sense, a computer can be thought of as a machine that transforms data numbers in memory under the control of code numbers in memory. The totality of code in memory is the computer program. All other addresses in memory can contain data for use by the program. *Very few programs use all of the available data memory,* however, and the unused data space contains whatever random bits (also known as *garbage*) are formed when the computer is powered up.

Computer Concepts: Starting Up

Since we, as programmers, are responsible for getting code and data properly organized in memory, how does the computer find the very first opcode, when it is first turned on? To find the first opcode, the CPU is *designed* to do so by the manufacturer.

When the CPU is first energized (also known as *resetting* or *booting*), it will *automatically* generate a *reset code memory address* that is designed into the CPU hardware and fetch the code bits at that address. The bits from the first address will be interpreted by the CPU as an instruction that the CPU is to perform. The CPU decodes the opcode, performs the instruction, and then obtains the next instruction from another code memory address.[3]

Computer Concepts: Machine Language

A computer instruction is a group of bits that controls the CPU so that it will perform a desired CPU operation. Most CPUs are designed to perform hundreds of different operations. Instruction codes are located in the code section of memory and are *placed there by the programmer.* Every CPU instruction requires a *unique* bit pattern to distinguish it from another instruction. The CPU instructions are said to be *coded,* because one unique instruction can be distinguished from another unique instruction only if each instruction is encoded as a different pattern of bits.

[3]Other microprocessor designs, notably those of Motorola, fetch the *address* of the first instruction from the reset address and *then* fetch and execute the first instruction from that address.

Another name for code is *machine language,* because the codes are the only bit patterns (language) to which the CPU (a machine) can respond. *Every* computer program must be coded, ultimately, in machine language. Machine language is understood perfectly by the CPU but is *not* suitable for human use.

Software Concepts: Assembly Language Mnemonics

Computer operations are directed by binary-coded machine language instructions fetched from code memory. Although the code bits fetched may make perfect sense to the CPU, they are not readily understandable by humans. The gap between humans and computers is bridged by human language. All computers, from calculators to supercomputers, are programmed using some sort of human *computer language.* A computer language is any human descriptive technique used by the programmer to get the computer to produce results. *Assembly language* is the human language most closely tied to machine code.

A human language that uses a *mnemonic* scheme is used for machine code programming. A mnemonic is defined[4] as some plan (a word, rhyme, or the like) designed to aid the (human) memory. As applied to assembly language programming, mnemonics are words that "sound like" an equivalent machine code operation.

Assembly language is usually written in instruction mnemonics that have been invented by the CPU *manufacturer.* Once a program has been written in mnemonics, it is translated into machine code by a process known as *assembling* the program. Assembling machine code, which was once done manually, is now done using a computer. Using the computer to assemble a program leads to the need for a text editor program, an assembler program, and a program that operates the computer.

4.2 The Mechanics of Programming

Getting Instructions into Code Memory

The task facing a computer programmer is, fundamentally, how to place the right set of instructions in code memory in order to get the CPU to solve the problem at hand. Some way must be found to translate human thoughts into bits inside of semiconductor code memory.

Personal computers are equipped with ROM code memory that has been formed and inserted permanently into the computer so that the computer is equipped with some minimum program when it is first turned on. The CPU, under the direction of ROM memory code, can then read the operating system program from a disk and place it in RAM. The operating system program then is given control of the computer and awaits orders from the computer program-

[4] *Computer Dictionary* (Redmond, WA: Microsoft Press, 1991), p. 229.

mer. Using the operating system program, the programmer can direct that an application program (*user code*) be placed in RAM and executed. The application program may help the programmer write programs, in which case it is often called a *utility program,* or it may be the programmer's own creation.

Many types of utility programs exist to help get human thoughts converted into code memory. The only fundamental difference between the humblest pocket calculator and the mightiest mainframe is the means by which their software and hardware convert thoughts, and language, into code bits.

Computer Languages

Humans think and communicate using languages (both verbal and more subtle means). Computers communicate using voltages and currents. An ideal way to convert from thoughts to voltages would be to issue instructions to the computer by speaking to it exactly the way one would speak to another person. A less desirable method, but the one most used today, is to type instructions to the computer on a keyboard.

The complexity of the language used to instruct the computer establishes the level of communication between human and computer. The simplest possible computer language is that used to get results from an inexpensive pocket calculator. We can program the calculator using no more than a few keys that have the programming language printed on separate keys (+, −, =, 0 – 9).

High-Level Languages

At a more advanced level than a pocket calculator, we can program a mainframe computer in a high-level language using a keyboard and a CRT. Interestingly, we need no knowledge of the internal workings of the calculator or the mainframe computer to use either one. High-level languages are said to be *machine independent* because the internal circuits of the computer used to program in a high-level language are not part of the high-level language. Examples of high-level languages include BASIC, FORTH, Pascal, C, and FORTRAN.

Note that the high-level language names are generic. Many versions of each high-level language exist, depending on what brand of computer they will program, but the language is the same. A high-level language is not tied to any particular processor type and no details of CPU construction are needed by the programmer. For instance, a Pascal program can execute on a mainframe, a PC, or an Apple with equal results. The Pascal programmer does not have to be concerned with what particular computer will be used to execute the Pascal program. Clearly, some utility programs must take the Pascal program and convert it into the machine language of the particular CPU in use, but such conversion is not the concern of the Pascal programmer.

High-level languages are often called *transportable,* that is, the same high-level language programs may be used on many makes of computer, all of which have different processors.

Assembly Languages

Assembly language programming, unlike high-level language programming, is tied very closely to the *physical* makeup of the CPU. Assembly language programs use the internal circuits (registers) of the computer as part of the assembly code language. The programmer must have a very *detailed* knowledge of the arrangement of a particular CPU's internal circuits in order to program that CPU in assembly language code. Examples of assembly-level languages include those associated with the 8051, 8086, 6502, 68000, and Z80 CPUs.

Assembly languages are *specific* to a unique microprocessor CPU type and are intimately bound up with the *exact construction* details for that particular microprocessor. An assembly language programmer *must know* the internal makeup of the CPU in order to program it. CPUs are not interchangeable. For instance, assembly language programs written for an Intel 8086 will not work for an Intel 8051. Assembly code programs are bound to a particular CPU type and cannot be used for a different CPU type.

Assembly-level languages are not transportable; programs written for one type cannot be used on another. In general, it is quicker to program in a high-level language than in assembly language because detailed knowledge concerning the hardware arrangement and operation of the processor is not needed for a high-level language. Detailed knowledge of the CPU, however, is essential for programming in assembly language for that processor.

Why Use Assembly Language?

There are at least five reasons to write computer instructions in assembly language:

1. To speed computer operation. Programs written in assembly language can be stored compactly in code memory, and less time is spent fetching the code. High-level languages are converted to code by utility programs named *compilers*. Because of the general nature of high-level languages, the compilers often produce excess or *overhead code*.
2. To reduce the size of the program. Assembly language requires no extra overhead code because the assembly language programmer is aware of the exact needs of the program for any given situation.
3. To write programs for special situations. Often, particularly when dealing with machine control, no standard programs (named *drivers*) exist. Robot arms and antilock brakes, for instance, have no standard drivers. It is generally more efficient to write nonstandard driver programs in assembly code. Also, when speed of response is critical, assembly-coded programs execute rapidly because of the exact fit of program code to task requirements.
4. To save money. Small computer systems, such as those that are embedded inside other machines, are often produced in large numbers. Reducing code size also reduces the cost of associated ROM chips.

5. To better understand how computers operate. In order to fully understand what is going on "under the hood" of the CPU, you should learn to program the CPU in assembly language.

Speed, size, and uniqueness are advantages assembly language programs offer over high-level languages.

Programs written in high-level languages are converted to machine code by a type of assembler named a compiler. Compilers take the high-level language and convert it to the machine code for the particular CPU in use. The resulting machine code program often requires large amounts of memory storage because each high-level instruction is compiled into many, often redundant, machine code instructions. High-level languages, after they are compiled into machine code, usually take longer to run and occupy more memory space than do assembly language programs. The excess high-level language machine code instructions take extra time to be fetched and executed by the processor. The time needed to run a high-level program is greater than the time needed to run a nonredundant assembly code program.

Execution time and program size may be of little interest if the application is, for instance, a word processing program. Word processing programs running on a personal computer have enormous amounts of memory at their disposal and the speed of program response far exceeds the response time of the user. It makes sense to write a word processing program in a high-level language.

Execution time and program size are of great interest, however, if the application is an automatic braking system on an automobile. Response time of an ABS program is in milliseconds, and the program must fit into a small ROM. Machine control programs are often written in assembly code where speed of response, and compact program size, are needed.

Cost can also be a factor in choosing assembly code over a high-level language. Pocket calculators do not require fast response time from internal code. Cost is crucial, however, in the calculator market, and the size of the program used to operate the calculator is costly in terms of memory chips. Program size restraints apply to any application in which cost is paramount, which includes most home appliances, computer peripherals, automotive electronics, and toys. Finally, size sometimes plays an important part in the decision to use assembly language rather than a high-level language. Applications that require the ultimate in miniaturization, such as portable phones, space vehicles, and many weapons, require that the controlling computer be as physically small as possible. Again, to shrink program size and memory chip requirements, assembly code is needed.

4.3 The Assembly Language Programming Process

Assembly language programming requires the use of application programs, known as *utilities* or *tools,* to get our assembly language programs converted into code memory bit charges. Assembly code utility programs can range from very crude board-level *monitors* to very sophisticated personal computer *assembler* programs. We shall assume that our assembly language programming is to be done using a personal computer. The personal computer enables us to speedily write and test our assembly language programs.

To get your thoughts converted into code memory charges, you must know how to use a set of programs that have been written to help ease the way. The following programs are essential to your success:

1. An *operating system* program, which controls the operation of the personal computer used for the entire programming process. Windows is the operating system program used by many PCs.
2. A *word processing,* often called a *text editor,* program. Programs in assembly language mnemonics are written using the text editor and stored on a disk as files that, normally, end in the extension *.A***. The .A** files are intended for the use of another program called the assembler program. Any text editor that can produce an ASCII (text) file is suitable for writing assembly language files.
3. An *assembler* program, which takes the assembly language program file and *converts* it to a machine code *.OBJ* file (the type we shall use most often). The assembler converts the ASCII mnemonic text file into an .OBJ file that contains machine code instructions to the CPU, in binary form.
4. A *testing* program, which lets you run and test your program under controlled conditions. Testing your program is the *most important step* of the programming process. To be able to test your program you must be able to execute each instruction and see the results. Utility programs that allow the user to test programs are called *debuggers,* or *simulators.*

4.4 The PAL Practice CPU

Let us, for the sake of practice, invent a very small computer CPU called the *Practice Assembly Language,* or PAL. We shall invent a set of binary-coded PAL instructions.

We shall use the PAL for practice programming by using only very crude (manual) means of converting PAL assembly language instructions into machine code. At the dawn of the computer age, program assembly was done by hand. Manual assembly methods have the distinct advantage of providing the user with a clear picture of exactly what happens when source (.asm) files become machine code. Manual assembly exercises usually provide the immediate desire to learn to use an assembler program such as our assembler, A51.

Figure 4.2 shows the details of the PAL *internal registers* and the RAM and ROM memory addresses that PAL may access. As shown in Figure 4.2, PAL contains three internal registers:

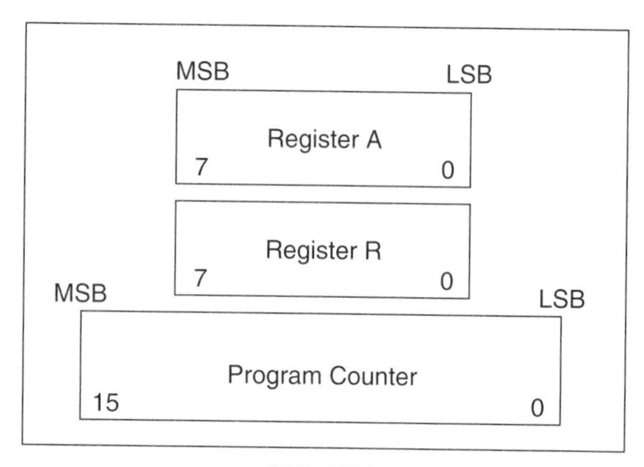

Address FFFFH	Contents of Address FFFFH
Address FFFEH	Contents of Address FFFEH
———	PAL 64K ROM Memory
Address 0001H	Second Code Byte
Address 0000H	First Code Byte

PAL CPU Fetches Code from Reset Address 0000H Up to End of Program

PAL CPU Resets to This Address

PAL ROM Code Memory

Address FFH	Contents of Address FFH
———	
Address 01 H	Contents of Address 01H
Address 00H	Contents of Address 00H

Program Data

PAL RAM Data Memory

FIGURE 4.2 ◆ PAL CPU and Memory

Register A: Performs all operations in the CPU

Register R: Stores numbers temporarily

Register PC: Holds the address of the next instruction to be executed in code memory.

Registers A and R can hold 8 bits (1 byte) of binary data, and register PC (the *Program Counter*) can hold a 16-bit address number. PAL is called an *8-bit computer* because the *working* registers A and R can hold a 1-byte number. PAL register PC limits the number of code byte addresses to 64K because it can hold code addresses from address 0000h to address FFFFh.

When PAL is reset, the PC register is set to 0000h by internal circuits. Every time an instruction is fetched from code memory, PC is indexed (counts up) to point to the address of the *next* instruction. Note that the PC points to the address of the next instruction *before* executing the current instruction it has just fetched. PAL fetches the first instruction at code memory address 0000h and proceeds exactly as directed by the programmer.

PAL Instructions

PAL has been equipped with nine instructions. These are the only instructions possible for PAL. The instruction mnemonics are shown in Table 4.2. In the table, the instruction mnemonic and data locations are listed on the left; the opcode for the mnemonic (in hexadecimal) with the instruction operand and the CPU action for the instruction are listed on the right.

TABLE 4.2

Instruction Mnemonic	Data Location	Machine Code (hex)	CPU Action
MOV	A, #num	3Dh, num	Move the 1-byte number num to A. num can be any constant number from 00h to FFh.
MOV	A, addr	3Eh, addr	Move the data byte contents of RAM address addr to A. addr is any 1-byte address number in RAM from 00h to FFh.
MOV	addr, A	3Fh, add	Move the data byte number in A to the contents of RAM address addr.
MOV	R, A	4Ch	Move the contents of A to R.
MOV	A, R	4Dh	Move the contents of R to A.
ANL	A, R	F3h	Logical AND the contents of A and R. Results of the AND operation in A.
ORL	A, R	9Ah	Logical OR the contents of A and R. Results of the OR operation in A.
CPL	A	67h	Logical complement of the contents of A.
GO	addr	77h, addr	Set the PC equal to the 2-byte number addr. Note that this action forces the PC to fetch the next instruction from code address addr.

TABLE 4.3

Instruction	Comment
mov a, #01h	;put the constant number 01h in A
mov 30h, a	;store contents of A in memory location 30h
mov a, #09h	;put the constant number 09h in A
mov 31h, a	;store contents of A in memory location 31h
mov a, #0Ch	;put the constant number 0Ch in A
mov 32h, a	;store contents of A in memory location 32h
go 0000h	;set PC back to start of program

Note that the *single* instruction mnemonic MOV can be encoded into one of *five* opcodes, depending on the *location of the data* to be moved. Note also that a constant number is indicated by the pound (#) sign, whereas an address number has no pound sign. (Once you have used assembler programs you will realize that there must always be some way for the assembler to be able to tell the difference between constant numbers and address numbers.)

We now have a set of instructions for PAL that are capable of performing some common CPU operations. Our example program will consist of placing the numbers 1, 9, and C in data memory locations 30h to 31h. Since we know that PAL resets to 0000h in memory, our program must start with the first opcode at memory location 0000h.

The first step is to write our program using the instruction mnemonics, with appropriate comments beside each instruction as a reminder of what each is to do. See Table 4.3.

Note that we do *not* have to write our program in capital letters. Instruction mnemonics are often printed as uppercase letters in manuals, but the assembler (in this case, ourselves) will accept lowercase letters. Numbers are all expressed in hexadecimal by using the letter *h*. If we do not use the *h* we may mistakenly use decimal numbers when we wish to use hex numbers. Our comments all *begin with a semicolon* to remind us that they are comments and not part of the instruction.

Now that the program is written in assembly language, we must assemble it into pure hexadecimal machine code and place each instruction in the right place in code memory. To assemble our program we must convert each line of our program from assembly language to its equivalent hex code. Our program is assembled as shown in Table 4.4.

We can see, from our manually assembled program, that a very short program takes 15 bytes of code memory. Several items of importance are illustrated by our example program:

- Manual and automated assembly (done using assembler programs) both use a text file as an input, and produce a hex file as an output.

TABLE 4.4

Code Memory Address	Contents of Code Memory Address	Instruction
0000h	3Dh	mov a, #01h
0001h	01h	
0002h	3Fh	mov 30h, a
0003h	30h	
0004h	3Dh	mov a, #09h
0005h	09h	
0006h	3Fh	mov 31h, a
0007h	31h	
0008h	3Dh	mov a, #0Ch
0009h	0Ch	
000Ah	3Fh	mov 32h, a
000Bh	32h	
000Ch	77h	go 0000h
000Dh	00h	
000Eh	00h	

- The program must begin at 0000h and proceed sequentially. We must make sure that the correct hex code byte is in each code address.
- Instructions are stored with the opcode first, then the data operands (constant numbers, or addresses) are stored next in code memory. Note that the register names do *not* have to be stored in code memory, because the opcode contains that information.
- Any 2-byte number, such as the one used with the GO instruction, is stored *low* byte first in code memory, then *high* byte next in code memory. This is not obvious from our example, because both bytes are 00h. If, however, the GO instructions had been:

```
go,1234h
```

the instruction would be stored in code memory as shown in Table 4.5.

TABLE 4.5

Code Memory Address	Contents of Code Memory Address	Instruction
000Ch	77h	go 1234h
000Dh	34h	
000Eh	12h	

TABLE 4.6

Instruction	Comment
mov a, #01h	;put the number 1 in A
mov r, a	;store the number in R *** change ***
mov 30h, a	;store in memory location 30h
mov a, #09h	;put the number 9 in A
mov 31h, a	;store in memory location 31h
mov a, #0Ch	;put the number C in A
mov 32h, a	;store in memory location 32h
go 0000h	;loop back to the beginning

There must be some instruction that lets the programmer limit the code addresses available to the program. PAL uses the GO instruction to limit the example program to occupy code addresses 0000h to 000Eh. If we did not have a GO instruction at locations 000Ch – 000Eh in code memory, the CPU would merrily keep on incrementing the PC by fetching every sequential byte in memory up to address FFFFh (and then *roll over* the PC back to 0000h.) The CPU has no way of knowing what is code, data, or just random garbage numbers. The *programmer* has the responsibility to make sure that the CPU stays *within code memory* and cannot possibly find its way to data or garbage.

Manually assembled programs suffer from two basic problems. First, the assembly process is very tedious and is prone to many mistakes when the human assembler puts in the wrong opcode, or forgets to number code memory addresses correctly. Second, should any changes be required in the program, the entire program must be re-assembled. To see the impact of the second point, take the original program and make the small change shown in Table 4.6.

Now, re-assemble the program. Every byte, after the first two, must be assigned new addresses in code memory!

Manual assembly, for even a small computer such as the PAL, is informative but soon becomes tedious. Manual assembly of programs for real CPUs, such as in the 8051, is totally impractical.

4.5 Programming Tools and Techniques

As noted earlier, a digital computer is a sequential circuit that fetches a program instruction from code memory, decodes the instruction, and then executes it. The computer instructions are placed in code memory by the programmer with the aid of utility programs that have been written by others to facilitate this chore.

TABLE 4.7

a. Learn about CPU architecture (internal register arrangement).
b. Learn a set of CPU assembly language instructions.
c. Learn how to use a personal computer and its host operating system.
d. Learn how to use a text editor program.
e. Learn how to use an assembler program.
f. Learn how to use a debugger program.
g. Practice writing and running small programs until proficient.
h. Add new assembly language instructions and master their use.
i. Continue to write larger and more complicated programs.

The challenges facing a beginning programmer are extensive because of a complete unfamiliarity with any of the procedures and tools that must be mastered before the first, simple program is run. A sequence of learning and action items found on the road to the first programs is listed in Table 4.7.

The list appears to be daunting, but the difficult parts are only the last three items.

Programming can be reduced to procedural steps that can be applied to any problem by a well-trained individual. There is still much art in the process though, and to master the art requires more than formal education and rote learning. Some individuals possess a unique programming talent that lets them "see" ways of doing things that are not obvious to most of their peers. People who possess extraordinary programming skills, whether through natural insight or (as is more often the case) by dint of long, hard work, are often referred to as *wizards.*[5] I cannot guarantee that you will become a wizard, but I do guarantee that you can become a competent programmer.

Understanding Assembly Language Syntax

8051 assembly language is a subset of the English language, but the rules of spelling, grammar, and punctuation are more *rigidly* defined: There are no alternate spellings in assembly language, and no leeway in how a comma is placed. Assembly language is rigid because *assembler programs* cannot tolerate any deviation from the way instructions are defined. The *rules* that apply to a particular assembly code language, and a particular assembler, are grouped under the heading *syntax.* Assemblers are very picky about syntax.

For instance, we might say to a friend, "I ain't got no money," or "I'm broke," or "I find myself temporarily financially embarrassed." In all cases we should be understood. Not so with assembly language. Each CPU assembly language,

[5]*Computer Dictionary* (Redmond, WA: Microsoft Press, 1991), p. 368.

TABLE 4.8

mov r0,r1	(Intel 8051)
ld a,b	(Zilog Z80)
move.w a1,d1	(Motorola 68000)
mov ds,ax	(Intel 8086)
txa	(Rockwell 6502)

and each assembler program for that language, admits only one way of expressing a particular action. For example, the assembly language instructions that move data from one internal register to another in several popular microprocessors are written as shown in Table 4.8.

Clearly, each instruction, which has the same result for each computer architecture used, is expressed in very different ways. One CPU uses the term mov and another uses the term ld to get the same result: copy data from one register to another. Each instruction listed in Table 4.8 uses a different mnemonic (memory aid) for the same CPU action. Each mnemonic reminds us of what the instruction does when it is executed by the computer CPU. For example, the Z80 "loads" (ld) data, the 68000 "moves" (move) data, and the 8086 "moves" data also (mov) but with a different mnemonic spelling.

Different manufacturers have adopted different mnemonics for the same CPU action for various competitive reasons. Originally, when almost all programs were written in assembly language, it was thought that the programmer might become "married" to the mnemonics, and thus would be loath to change over to a competitor. Legal questions also arose, because instruction mnemonics could be copyrighted and could not be used by another company without permission. The industry has matured. Instruction mnemonics are no longer subject to copyright law, and most programming is now done in semi-generic, high-level languages. Microprocessor manufacturers have remained with the original forms of their early mnemonics, however, and thus we have 8051 mnemonics that can trace their roots to the first Intel primitive 4-bit microprocessor, the 4004.

Understanding the Assembler Program

Assembly language programming, in the bad old days, was done entirely in a manual fashion, as we saw earlier for the fictitious PAL CPU. An outline of a typical manual assembly code process is listed in Table 4.9.

The process listed in Table 4.9 is very tedious, especially when new lines of code have to be inserted into the program and all of the addresses of each line of code reassigned. Today, with the availability of inexpensive personal

TABLE 4.9

a. Write the program, on a pad of paper, using the instruction mnemonics.
b. Convert each instruction to its hexadecimal code, using a code book that lists each instruction mnemonic and its hexadecimal code.
c. Assign each code byte an address in code memory.
d. Use a paper tape punch to record the program code bytes.
e. Enter the program into computer code memory via a paper tape reader under the control of a reader utility program.
f. Find the mistakes, re-write and re-code the program, and re-enter it.

computers and computer software, the process has been greatly automated. The process for writing assembly language programs is now as shown in Table 4.10.

Writing the program is no easier using automated means. But the process of converting assembly language instructions to hex codes, assigning addresses, loading the program, and changing the program is much, much faster using automated assembly.

The increase in the pace at which programs can be edited and assembled has not come without a price, however. Now, in addition to having to learn the instruction mnemonics for a CPU, the budding programmer must also learn to use a personal computer, a text editor, and an assembler program.

Assembler Directives

An assembler program is a program like any other, one that has its own programming language instructions that must be understood before any programs can be written.

Assembler program concerns deal with such things as setting the PC contents, assigning names to numbers, and defining fixed numbers. The title given to an assembler program language instruction is a *directive*.

TABLE 4.10

a. Write the program in instruction mnemonics, using a computer and text editor.
b. Store the edited assembly language program in a disk file.
c. Use a program, called an *assembler*, to convert the assembly language file stored on disk to a second disk file. The second disk file contains binary equivalents of the assembly language file that the computer can execute (machine language instructions).
d. Load the assembled machine-language program from disk into computer RAM and test the program using a debugger program.
e. Find the mistakes, re-edit the program, re-assemble, load, and test again.

Appendix B discusses the assembler used in this book (A51) and some of the more commonly used A51 directives.

Understanding the Problem to Be Solved

The problem at hand, whether assigned by the teacher as a classroom exercise or presented by an actual application, must be *fully* understood by the programmer before any programs are written. At the problem *definition* stage, it is not important to know which computer is to be used; any computer type could do the job. But the most powerful, multi-byte, multi-megahertz computing wonder will do a poor job if the program does *not* solve the problem. There are no shortcuts here; you must persist in asking questions of the teacher or the customer, or nature, until both you and the end user are in agreement as to *what* is to be done.

Many times, especially for complex and expensive applications, a formal, written definition of the problem (a contract) is formulated and agreed on by all parties. The problem definition phase is the *most important* part of the programming process.

Designing the Program

Once the problem is known, then the programmer can begin to lay out a sketch or overall plan of how to solve the problem. The plan is called an *algorithm*. An algorithm is any scheme, such as a list of actions or a diagram, by which the programmer is guided in solving the problem. It can be very precise or quite general. For instance, the algorithm for finding the square root of a 4-word number can be exactly specified, and in great detail. The algorithm for adding together several sets of numbers in memory can be quite general.

A common technique used to *document* an algorithm are diagrams called *flow charts*. The history of flow charts goes back to the dawn of the computer age, when they were considered essential in the programming process. Flow charting very large programs soon became unproductive, however, because of their ever-increasing complexity, so large programs were divided into independent pieces called *modules*. Each module is then written using standard programming techniques named *structures* and is documented using written *comments*. Almost all of the example programs in this book are short, and comments will be used to document them.

For those programmers who wish to review flow chart elements, the next section explains flow chart basics.

Flow Charts

Diagrams of programs are called *flow charts* because they visually show the way a program operates or "flows" as it runs. These charts are not unique to computer programming and are used in many other fields, such as business and construction planning.

Flow charts can be a valuable aid in visualizing programs. Let us concentrate on concepts before becoming engrossed in opcode details. As mentioned previously, flow charts have been supplanted by modular programming as a technique for organizing and documenting programs. Flow charts are, however, very useful learning tools for the beginning programmer.

Flow Chart Elements

For our purposes, we shall use five elements, or parts, to draw a flow chart. These are: the box, the bubble, the diamond, the line, and the arrow. The five flow chart elements are shown in Figure 4.3.

Actions

The *action box* is used to contain statements of actions that the computer program will execute. The box is entered from the top and exited from the bottom. The action statements may be data moves, math operations, or any other instructions that denote a program action. Figure 4.4 shows several boxes that contain action instructions for the 8051 microprocessor.

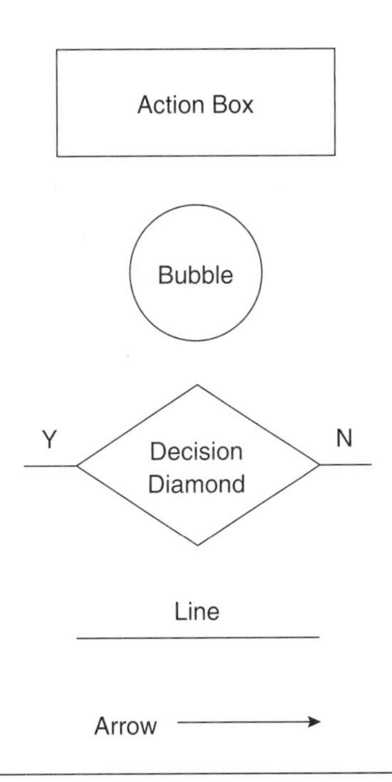

FIGURE 4.3 ◆ Flow Chart Elements

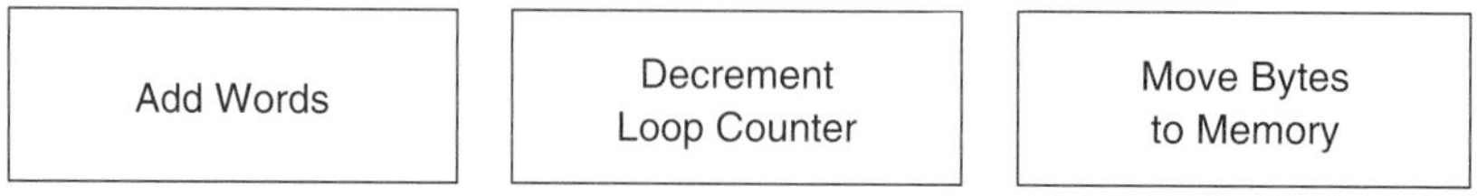

Add Words

Decrement Loop Counter

Move Bytes to Memory

FIGURE 4.4 ♦ Action Box Elements

Decisions

Decisions are shown on a flow chart by using the *decision diamond* element. The diamond is entered from the top corner and exited through any one of the other three points of the diamond: the left, right, or bottom corner. Decisions are indicated by a flag or condition statement. The usual answer to a decision is yes (Y) or no (N). Normally only one decision per diamond is shown. Figure 4.5 shows several diamonds that indicate 8051 decision instructions.

Flow

Program flow is indicated by connecting the boxes and diamonds of a program together with *lines* and *arrows.* Each line begins at a box, or diamond, and has an arrow at the termination end that shows the direction the program is to go. Traditionally, the flow of a program begins at the top of the page and flows down.

Bubbles

Bubbles, or small circles, are used to begin and end programs. They are also used as markers to indicate where a line leaves an area and where it re-enters an area. Using bubbles as markers tends to make the diagram more readable when there are many lines of flow, or when the program leaves one page and resumes on another. Figure 4.6 shows a complete flow chart for a simple program.

We should again note here that large programs are not usually put into flow chart form. The advantage of flow charting is to show, hopefully on a single page, a single simple program. Flow charts are also useful for diagramming concepts within a larger program. The trend in programming today is to write programs in modular form and to copiously document and comment those modules.

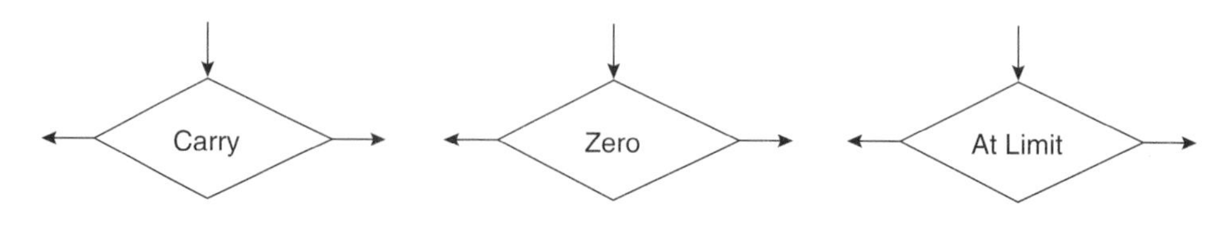

FIGURE 4.5 ♦ Diamond Decision Elements

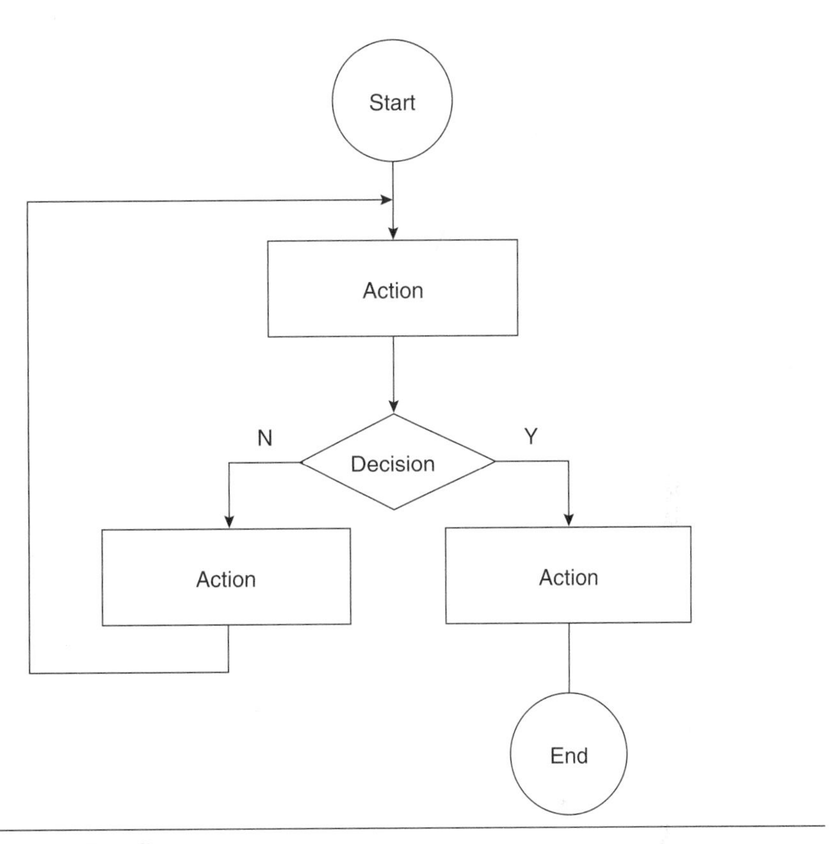

FIGURE 4.6 ◆ A Flow Chart

Writing and Testing the Program

Once the program algorithm has been designed, the actual assembly language instructions for the type of CPU that is to be used are written, assembled into instruction codes, and placed in the computer. The means by which the code is placed in the computer, and tested, varies with the computer hardware and software.

Testing Programs Using a Personal Computer

If the program is to be run on a personal computer, then the text editing, assembly, and testing of the program proceeds without leaving the keyboard. Personal computers are equipped with a master program, the operating system, which operates the assets of the computer under user command. Personal computers based on the 8686 family of processors generally use an operating system called Windows. The programmer, using the operating system as a tool to load and run programs, proceeds to test and debug an 8051 program as shown next:

1. A text editor is used to edit the program, saving it on a disk as a pure ASCII file. The saved program is referred to as the *source* file.
2. The assembler program assembles the source program as directed by the programmer. The assembler saves the assembled program on disk as a machine-language file. Any errors found during assembly are reported by the assembler program to the programmer on the monitor.
3. The syntax errors in the source file are corrected by the programmer and the program re-assembled. A debugging or simulation program tests the machine-language file under user control.
4. Any conceptual errors (also called *run-time* errors) are found during debugging, and the entire process repeated until the program is seen to run as intended.

Debugging programs usually permits the programmer to execute the test program one instruction at a time. After each step, the programmer can visually inspect CPU register and system memory contents for proper test program operation. The debugger can also run the program full speed, stopping only at programmer-specified places in the program, called *breakpoints.*

Testing Programs on a Single-Board Computer

If the test program is to be exercised on a small, board-level computer that has no disk drives or operating system, then the debugging procedure is more cumbersome. Programs are written and assembled using the PC, and then transmitted, via a serial data link, to the computer board using the PC as a terminal. The process of transmitting the program to the board is called *downloading* the program.

The board will have a small debugging program, called a *monitor* program, in a ROM memory on the board, which controls the downloading and subsequent communications with the PC. The monitor will generally have the same general debugging abilities as the PC-based debugging program. Debuggers tend to be menu-driven and easy to use because the instructions are always shown on the screen. Monitors tend to require that the user remember certain keystrokes for each desired result. Board-level monitors are very inexpensive and are usually supplied by the board manufacturer.

More advanced board-level debuggers are called *in-circuit emulators* or ICE debuggers. ICE debuggers involve plugging special hardware into the PC and also into the CPU socket on the board. ICE debuggers tend to be costly, in the range of $800 to $15,000.

Using a simulator, such as the one used in this book, becomes natural after a short exposure to its command structure. Testing for (and finding) all of the errors in the program is generally the *real* challenge to the programmer. Small programs, such as are found in this and other books, are relatively easy to test with some assurance that the programs really do work.

The Importance of Testing Programs

Testing a program is the most *crucial* programming step after defining the problem and causes the most grief to novice and experienced programmers alike. Simple programs can be tested by exhaustive means, that is, all possible combinations of conditions can be put into the computer and the simple program run for each condition. Large, complex programs cannot be exhaustively tested, because the combinations of possible inputs are so numerous that they cannot be known. Large, complex programs can sometimes be tested using test programs that are specifically written to simulate actual use. Then the final bugs are found by using the program in actual situations until, hopefully, all of the problems have been identified. Even then, it may be that in fixing one bug, another is inadvertently introduced. This has led to the cynical observation by jaded programmers that one has "the latest fix—and the latest bug."

4.6 Programming the 8051

We begin our study of the 8051 family of microcontrollers by investigating the general structure of 8051 syntax.

Lines of Code

Assembly language programs are written as a sequence of text *lines.* A text line is nothing more than a line of alphanumeric characters that are stored in a disk file. Each line ends when a carriage return character (or carriage return–line feed) is stored by the text editor on the disk. The file is created by using a word processor or text editor. The only condition placed on the file is that it must be stored on the disk in pure ASCII (American Standard Code for Information Interchange) format. Using ASCII format for the file ensures that unusual characters, used by most word processors for special functions (such as underlining or tabs), do not appear in the final file.

Each line of text is an instruction to the CPU, a directive to the assembler, or a combination of the two. Many names have been used for a single text line in a source file, such as a *statement*, or a *line of code*, or an *instruction.*

8051 Instruction Syntax

Almost all computer instructions involve taking one piece of data from a location somewhere inside the computer and "operating" on that data together with another piece of data located somewhere else inside the computer. Data originates at the *source* location and ends up at the *destination* location. (We will use the terms *location* and *address* interchangeably when discussing the physical place that binary data is stored.)

Instructions commonly start with an action mnemonic that reminds us of what the instruction does. The instruction mnemonic[6] is the *first* word of the instruction and indicates in what manner, or how, the data are to be combined or otherwise manipulated by the CPU.

Each code line may also contain comments and a name for the instruction address in code memory, called the *label* of the code address. A label is a *name, or symbol,* given to the address *number* where the first byte of the labeled instruction is stored in code memory. We have seen that instructions are assembled into code memory with each byte of code located at a unique address number. *Labels* let us *convert* address *numbers in code memory* into *names* of our choosing and let the assembler worry about exactly which address number has each name. If the address number of a line of code should happen to change as a result of adding a new instruction, for instance, the name remains the same, only the address number for that name will change. The assembler can keep track of address numbers for each name, and we can keep track of the names.

The overall syntax of a line of 8051 program code is shown in Figure 4.7.

Labels

A label can be any combination of up to 128 letters (A–Z), numbers (0 – 9), underline (_), and question mark (?).

Labels *must* begin with a letter from A(a) to Z(z) or ? or _ . Do *not* begin a label with any decimal number from 0 to 9, or the assembler may assume it is a *number*. Examples of code address labels are the following:

```
fred.2:
square:
tabletwo:
curses:
setmode:
transmit:
receive:
ad1234h:
```

Although you may legally use labels up to 128 characters in length, experienced programmers recommend names of 7 or fewer characters. Try to keep label

[6]The instruction mnemonic is sometimes called an *opcode* because early microprocessors had one *unique* code per instruction type. Contemporary microprocessors may have many different codes for the same mnemonic.

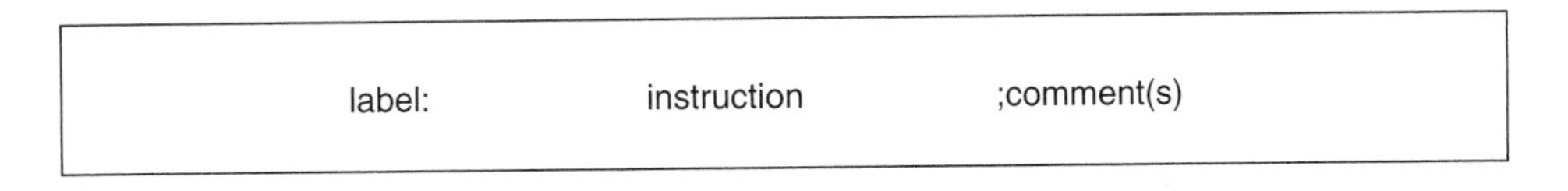

FIGURE 4.7 ◆ An Assembly Language Instruction Line

names *short* and *related* to the program. Do *not* use label names that could be considered *offensive* to a reasonable person (such as your instructor).

Labels should be placed in the first text column, on the far left of the source file page, so that we can see, quickly, that it is a label and not an instruction. Labels may also be used alone on a line, with no following instruction. The assembler will associate the label with the first instruction line *after* the label.

Labels, *when used to assign a name to a location in the program,* always end with a colon (:) to indicate they are labels and not instructions. Any other use of a label, such as part of an instruction operand, *always omits* the colon.

Labels can*not* be the name of any register or instruction used by the 8051 such as DPH, mov, or djnz. Such names are called *reserved words*, or *keywords*, because they have *already been defined and reserved for assembler* use by the A51 assembler author. Appendix B has a list of A51 assembler program reserved words, at the end of the appendix.

Instructions, without a label preceding them on a line, can begin in any text column. Programs are more readable, however, if instructions are indented 10 spaces from the left margin.

Instructions

An 8051 instruction is any of the *coded set that have been defined by the manufacturer* of the 8051. Every instruction can be converted into a unique machine-language binary code that can be acted on by the 8051 internal circuitry. Undefined binary codes will cause the CPU to perform erratically.

Each instruction is generally made up of three distinct parts, as shown in Figure 4.8.

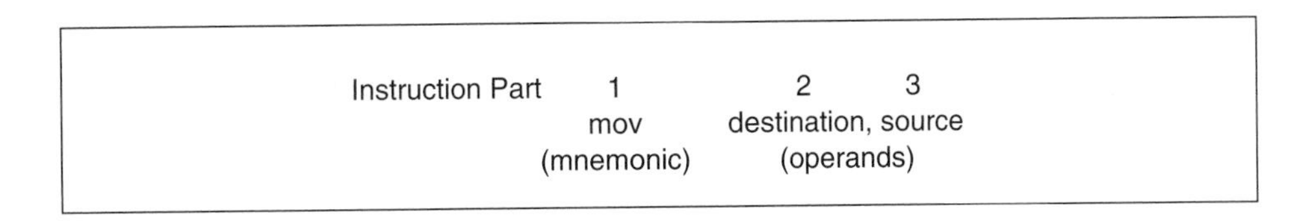

FIGURE 4.8 ◆ An 8051 Instruction

Part 1, the instruction mnemonic, is intended to jog our memories by sounding like the operation to be performed by the CPU. For instance, the instruction shown in Figure 4.8 will move data from one location in the computer (the source) to another location (the destination).

Part 2 is called an *operand* or *data address,* because it specifies the destination for the data that is being copied from the source.

Part 3 is also an operand and contains the address of the source location in the computer that is providing data to be copied to the destination.

The names of the destination and source address are separated by a comma (,) so that the assembler will know when one operand name ends and the other begins. Do not leave out the comma, for it is as important as any other part of the instruction.

Note that I have used the term *copied* rather than *moved* in the description of a data mov instruction. This is usually the case for most computer data transfer operations. Data is copied from the source to the destination, not physically moved and nothing left behind.

There is no reason that the destination address should be listed before the source address except that doing so is a *convention* established for the 8051 assembler. A programming convention, unlike a law of physics, is purely a human invention.

Comments

Comments are included by the programmer as simple, terse explanations of exactly what the instruction is doing to make the program function. Beginning programmers often omit comments, a very *dangerous* practice that can lead to sloppy and erroneous programs. Commenting each program line is encouraged, even when the meaning of the line is obvious. Comments *begin with a semicolon* (;) to indicate that they are not one of the operands. Anything can be typed after the semicolon with impunity, even reserved words. Examples of comments include the following:

```
;this is a comment
;and so is this
;mov a, b
;hello mom!
;and so on
```

One rather handy way to use the semicolon when debugging a program is to place a semicolon in front of a line of code that you may think (but are not absolutely sure) needs to be deleted. Placing a semicolon in front of the line of suspect code will ensure it is not part of the assembled program. You can easily restore the line in your program by removing the semicolon, should your suspicions prove wrong.

4.7 Summary

A fundamental computer model can be divided into two parts, a central processing unit (CPU) and memory. The CPU fetches binary instruction numbers from memory and performs binary operations on other numbers as directed by the opcodes. The sequence of CPU opcodes in memory is called a computer program. Results of the binary operations are stored in memory as data. Fetching opcodes is an automatic function of the CPU; all other CPU actions are controlled by the program in memory.

Assembly language programming consists of writing programs using instruction mnemonics that are specified by the manufacturer of the computer. An instruction consists of a label, an instruction mnemonic, operands, and optional comments. The instruction mnemonic specifies the action to be taken by the CPU, and the operands specify the addresses of data used in the action. Labels are names given to address numbers in program code memory.

The mechanics of programming involve using a text editor program to prepare the 8051 program in mnemonic form, an assembler program to convert the 8051 instructions into hexadecimal machine code, and a debug or simulator program to test the program. Personal computers are often used to run the various utility programs.

Programs may be documented using flow charts, copious comments, or both. Modern programming techniques emphasize using small program modules, which are collected into one main program.

4.8 Questions and Problems

1. Name two basic components of a computer.
2. Computers deal only with binary numbers. Name two types of computer numbers.
3. Identify three binary number lengths, starting with the smallest.
4. What is the function of an address?
5. What is the difference between fetching a number from memory and reading a number from memory?
6. What is the function of an operating system program?
7. What is software?
8. What is hardware?
9. Name three high-level languages.
10. Why are high-level languages machine-independent?

11. Why must the programmer know about the CPU in order to program in assembly language?
12. Name three assembly languages.
13. Identify two reasons to program in a high-level language.
14. In general, which takes more space in memory, an assembly-coded program or a high-level language program? Why?
15. List four reasons to program a CPU in assembly language.
16. List four types of utility programs.
17. What type of file goes into an assembler program?
18. What type of file comes out of an assembler program?
19. What utility is used to test a program?
20. Name the programs that are used to convert from human thoughts to binary-coded computer instructions.

Write, and manually assemble, the following programs for the PAL CPU. All memory addresses include the starting and ending addresses. Begin your code at memory address 0000h.

21. Invert all the data from RAM memory locations A0h to A3h.
22. Place 00h in all memory locations from 20h to 2Ah.
23. Move the data in addresses 10h to 1Ah to addresses 20h to 2Ah. For example, move the data at address 10h to address 20h, and so on until data at address 1Ah is moved to address 2Ah.
24. Make every odd bit (bit 1, 3, 5, and 7) in memory addresses C0h and D0h a 0. Do not change any other bit at each address.
25. Make the low-order nibble of the byte at address 34h a 1. Do not change the high-order nibble.
26. Determine what you must do to assemble the program of Problem 23 beginning at code memory address 030h.
27. Add the instruction `mov r,a` to the beginning of the program of Problem 23 and re-assemble the program.
28. What determines the address of the first instruction in memory?
29. Explain why you cannot tell the difference between a code number and a data number, if viewed randomly in memory.
30. Name two automatic (not the result of a program code instruction) actions a computer performs.

5

Moving Data

Chapter Outline

CHAPTER OBJECTIVES

On successful completion of this chapter you will be able to:

- Use commands that place data in registers, internal memory, and external memory.
- List the data addressing modes.
- Describe how data may be pushed and popped using a stack.
- Use commands that get data from ROM addresses.
- Use commands that exchange data.
- Write simple data movement programs.

5.0 Introduction

A computer typically spends more time moving data from one location to another than it spends on any other operation. It is not surprising, therefore, to find that more instructions are provided for moving data than for any other type of operation.

Data is stored at a *source* address and moved (actually, the data is *copied*) to a *destination* address. The ways by which these addresses are specified are called the *addressing modes.* The 8051 mnemonics are written with the *destination* address named *first,* followed by the source address.

A detailed study of the operational codes (opcodes) of the 8051 begins in this chapter. Although there are 28 distinct mnemonics that copy data from a source to a destination, they may be divided into the following three main types:

1. MOV destination, source
2. PUSH source or POP destination
3. XCH destination, source

The following four addressing modes are used to access data:

1. Immediate addressing mode
2. Register addressing mode
3. Direct addressing mode
4. Indirect addressing mode

The MOV opcodes involve data transfers within the 8051 memory. This memory is divided into the following four distinct physical parts:

1. Internal RAM
2. Internal special-function registers
3. External RAM
4. Internal and external ROM

Finally, the following five types of opcodes are used to move data:

1. MOV
2. MOVX
3. MOVC
4. PUSH and POP
5. XCH

◆ COMMENT ◆

All of the following opcode examples may be converted into operating programs that may be assembled and debugged by adding these code lines to the code found in each example:

```
        org 0000h       ; start code at 0000h
loop:                   ; a label to jump to (Chapter 8)
    (insert code here)
        sjmp loop       ; jump back to beginning
        end             ; end the program
```

5.1 Addressing Modes

The way the data sources or destination addresses are specified in the mnemonic that moves that data determines the addressing mode. Figure 5.1 diagrams the four addressing modes: immediate, register, direct, and indirect.

Immediate Addressing Mode

The simplest way to get data to a destination is to make the source of the data part of the opcode. The data source is then immediately available as part of the instruction itself.

When the 8051 executes an immediate data move, the program counter is automatically incremented to point to the byte(s) following the opcode byte in the program memory. Whatever data is found there is copied to the destination address.

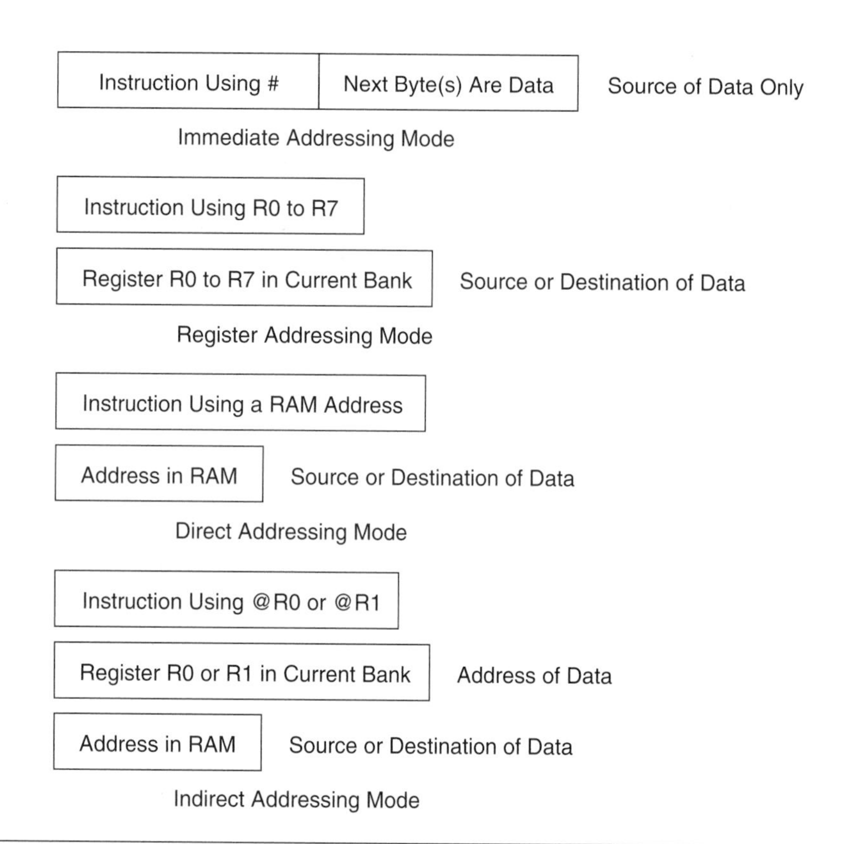

FIGURE 5.1 ◆ Addressing Modes

The mnemonic for immediate data is the pound sign (#). Occasionally, in the rush to meet a deadline, we might forget to use the # for immediate data. The resulting opcode is often a legal command that is assembled with no objections by the assembler. This omission guarantees that the rush will continue.

Three mnemonics can copy immediate numbers from the opcode into registers R0–R7 (of the currently selected register bank), A, and DPTR:

Mnemonic	Operation
MOV Rr,#n	Copy the 8-bit number n into register Rr (of the current register bank)
MOV A,#n	Copy the 8-bit number n into the Accumulator register
MOV DPTR,#nn	Copy the 16-bit number nn into the DPTR register

Immediate addressing modes, using some of the registers from R0 to R7 in the currently selected register bank, register A, and register DPTR, are shown in the following list:

Mnemonic	Operation
MOV R0,#00h	Put the immediate 8-bit number 00h in register R0
MOV R1,#01h	Put the immediate 8-bit number 01h in register R1
MOV R4,#04h	Put the immediate 8-bit number 04h in register R4
MOV R7,#07h	Put the immediate 8-bit number 07 in register R7
MOV A,#0AAh	Put the immediate 8-bit number AAh in register A
MOV DPTR,#1234h	Put the immediate 16-bit number 1234h in register DPTR

Register Addressing Mode

Certain register names may be used as part of the opcode mnemonic as sources or destinations of data. Registers A, DPTR, and R0 to R7 may be named as part of the opcode mnemonic. Other registers in the 8051 may be addressed using the direct addressing mode. Some assemblers can equate many of the direct addresses to the register name (as is the case with the assembler discussed in this book) so that register names may be used in lieu of register addresses. Remember that the registers used in the opcode as R0 to R7 are the ones that are *currently* chosen by the bank-select bits, RS0 and RS1 in the PSW.

Register-to-register moves are as follows:

Mnemonic	Operation
MOV A,Rr	Copy data from register Rr to register A
MOV Rr,A	Copy data from register A to register Rr

A data MOV does not alter the contents of the data source address. A *copy* of the data is made from the source and moved to the destination address. The contents of the destination address are replaced by the source address contents. The following list shows examples of MOV opcodes with immediate and register addressing modes:

Mnemonic	Operation
MOV A,#0F1h	Move the immediate data byte F1h to the A register
MOV A,R0	Copy the data in register R0 to register A
MOV DPTR,#0ABCDh	Move the immediate data bytes ABCDh to the DPTR
MOV R5,A	Copy the data in register A to register R5
MOV R3,#1Ch	Move the immediate data byte 1Ch to register R3

◆ CAUTION ◆

- It is impossible to have immediate data as a destination.
- All numbers *must* start with a decimal number (0 – 9), or the assembler assumes the number is a *label*.
- Register-to-register moves using the register addressing mode occur between registers A and R0 to R7.

Direct Addressing Mode

All 128 bytes of internal RAM and the SFRs may be addressed directly using the single-byte address assigned to each RAM location and each special-function register. See Appendix F for an overall SFR/memory map.

Internal RAM uses addresses from 00h to 7Fh to address each byte. The SFR addresses exist from 80h to FFh at the locations shown in Table 5.1.

◆ CAUTION ◆

- Note that there are "gaps" in the addresses of the SFRs; the addresses are not in order.
- Note the use of a leading 0 for all numbers that begin with an alphabetic (alpha) character.

RAM addresses 00 to 1Fh are *also* the locations assigned to the four banks of eight working registers, R0 to R7. This assignment means that R2 of register

TABLE 5.1

SFR	Address (hex)
A	0E0
B	0F0
DPL	82
DPH	83
IE	0A8
IP	0B8
P0	80
P1	90
P2	0A0
P3	0B0
PCON	87
PSW	0D0
SBUF	99
SCON	98
SP	81
TCON	88
TMOD	89
TH0	8C
TL0	8A
TH1	8D
TL1	8B

TABLE 5.2

Bank	Register	Address (hex)	Bank	Register	Address (hex)
0	R0	00	2	R0	10
0	R1	01	2	R1	11
0	R2	02	2	R2	12
0	R3	03	2	R3	13
0	R4	04	2	R4	14
0	R5	05	2	R5	15
0	R6	06	2	R6	16
0	R7	07	2	R7	17
1	R0	08	3	R0	18
1	R1	09	3	R1	19
1	R2	0A	3	R2	1A
1	R3	0B	3	R3	1B
1	R4	0C	3	R4	1C
1	R5	0D	3	R5	1D
1	R6	0E	3	R6	1E
1	R7	0F	3	R7	1F

bank 0 can be addressed in the register mode as R2 or in the direct mode as 02h. The direct addresses of the working registers are shown in Table 5.2.

Only one bank of working registers is *active* at any given time. The PSW special-function register holds the bank-select bits, RS0 and RS1, which determine which register bank is in use.

When the 8051 is reset, RS0 and RS1 are set to 00b to select the working registers in bank 0, located from 00h to 07h in internal RAM. Reset also sets SP to 07h, and the stack will grow *up* as it is used. This growing stack will overwrite the register banks above bank 0. Be *sure* to set the SP to a number above those of any working registers the program may use.

The programmer may choose any other bank by setting RS0 and RS1 as desired; this bank change is often done to "save" one bank and choose another when servicing an interrupt or using a subroutine.

The programmer may elect to use the *absolute* numeric address number for an SFR or may use a *symbol* (name) for the SFR. For example, the following instructions both move a *constant* number into port 1:

```
mov 90h,#0a5h
mov p1,#0a5h
```

The A51 assembler, supplied with this book, "looks up" the actual address of an SFR when the programmer uses an SFR symbol. Please refer to the end of Appendix B for a list of SFR symbols.

We shall use both methods of specifying SFRs in this book to emphasize the fact that SFRs are internal RAM addresses.

The moves made possible using direct, immediate, and register addressing modes are as follows:

Mnemonic	Operation
MOV A,add	Copy data from direct address add to register A
MOV add,A	Copy data from register A to direct address add
MOV Rr,add	Copy data from direct address add to register Rr
MOV add,Rr	Copy data from register Rr to direct address add
MOV add,#n	Copy immediate data byte n to direct address add
MOV add1,add2	Copy data from direct address add2 to direct address add1

The following list shows examples of MOV opcodes using direct, immediate, and register addressing modes:

Mnemonic	Operation
MOV A,80h	Copy data from the port 0 pins to register A
MOV 80h,A	Copy data from register A to the port 0 latch
MOV 3Ah,#3Ah	Copy immediate data byte 3Ah to RAM location 3Ah
MOV R0,12h	Copy data from RAM location 12h to register R0
MOV 8Ch,R7	Copy data from register R7 to timer 0 high byte
MOV 5Ch,A	Copy data from register A to RAM location 5Ch
MOV 0A8h,77h	Copy data from RAM location 77h to IE register

- MOV instructions that refer to direct addresses above 7Fh that are not SFRs will result in errors. The SFRs are physically on the chip; all other addresses above 7Fh do not physically exist.
- Moving data to a port changes the port *latch;* moving data from a port gets data from the port *pins.*
- Moving data from a direct address to itself is not predictable and could lead to errors.

Indirect Addressing Mode

For all the addressing modes covered to this point, the source or destination of the data is an absolute number or a name. Inspection of the opcode reveals exactly what the addresses are of the destination and source. For example, the opcode MOV A,R7 says that the A register will get a copy of whatever data is in register R7; MOV 33h,#32h moves the hex number 32 to hex RAM address 33.

The indirect addressing mode uses a register to *hold* the actual address that will finally be used in the data move; the register itself is *not* the address, but rather the number *in* the register. Indirect addressing for MOV opcodes uses register R0 or R1, often called a *data pointer,* to hold the address of one of the data locations in RAM from address 00h to 7Fh. The number that is in the pointing register (Rp) cannot be known unless the history of the register is known. The mnemonic symbol used for indirect addressing is the "at" sign, which is printed as @.

The moves made possible using immediate, direct, register, and indirect addressing modes are as follows:

Mnemonic	Operation
MOV @Rp,#n	Copy the immediate byte n to the address in Rp
MOV @Rp,add	Copy the contents of add to the address in Rp
MOV @Rp,A	Copy the data in A to the address in Rp
MOV add,@Rp	Copy the contents of the address in Rp to add
MOV A,@Rp	Copy the contents of the address in Rp to A

The following list shows examples of MOV opcodes, using immediate, register, direct, and indirect modes

Mnemonic	Operation
MOV A,@R0	Copy the contents of the address in R0 to the A register
MOV @R1,#35h	Copy the number 35h to the address in R1
MOV add,@R0	Copy the contents of the address in R0 to add
MOV @R1,A	Copy the contents of A to the address in R1
MOV @R0,80h	Copy the contents of the port 0 pins to the address in R0

- The number in register Rp must be a RAM address.
- *Only* registers R0 or R1 may be used for indirect addressing.

5.2 External Data Moves

As discussed in Chapter 3, it is possible to expand RAM and ROM memory space by adding external memory chips to the 8051 microcontroller. The external memory can be as large as 64K for each of the RAM and ROM memory areas. Opcodes that access this external memory *always* use indirect addressing to specify the external memory.

Figure 5.2 shows that registers R0, R1, and the aptly named DPTR can be used to hold the address of the data byte in external RAM. R0 and R1 are lim-

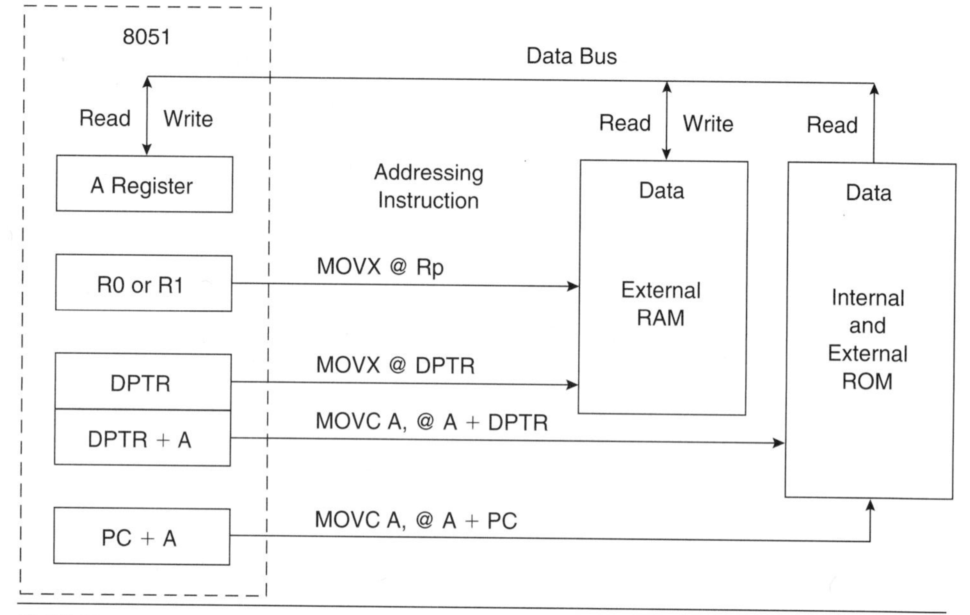

FIGURE 5.2 ◆ External Addressing Using MOVX and MOVC

ited to external RAM address ranges of 00h to 0FFh, while the DPTR register can address the maximum RAM space of 0000h to 0FFFFh.

An X is added to the MOV mnemonics to serve as a reminder that the data move is external to the 8051, as shown in the following list:

Mnemonic	Operation
MOVX A,@Rp	Copy the contents of the external address in Rp to A
MOVX A,@DPTR	Copy the contents of the external address in DPTR to A
MOVX @Rp,A	Copy data from A to the external address in Rp
MOVX @DPTR,A	Copy data from A to the external address in DPTR

The following list shows examples of external moves using register and indirect addressing modes:

Mnemonic	Operation
MOVX @DPTR,A	Copy data from A to the 16-bit address in DPTR
MOVX @R0,A	Copy data from A to the 8-bit address in R0
MOVX A,@R1	Copy data from the 8-bit address in R1 to A
MOVX A,@DPTR	Copy data from the 16-bit address in DPTR to A

◆ CAUTION ◆

- All external data moves must involve the A register.
- Rp can address 256 bytes; DPTR can address 64K bytes.
- MOVX is normally used with external RAM or I/O addresses.
- Note that there are two sets of RAM addresses between 00h and 0FFh: one internal and one external to the 8051.

5.3 Code Memory Read-Only Data Moves

Data moves between RAM locations and 8051 registers are made by using MOV and MOVX opcodes. The data is usually of a temporary or "scratch pad" nature and disappears when the system is powered down.

There are times when access to a preprogrammed mass of data is needed, such as when using tables of predefined bytes. This data must be permanent to be of repeated use and is stored in the program ROM using assembler directives that store programmed data anywhere in ROM that the programmer wishes.

Access to this data is made possible by using indirect addressing and the A register in conjunction with either the PC or the DPTR, as shown in Figure 5.2. In both cases, the number in register A is *added* to the pointing register to form the address in ROM where the desired data is to be found. The data is then fetched from the ROM address so formed and placed in the A register. The original data in A is lost, and the addressed data takes its place.

As shown in the following list, the letter C is added to the MOV mnemonic to highlight the use of the opcodes for moving data from the source address in the code ROM to the A register in the 8051:

Mnemonic	Operation
MOVC A,@A+DPTR	Copy the code byte, found at the ROM address formed by adding A and the DPTR, to A
MOVC A,@A+PC	Copy the code byte, found at the ROM address formed by adding A and the PC, to A

Note that the DPTR and the PC are not changed; the A register contains the ROM byte found at the address formed.

The following list shows examples of code ROM moves using register and indirect addressing modes:

Mnemonic	Operation
MOV DPTR,#1234h	Copy the immediate number 1234h to the DPTR
MOV A,#56h	Copy the immediate number 56h to A
MOVC A,@A+DPTR	Copy the contents of address 128Ah to A
MOVC A,@A+PC	Copies the contents of address 4059h to A if the PC contains 4000h and A contains 58h when the opcode is executed

- The PC is incremented by 1 (to point to the next instruction) *before* it is added to A to form the final address of the code byte.
- All data is moved *from* the code memory to the A register.
- MOVC is normally used with internal or external ROM and can address 4K of internal or 64K of external code.

5.4 Push and Pop Opcodes

The PUSH and POP opcodes specify the direct address of the data. The data moves between an area of internal RAM, known as the stack, and the specified direct address. The stack pointer special-function register (SP) contains the address in RAM where data *from* the source address will be PUSHed, or where data to be POPed *to* the destination address is found. The SP register actually is used in the indirect addressing mode but is *not* named in the mnemonic. It is *implied* that the SP holds the indirect address whenever PUSHing or POPing. Figure 5.3 shows the operation of the stack pointer as data is PUSHed or POPed to the stack area in internal RAM.

A PUSH opcode copies data from the source address to the stack. SP is *incremented* by 1 *before* the data is copied to the internal RAM location contained in SP so that the data is stored from low addresses to high addresses in the internal RAM. The stack grows *up* in memory as it is PUSHed. Excessive PUSHing can make the stack exceed 7Fh (the top of internal RAM), after which PUSHed data is lost.

A POP opcode copies data from the stack to the destination address. SP is *decremented* by 1 *after* data is copied from the stack RAM address to the direct destination to ensure that data placed on the stack is retrieved in the same order as it was stored.

The PUSH and POP opcodes behave as explained in the following list:

Mnemonic	Operation
PUSH add	Increment SP; copy the data in add to the internal RAM address contained in SP
POP add	Copy the data from the internal RAM address contained in SP to add; decrement the SP

The SP register is set to 07h when the 8051 is reset, which is the same direct address in internal RAM as register R7 in bank 0. The first PUSH opcode would write data to R0 of bank 1. The SP should be initialized by the programmer to point to an internal RAM address above the highest address likely to be used by the program.

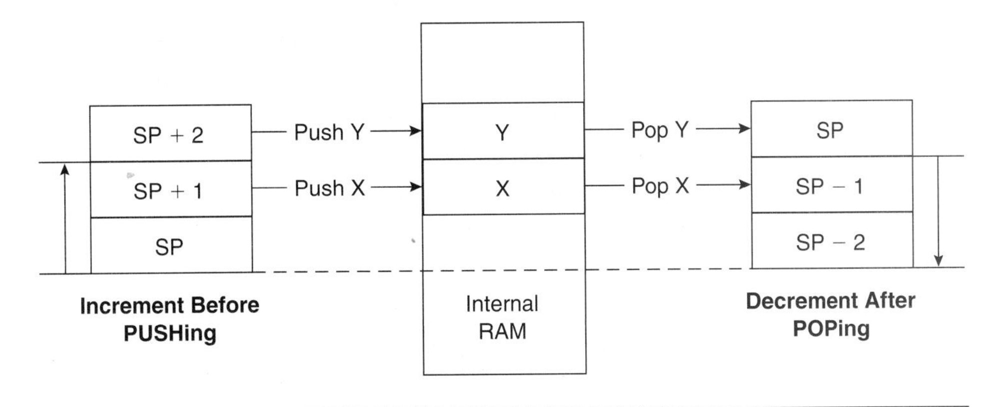

FIGURE 5.3 ◆ PUSH and POP the Stack

The following list shows examples of PUSH and POP opcodes:

Mnemonic	Operation
MOV 81h,#30h	Copy the immediate data 30h to the SP
MOV R0, #0ACh	Copy the immediate data ACh to R0
PUSH 00h	SP = 31h; address 31h contains the number ACh
PUSH 00h	SP = 32h; address 32h contains the number ACh
POP 01h	SP = 31h; register R1 now contains the number ACh
POP 80h	SP = 30h; port 0 latch now contains the number ACh

- When the SP reaches FFh it "rolls over" to 00h (R0).
- RAM ends at address 7Fh; PUSHes above 7Fh result in errors.
- The SP is usually set at addresses above the register banks.
- The SP may be PUSHed and POPed to the stack.
- Note that direct addresses, *not* register names, must be used for R0–R7. The stack mnemonics have no way of knowing which bank is in use.

5.5 Data Exchanges

MOV, PUSH, and POP instructions all involve copying the data found in the source address to the destination address; the original data in the source is not changed. Exchange instructions actually move data in two directions: from source to destination and from destination to source. All addressing modes except immediate may be used in the XCH (exchange) instructions:

Mnemonic	Operation
XCH A,Rr	Exchange data bytes between register Rr and A
XCH A,add	Exchange data bytes between add and A
XCH A,@Rp	Exchange data bytes between A and address in Rp
XCHD A,@Rp	Exchange lower nibble between A and address in Rp

Exchanges between A and any port location copy the data on the port *pins* to A, whereas the data in A is copied to the port *latch*. Register A is used for so many instructions that the XCH opcode provides a very convenient way to "save" the contents of A without the necessity of using a PUSH opcode and then a POP opcode.

The following list shows examples of data moves using exchange opcodes:

Mnemonic	Operation
XCH A,R7	Exchange bytes between register A and register R7
XCH A,0F0h	Exchange bytes between register A and register B
XCH A,@R1	Exchange bytes between register A and address in R1
XCHD A,@R1	Exchange lower nibble in A and the address in R1

- All exchanges are *internal* to the 8051.
- All exchanges use register A.
- When using XCHD, the upper nibble of A and the upper nibble of the address location in Rp do not change.

This section concludes the listing of the various data moving instructions; the remaining sections will concentrate on using these opcodes to write short programs.

5.6 Example Problems

Programming is at once a skill and an art. Just as anyone may learn to play a musical instrument after sufficient instruction and practice, so may anyone learn to program a computer. Some individuals, however, have a gift for programming that sets them apart from their peers with the same level of experience, just as some musicians are more talented than their contemporaries.

Gifted or not, you will not become adept at programming until you have written and rewritten many programs. The emphasis here is on practice; you can read many books on how to ride a bicycle, but you do not know how to ride until you do it.

If some of the examples and problems seem trivial or without any "real-world" application, remember the playing of scales on a piano by a budding musician. Each example will be done using several methods; the best method depends on what resource is in short supply. If programming time is valuable, then the best program is the one that uses the fewest lines of code; if either ROM or execution time is limited, then the program that uses the fewest code bytes is best.

EXAMPLE PROBLEM 5.1

Copy the byte in TCON to register R2 using at least four different methods.

- **Method 1:** Use the direct address for TCON (88h) and register R2:

Mnemonic	Operation
MOV R2,88h	Copy TCON to R2

- **Method 2:** Use the direct addresses for TCON and R2:

Mnemonic	Operation
MOV 02h,88h	Copy TCON to direct address 02h (R2)

- **Method 3:** Use R1 as a pointer to R2 and use the address of TCON:

Mnemonic	Operation
MOV R1,#02h	Use R1 as a pointer to R2
MOV @R1,88h	Copy TCON byte to address in R1 (02h = R2)

- **Method 4:** Push the contents of TCON into direct address 02h (R2):

Mnemonic	Operation
MOV 81h,#01h	Set the SP to address 01h in RAM
PUSH 88h	Push TCON (88h) to address 02h (R2)

◆

EXAMPLE PROBLEM 5.2

Set timer T0 to an initial setting of 1234h.

Use the direct address with an immediate number to set TH0 and TL0:

Mnemonic	Operation
MOV 8Ch,#12h	Set TH0 to 12h
MOV 8Ah,#34h	Set TL0 to 34h
	Totals: 6 bytes, 2 lines

◆

EXAMPLE PROBLEM 5.3

Put the number 34h in registers R5, R6, and R7.

- **Method 1:** Use an immediate number and register addressing:

Mnemonic	Operation
MOV R5,#34h	Move 34h to R5
MOV R6,#34h	Move 34h to R6
MOV R7,#34h	Move 34h to R7
	Totals: 6 bytes, 3 lines

- **Method 2:** Since the number is the same for each register, put the number in A and MOV A to each register:

Mnemonic	Operation
MOV A,#34h	Move a 34h to A
MOV R5,A	Copy A to R5
MOV R6,A	Copy A to R6
MOV R7,A	Copy A to R7
	Totals: 5 bytes, 4 lines

- **Method 3:** Copy one direct address to another:

Mnemonic	Operation
MOV R5,#34h	Move 34h to register R5
MOV 06h,05h	Copy R5 (add 05) to R6 (add 06)
MOV 07h,06h	Copy R6 to R7
	Totals: 8 bytes, 3 lines

◆

EXAMPLE PROBLEM 5.4

Put the number 8Dh in RAM locations 30h to 34h.

- **Method 1:** Use the immediate number to a direct address:

Mnemonic	Operation
MOV 30h,#8Dh	Copy the number 8Dh to RAM address 30h
MOV 31h,#8Dh	Copy the number 8Dh to RAM address 31h
MOV 32h,#8Dh	Copy the number 8Dh to RAM address 32h
MOV 33h,#8Dh	Copy the number 8Dh to RAM address 33h
MOV 34h,#8Dh	Copy the number 8Dh to RAM address 34h
	Totals: 15 bytes, 5 lines

- **Method 2:** Using the immediate number in each instruction uses bytes; use a register to hold the number:

Mnemonic	Operation
MOV A,#8Dh	Copy the number 8Dh to the A register
MOV 30h,A	Copy the contents of A to RAM location 30h
MOV 31h,A	Copy the contents of A to the remaining addresses
MOV 32h,A	
MOV 33h,A	
MOV 34h,A	Totals: 12 bytes, 6 lines

- **Method 3:** There must be a way to avoid naming each address; the PUSH opcode can increment to each address:

Mnemonic	Operation
MOV 30h,#8Dh	Copy the number 8Dh to RAM address 30h
MOV 81h,#30h	Set the SP to 30h
PUSH 30h	Push the contents of 30h (=8Dh) to address 31h
PUSH 30h	Continue pushing to address 34h
PUSH 30h	
PUSH 30h	Totals: 14 bytes, 6 lines

◆

EXAMPLE PROBLEM 5.5 Swap the contents of registers R7 and R6 in register bank 0.

- **Method 1:** Use a series of register address mode MOVes:

Mnemonic	Operation
MOV A, R6	Copy the contents of R6 to A
MOV R5, A	Save contents of R6 in R5
MOV A, R7	Copy contents of R7 to A
MOV R6, A	Contents of R7 now in R6
MOV A, R5	Retrieve contents of R6
MOV R7, A	Contents of R6 now in R7
	Totals: 6 bytes, 6 lines

- **Method 2:** Use a series of direct address mode MOVes:

Mnemonic	Operation
MOV 10h, 06h	Copy the contents of R6 to RAM address 10h
MOV 06h, 07h	Copy the contents of R7 to R6
MOV 07h,10h	Copy the saved contents of R6 to R7
	Totals: 9 bytes, 3 lines

- **Method 3:** Use a series of PUSHes and POPs:

Mnemonic	Operation
PUSH 07h	Push contents of R7 on the stack
PUSH 06h	Push contents of R6 on the stack
POP 07h	POP contents of R6 to R7
POP 06h	POP contents of R7 to R6
	Totals: 8 bytes, 4 lines

◆ **Method 4:** Use a series of XCHes:

Mnemonic	Operation
XCH A, R6	Exchange contents of A and R6
XCH A, R7	Contents of R6 now in R7; R7 in A
XCH A, R6	Contents of R7 now in R6; A same
	Totals: 3 bytes, 3 lines

◆

◆ COMMENT ◆

Indirect addressing with the number in A and the indirect address in R1 could be done; however, R1 would have to be loaded with each address from 30h to 34h. Loading R1 would take a total of 17 bytes and 11 lines of code. Indirect addressing is advantageous when we have opcodes that can change the contents of the pointing registers automatically.

EXAMPLE PROBLEM 5.6

Square any number in register A, from 0 to 5, and put the result in register A.

As shown in Figure 5.2, MOVC instructions can move code bytes from both internal and external ROM. Internal ROM is used for those systems that are factory-programmed, or where an EPROM version of the 8051, an 8751, is used. All code bytes are stored beginning at code address 0000h, because the PC is set to 0000h at start-up.

This example makes use of a "look-up" table, which is constructed as part of program code. The number in A is used to locate the address of the square of the number in A in the code bytes which follow the MOVC opcode. The debugger should be used to assemble and test the program; the programmer may place numbers from 0 to 5 in A, and verify that the square, from 00h to 19h, is obtained. Reset the debugger after executing the MOVC opcode, or the debugger will try to execute the table byte at address 0001h (an NOP instruction). Working programs would jump over the table, using opcodes discussed in Chapter 8, to prevent executing table bytes.

Code Address	Mnemonic	Operation
0000h	MOVC A,@A+PC	A is 0 to 5; PC is 0001h
0001h	DB 00h	A = 0, square = 00h, A+PC = 0001h
0002h	DB 01h	A = 1, square = 01h, A+PC = 0002h
0003h	DB 04h	A = 2, square = 04h, A+PC = 0003h
0004h	DB 09h	A = 3, square = 09h, A+PC = 0004h
0005h	DB 10h	A = 4, square = 10h, A+PC = 0005h
0006h	DB 19h	A = 5, square = 19h, A+PC = 0006h

◆

TABLE 5.3

Instruction Type	Result
MOV destination,source	Copy data from the internal RAM source address to the internal RAM destination address
MOVC A,source	Copy internal or external program memory byte from the source to register A
MOVX destination,source	Copy byte to or from external RAM to register A
PUSH source	Copy byte to internal RAM stack from internal RAM source
POP destination	Copy byte from internal RAM stack to internal RAM destination
XCH A,source	Exchange data between register A and the internal RAM source
XCHD A,source	Exchange lower nibble between register A and the internal RAM source

5.7 Summary

The opcodes that move data between locations within the 8051 and between the 8051 and external memory have been discussed. The general form and results of these instructions are given in Table 5.3.

There are four addressing modes: an immediate number, a register name, a direct internal RAM address, and an indirect address contained in a register.

5.8 Problems

Write programs that will accomplish the desired tasks listed below, using as few lines of code as possible. Use only opcodes that have been covered up to this chapter. Comment on each line of code.

1. Place the number 3Bh in internal RAM locations 30h to 32h.
2. Copy the data at internal RAM location 70h to R0 and R3.
3. Set the SP at the byte address just above the last working register address.
4. Exchange the contents of the SP and the PSW.
5. Copy the byte at internal RAM address 27h to external RAM address 27h.
6. Set timer 1 to A23Dh.
7. Copy the contents of DPTR to registers R0 (DPL) and R1 (DPH).
8. Copy the data in external RAM location 0123h to TL0 and the data in external RAM location 0234h to TH0.
9. Copy the data in internal RAM locations 12h to 15h to internal RAM locations 20h to 23h: Copy 12h to 20h, 13h to 21h, etc.
10. Set the SP register to 07h and PUSH the SP register on the stack; predict what number is PUSHed to address 08h.

11. Exchange the contents of the B register and external RAM address 02CFh.
12. Rotate the bytes in registers R0 to R3: copy the data in R0 to R1, R1 to R2, R2 to R3, and R3 to R0.
13. Copy the external code byte at address 007Dh to the SP.
14. Copy the data in register R5 to external RAM address 032Fh.
15. Copy the internal code byte at address 0300h to external RAM address 0300h.
16. Swap the bytes in timer 0: put TL0 in TH0 and TH0 in TL0.
17. Store DPTR in external RAM locations 0123h (DPL) and 02BCh (DPH).
18. Exchange both low nibbles of registers R0 and R1: put the low nibble of R0 in R1, and the low nibble of R1 in R0.
19. Store the contents of register R3 at the internal RAM address contained in R2. (Be sure the address in R2 is legal.)
20. Store the contents of RAM location 20h at the address contained in RAM location 08h.
21. Store register A at the internal RAM location address in register A.
22. Copy program bytes 0100h to 0102h to internal RAM locations 20h to 22h.
23. Copy the data on the pins of port 2 to the port 2 latch.
24. PUSH the contents of the B register to TMOD.
25. Copy the contents of external code memory address 0040h to IE.
26. Show that a set of XCH instructions executes faster than a PUSH and POP when saving the contents of the A register.
27. Use the debugger to determine if pin data or latch data is POPed from a port.
28. Swap the contents of register bank 3 with register bank 0 as follows: R0 of bank 3 with R0 of bank 0, R1 of bank 3 with R1 of bank 0, and so on, until R7 of bank 3 is swapped with R7 of bank 0.
29. Rotate register bank 0 upward: the contents of R0 to R1, R1 to R2, and so on, until the contents of R7 are placed in R0.
30. Rewrite Example Problem 5.6 using the DPTR instead of the PC.

6

Logical Operations

Chapter Outline

CHAPTER OBJECTIVES

On successful completion of this chapter you will be able to:

- Use byte-level AND, OR, XOR, and NOT Boolean instructions.
- Use bit-level AND, OR, and NOT Boolean instructions.
- Use bit-level set, clear, and data-moving instructions.
- Use 8- and 9-bit rotate instructions.
- Use the A register nibble-swapping instruction.

6.0 Introduction

One application area the 8051 is designed to fill is that of machine control. A large part of machine control concerns sensing the on/off states of external switches, making decisions based on the switch states, and then turning external circuits on or off.

Single point sensing and control implies a need for *byte* and *bit* opcodes that operate on data using Boolean operators. All 8051 RAM areas, both data and SFRs, may be manipulated using byte opcodes. Many of the SFRs, and a unique internal RAM area that is bit addressable, may be operated on at the *individual* bit level. Bit operators are notably efficient when speed of response is needed. Bit operators yield compact program code that enhances program execution speed.

The two data levels, byte or bit, at which the Boolean instructions operate are shown in Table 6.1.

There are also rotate opcodes that operate only on a byte, or a byte and the Carry flag, to permit limited 8- and 9-bit shift-register operations. The following list shows the rotate opcodes:

Mnemonic	Operation
RL	Rotate a byte to the left; the Most Significant Bit (MSB) becomes the Least Significant Bit (LSB)
RLC	Rotate a byte and the carry bit left; the carry becomes the LSB, the MSB becomes the carry
RR	Rotate a byte to the right; the LSB becomes the MSB
RRC	Rotate a byte and the carry to the right; the LSB becomes the carry, and the carry the MSB
SWAP	Exchange the low and high nibbles in a byte

Refer to Figure 6.2 to better visualize the rotate instructions.

TABLE 6.1

Boolean Operator	8051 Mnemonic
AND	ANL (AND logical)
OR	ORL (OR logical)
XOR	XRL (exclusive OR logical)
NOT	CPL (complement)

6.1 Byte-Level Logical Operations

The byte-level logical operations use all four addressing modes for the source of a data byte. The A register or a direct address in internal RAM is the destination of the logical operation result.

Keep in mind that all such operations are done using each individual bit of the destination and source bytes. These operations, called *byte-level Boolean operations* because the entire byte is affected, are as follows.

Mnemonic	Operation
ANL A,#n	AND each bit of A with the same bit of immediate number n; put the results in A
ANL A,add	AND each bit of A with the same bit of the direct RAM address; put the results in A
ANL A,Rr	AND each bit of A with the same bit of register Rr; put the results in A
ANL A,@Rp	AND each bit of A with the same bit of the contents of the RAM address contained in Rp; put the results in A
ANL add,A	AND each bit of A with the direct RAM address; put the results in the direct RAM address
ANL add,#n	AND each bit of the RAM address with the same bit in the number n; put the result in the RAM address
ORL A,#n	OR each bit of A with the same bit of n; put the results in A
ORL A,add	OR each bit of A with the same bit of the direct RAM address; put the results in A
ORL A,Rr	OR each bit of A with the same bit of register Rr; put the results in A
ORL A,@Rp	OR each bit of A with the same bit of the contents of the RAM address contained in Rp; put the results in A
ORL add,A	OR each bit of A with the direct RAM address; put the results in the direct RAM address
ORL add,#n	OR each bit of the RAM address with the same bit in the number n; put the result in the RAM address
XRL A,#n	XOR each bit of A with the same bit of n; put the results in A
XRL A,add	XOR each bit of A with the same bit of the direct RAM address; put the results in A
XRL A,Rr	XOR each bit of A with the same bit of register Rr; put the results in A
XRL A,@Rp	XOR each bit of A with the same bit of the contents of the RAM address contained in Rp; put the results in A
XRL add,A	XOR each bit of A with the direct RAM address; put the results in the direct RAM address
XRL add,#n	XOR each bit of the RAM address with the same bit in the number n; put the result in the RAM address
CLR A	Clear each bit of the A register to 0
CPL A	Complement each bit of A: every 1 becomes a 0, and each 0 becomes a 1

Note that no flags are affected by the byte-level logical operations unless the direct RAM address is the PSW.

Many of these byte-level operations use a direct address, which can include the port SFR addresses, as a destination. The normal source of data from a port are the port pins; the normal destination for port data is the port latch. When the *destination* of a logical operation is the direct address of a port, the *latch* register, *not* the pins, is used *both* as the *source* for the original data and then the *destination* for the altered byte of data. *Any* port operation that must first *read* the source data, logically operate on it, and then *write* it back to the source (now the destination) must use the *latch*. Logical operations that use the port as a source, but *not* as a destination, use the *pins* of the port as the source of the data.

For example, the port 0 latch contains FFh, but the pins are all driving transistor bases and are close to ground level. The logical operation

```
ANL P0,#0Fh
```

which is designed to turn the upper nibble transistors off, reads FFh from the latch, ANDs it with 0Fh to produce 0Fh as a result, and then writes it back to the latch to turn these transistors off. Reading the pins produces the result 00h, turning all transistors off, in error. But, the operation

```
ANL A,P0
```

produces A = 00h by using the port 0 pin data, which is 00h.

The following list shows byte-level logical operation examples:

Mnemonic	Operation
MOV A,#0FFh	A = FFh
MOV R0,#77h	R0 = 77h
ANL A,R0	A = 77h
MOV 15h,A	15h = 77h
CPL A	A = 88h
ORL 15h,#88h	15h = FFh
XRL A,15h	A = 77h
XRL A,R0	A = 00h
ANL A,15h	A = 00h
ORL A,R0	A = 77h
CLR A	A = 00h
XRL 15h,A	15h = FFh
XRL A,R0	A = 77h

Note that instructions that can use the SFR port latches as destinations are ANL, ORL, and XRL.

◆ CAUTION ◆

- If the direct address destination is one of the port SFRs, the data latched in the SFR, not the pin data, is used.
- No flags are affected unless the direct address is the PSW.
- *Only* internal RAM or SFRs may be logically manipulated.

6.2 Bit-Level Logical Operations

Certain internal RAM and SFRs can be addressed by their byte addresses or by the address of each bit within a byte. Bit addressing is *very* convenient when you wish to alter a single bit of a byte, in a control register for instance, without having to wonder what you need to do to avoid altering some other crucial bit of the same byte. The assembler can also equate bit addresses to labels that make the program more readable. For example, bit 4 of TCON can become TR0, a label for the timer 0 run bit.

The ability to operate on individual bits creates the need for an area of RAM that contains data addresses that hold a single bit. Internal RAM byte addresses 20h to 2Fh serve this need and are both byte and bit addressable. The bit addresses are numbered from 00h to 7Fh to represent the 128d bit addresses (16d bytes × 8 bits) that exist from byte addresses 20h to 2Fh. Bit 0 of *byte* address 20h is *bit* address 00h, and bit 7 of *byte* address 2Fh is *bit* address 7Fh. You must know your bits from your bytes to take advantage of this RAM area.

Internal RAM Bit Addresses

The availability of individual bit addresses in internal RAM makes the use of the RAM very efficient when storing bit information. Whole bytes do not have to be used up to store one or two bits of data.

The correspondence between byte and bit addresses is shown in Table 6.2. Interpolation of the table shows, for example, the address of bit 3 of internal RAM byte address 2Ch is 63h, the bit address of bit 5 of RAM address 21h is 0Dh, and bit address 47h is bit 7 of RAM byte address 28h. A complete listing of the addresses for these bits may be found in Appendix F.

SFR Bit Addresses

All SFRs may be addressed at the byte level by using the direct address assigned to it, but not all of the SFRs are addressable at the bit level. The SFRs that are also bit addressable form the bit address by using the five most significant bits of the direct address for that SFR, together with the three least significant bits that identify the bit position from position 0 (LSB) to 7 (MSB).

TABLE 6.2

Byte Address (hex)	Bit Addresses (hex)
20	00–07
21	08–0F
22	10–17
23	18–1F
24	20–27
25	28–2F
26	30–37
27	38–3F
28	40–47
29	48–4F
2A	50–57
2B	58–5F
2C	60–67
2D	68–6F
2E	70–77
2F	78–7F

The bit-addressable SFR and the corresponding bit addresses are given in Table 6.3.

The patterns in Table 6.3 show the direct addresses assigned to the SFR bytes all have bits 0–3 equal to zero so that the address of the byte is also the address of the LSB. For example, bit 0E3h is bit 3 of the A register. The Carry flag, which is bit 7 of the PSW, is bit addressable as 0D7h. The assembler can also "understand" more descriptive mnemonics, such as P0.5 for bit 5 of port 0, which is more formally addressed as 85h.

TABLE 6.3

SFR	Direct Address (hex)	Bit Addresses (hex)
A	0E0	0E0–0E7
B	0F0	0F0–0F7
IE	0A8	0A8–0AF
IP	0B8	0B8–0BF
P0	80	80–87
P1	90	90–97
P2	0A0	0A0–0A7
P3	0B0	0B0–0B7
PSW	0D0	0D0–0D7
TCON	88	88–8F
SCON	98	98–9F

For example, to clear the A register to 00h, we write:

```
CLR A
```

whereas to clear bit 5 of the A register, the instruction is:

```
CLR ACC.5
```

It is unfortunate that Intel chose to use the same mnemonic to clear both the A register and to clear an A register addressable bit. Moreover, the instructions CLR ACC and CLR ACC.0 accomplish the same thing: clearing bit 0 of the A register. The SETB mnemonic is much less ambiguous. The other byte and bit mnemonic is CPL (complement).

Figure 6.1 shows all the bit-addressable SFRs and the function of each addressable bit. (Refer to Chapter 3 for more detailed descriptions of the SFR bit functions.)

Bit-Level Boolean Operations

The bit-level Boolean logical opcodes operate on any addressable RAM or SFR bit. The Carry flag (C) in the PSW special-function register is the destination for most of the opcodes because the flag can be tested and the program flow changed using instructions covered in Chapter 8.

The Boolean bit-level operations are as follows:

Mnemonic	Operation
ANL C,b	AND C and the addressed bit; put the result in C
ANL C,/b	AND C and the complement of the addressed bit; put the result in C; the addressed bit is not altered
ORL C,b	OR C and the addressed bit; put the result in C
ORL C,/b	OR C and the complement of the addressed bit; put the result in C; the addressed bit is not altered
CPL C	Complement the C flag
CPL b	Complement the addressed bit
CLR C	Clear the C flag to 0
CLR b	Clear the addressed bit to 0
MOV C,b	Copy the addressed bit to the C flag
MOV b,C	Copy the C flag to the addressed bit
SETB C	Set the C flag to 1
SETB b	Set the addressed bit to 1

Note that no flags, other than the C flag, are affected, unless the flag is an addressed bit.

As is the case for byte-logical operations when addressing ports as destinations, a port bit used as a destination for a logical operation is part of the SFR latch, not the pin. A port bit used as a source *only* is a pin, not the latch. The bit instructions that can use a SFR latch bit are: CLR, CPL, MOV, and SETB.

7	6	5	4	3	2	1	0
CY	AC	F0	RS1	RS0	OV	Reserved	P

Program Status Word (PSW) Special Function Register. Bit Addresses D0h to D7h.

Bit	Function	Bit	Function
7	Carry flag	3	Register bank select bit 0
6	Auxiliary Carry flag	2	Overflow flag
5	User flag 0	1	Not used (reserved for future)
4	Register bank select bit 1	0	Parity flag

7	6	5	4	3	2	1	0
EA	Reserved	Reserved	ES	ET1	EX1	ET0	EX0

Interrupt Enable (IE) Special Function Register. Bit Addresses A8h to AFh.

Bit	Function	Bit	Function
7	Disables all interrupts	3	Timer 1 overflow interrupt enable
6	Not used (reserved for future)	2	External interrupt 1 enable
5	Not used (reserved for future)	1	Timer 0 interrupt enable
4	Serial port interrupt enable	0	External interrupt 0 enable

EA disables all interrupts when cleared to 0; if EA = 1 then each individual interrupt will be enabled if 1, and disabled if 0.

7	6	5	4	3	2	1	0
*	*	Reserved	PS	PT1	PX1	PT0	PX0

Interrupt Priority (IP) Special Function Register. Bit Addresses B8h to BFh.

Bit	Function	Bit	Function
7	Not implemented	3	Timer 1 interrupt priority
6	Not implemented	2	External interrupt 1 priority
5	Not used (reserved for future)	1	Timer 0 interrupt priority
4	Serial port interrupt priority	0	External interrupt 0 priority

The priority bit may be set to 1 (highest) or 0 (lowest).

7	6	5	4	3	2	1	0
TF1	TR1	TF0	TR0	IE1	IT1	IE0	IT0

Timer/Counter Control (TCON) Special Function Register. Bit Addresses 88h to 8Fh.

Bit	Function	Bit	Function
7	Timer 1 Overflow flag	3	External interrupt 1 Edge flag
6	Timer run control	2	External interrupt 1 mode control
5	Timer 0 Overflow flag	1	External interrupt 0 Edge flag
4	Timer 0 run control	0	External interrupt 0 mode control

All flags can be set by the indicated hardware action; the flags are cleared when interrupt is serviced by the processor.

7	6	5	4	3	2	1	0
SM0	SM1	SM2	REN	TB8	RB8	TI	RI

Serial Port Control (SCON) Special Function Register. Bit Addresses 98h to 9Fh.

Bit	Function	Bit	Function
7	Serial port mode bit 0	3	Transmitted bit in modes 2 and 3
6	Serial port mode bit 1	2	Received bit in modes 2 and 3
5	Multiprocessor communications enable	1	Transmit Interrupt flag
4	Receive enable	0	Receive Interrupt flag

FIGURE 6.1 ◆ Bit-Addressable Control Registers

Bit-level logical operation examples are shown in the following list:

Mnemonic	**Operation**
SETB 00h	Bit 0 of RAM byte 20h = 1
MOV C,00h	C = 1
MOV 7Fh,C	Bit 7 of RAM byte 2Fh = 1
ANL C,/00h	C = 0; bit 0 of RAM byte 20h = 1
ORL C,00h	C = 1
CPL 7Fh	Bit 7 of RAM byte 2Fh = 0
CLR C	C = 0
ORL C,/7Fh	C = 1; bit 7 of RAM byte 2Fh = 0

◆ CAUTION ◆

- Only the SFRs that have been identified as bit addressable may be used in bit operations.
- If the destination bit is a port bit, the SFR latch bit is affected, not the pin.
- ANL C,/b and ORL C,/b do not alter the addressed bit b.

6.3 Rotate and Swap Operations

The ability to rotate data is useful for inspecting bits of a byte without using individual bit opcodes. The A register can be rotated one bit position to the left or right with or without including the C flag in the rotation. If the C flag is not included, then the rotation involves the eight bits of the A register. If the C flag is included, then nine bits are involved in the rotation. Including the C flag enables the programmer to construct rotate operations involving any number of bytes.

The SWAP instruction can be thought of as a rotation of nibbles in the A register. Figure 6.2 diagrams the rotate and swap operations, which are given in the following list:

Mnemonic	**Operation**
RL A	Rotate the A register one bit position to the left; bit A0 to bit A1, A1 to A2, A2 to A3, A3 to A4, A4 to A5, A5 to A6, A6 to A7, and A7 to A0
RLC A	Rotate the A register and the Carry flag, as a ninth bit, one bit position to the left; bit A0 to bit A1, A1 to A2, A2 to A3, A3 to A4, A4 to A5, A5 to A6, A6 to A7, A7 to the Carry flag, and the Carry flag to A0
RR A	Rotate the A register one bit position to the right; bit A0 to bit A7, A7 to A6, A6 to A5, A5 to A4, A4 to A3, A3 to A2, A2 to A1, and A1 to A0
RRC A	Rotate the A register and the Carry flag, as a ninth bit, one bit position to the right; bit A0 to the Carry flag, Carry flag to A7, A7 to A6, A6 to A5, A5 to A4, A4 to A3, A3 to A2, A2 to A1, and A1 to A0
SWAP A	Interchange the nibbles of register A; put the high nibble in the low nibble position and the low nibble in the high nibble position

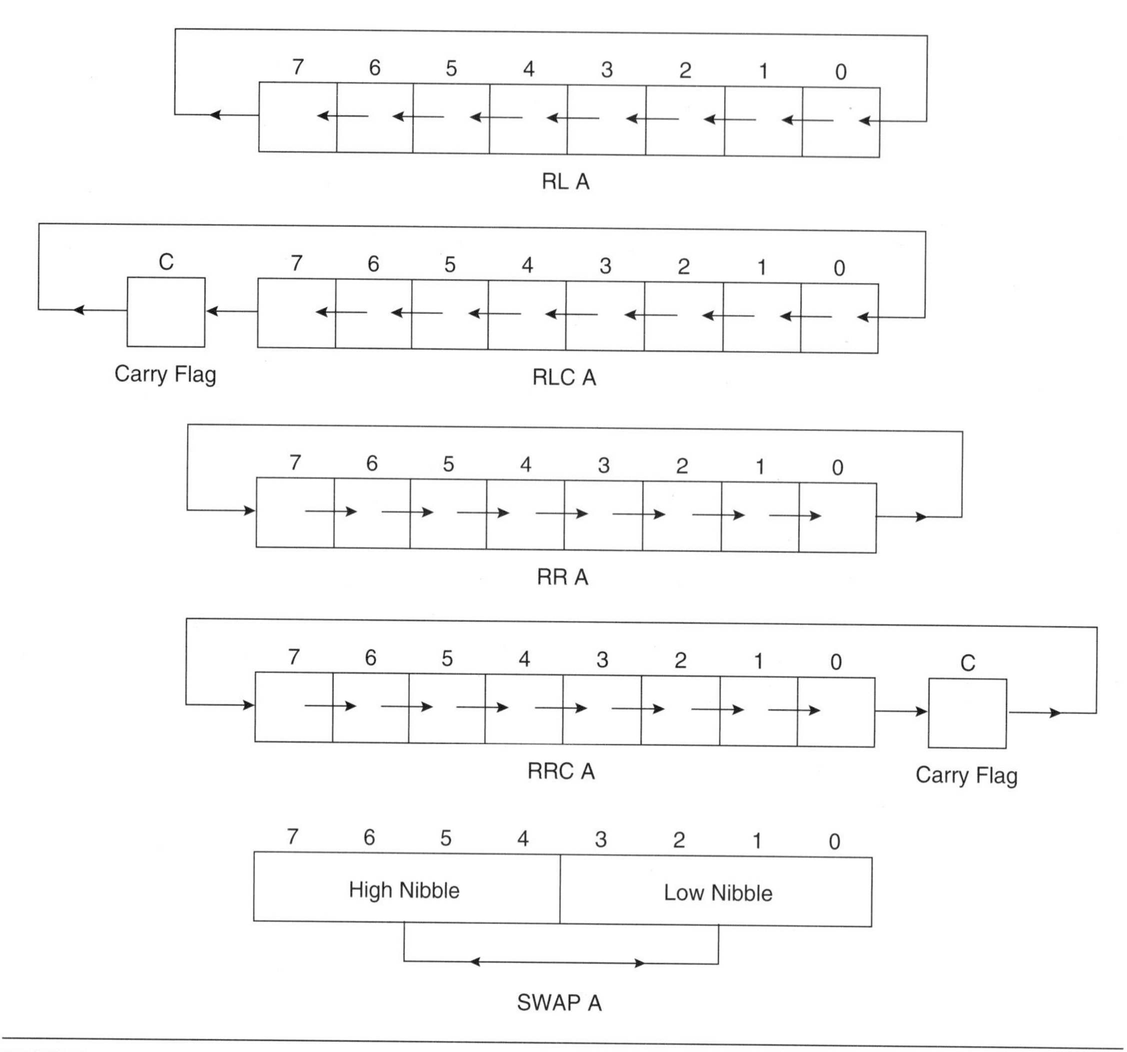

FIGURE 6.2 ◆ Register A Rotate Operations

Note that no flags, other than the Carry flag in RRC and RLC, are affected. If the carry is used as part of a rotate instruction, the state of the Carry flag should be known before the rotate is done.

The following list shows examples of rotate and swap operations:

Mnemonic	Operation Result
MOV A,#0A5h	A = 10100101b = A5h
RR A	A = 11010010b = D2h

RR A	A = 01101001b = 69h
RR A	A = 10110100b = B4h
RR A	A = 01011010b = 5Ah
SWAP A	A = 10100101b = A5h
CLR C	C = 0; A = 10100101b = A5h
RRC A	C = 1; A = 01010010b = 52h
RRC A	C = 0; A = 10101001b = A9h
RL A	A = 01010011b = 53h
RL A	A = 10100110b = A6h
SWAP A	C = 0; A = 01101010b = 6Ah
RLC A	C = 0; A = 11010100b = D4h
RLC A	C = 1; A = 10101000b = A8h
SWAP A	C = 1; A = 10001010b = 8Ah

◆ CAUTION ◆

- Know the state of the Carry flag when using RRC or RLC.
- Rotation and swap operations are limited to the A register.

6.4 Example Problems

The programs in this section are written using only opcodes covered to this point in the book. The challenge is to minimize the number of lines of code.

EXAMPLE PROBLEM 6.1

Use three different instructions to clear the contents of the A register.

- **Method 1:** A MOV instruction:

Mnemonic	Operation
MOV A, #00h	Move the immediate number 00h to A

- **Method 2:** A CLR instruction:

Mnemonic	Operation
CLR A	Clear register A to 00h

- **Method 3:** ANL the contents of A with 00h:

Mnemonic	Operation
ANL A,#00h	AND the contents of A with 00h

◆

EXAMPLE PROBLEM 6.2

Write programs that will invert (complement) every bit in register R6 of bank 0.

- **Method 1:** Move the contents of R6 to A, complement them, and move them back:

Mnemonic	Operation
MOV A,R6	Move contents of R6 to A
CPL A	Complement the contents of A
MOV R6,A	Move contents of A back to R6

- **Method 2:** Make all bits in A one, XRL with R6, then move the results back to R6:

Mnemonic	Operation
MOV A,#0ffh	A = ffh
XRL, A,R6	A = complement of R6
MOV R6,A	R6 = A

- **Method 3:** XRL R6 with FFh using the direct address of R6 in bank 0:

Mnemonic	Operation
XRL 06h,#0ffh	XRL R6 with ffh

◆

EXAMPLE PROBLEM 6.3

Configure Timer T0 to act as a 16-bit timer and Timer T1 to act as a 13-bit counter, and then start them.

Timer configuration is programmed in the TMOD register (see Chapter 3), which is *not* bit addressable. This example will "build" a control byte in register A using bit addressing and then move it to TMOD, in order to show step by step how timers are programmed. Bit-addressable bits in TCON then start each timer. (It is more efficient to just move a control byte to TMOD, as is shown in the last step of the example.)

Mnemonic	Operation
SETB ACC.0	T0 timer mode is 1; M0(T0) = 1
CLR ACC.1	M1(T0) = 0
CLR ACC.2	C/T(T0) = 0 to time
CLR ACC.3	Disable T0 gate bit
CLR ACC.4	T1 timer mode is 0; M0(T1) = 0
CLR ACC.5	M1(T1) = 0
SETB ACC.6	C/T(T1) = 1 to count

CLR ACC.7	Disable T1 gate bit
MOV TMOD,A	Configure timers
SETB TR1	Set T1 run bit
SETB TR0	Set T0 run bit
MOV TMOD, #41h	Also configures the timers

◆

EXAMPLE PROBLEM 6.4

Set and then clear bit 7 of internal RAM address 28h without changing any other bits in the byte.

◆ **Thoughts on the Problem** Internal RAM address 28h may be directly addressed using byte-level Boolean mnemonics, or, since byte 28h is in the bit-addressable area, bit-level operators may be used on bit 7 of byte 28h. Bit 7 of byte 28h is direct bit address 47h.

Mnemonic	Operation
ORL 28h,#80h	1 OR X = 1; set bit 7 only
ANL 28h,#7fh	0 AND X = 0; clear bit 7 only
SETB 47h	Set the bit
CLR 47h	Clear the bit

◆

EXAMPLE PROBLEM 6.5

Double the number in register R2, and put the result in registers R3 (high byte) and R4 (low byte).

◆ **Thoughts on the Problem** The largest number in R2 is FFh; the largest result is 1FEh. There are at least three ways to solve this problem: Use the MUL instruction (multiply, covered in Chapter 7), add R2 to itself, or shift R2 left one time. The solution that shifts R2 left is as follows:

Mnemonic	Operation
CLR C	Clear the carry to rotate as MSB of byte
MOV A,R2	Get R2 to A
RLC A	Rotate left, which doubles the number
MOV R4,A	Put low byte of result in R4
CLR A	Clear A to receive carry
RLC A	The carry bit is now bit 0 of A
MOV R3,A	Transfer any carry bit to R3

◆

◆ COMMENT ◆

Note how the Carry flag has to be cleared to a known state before being used in a rotate operation.

EXAMPLE PROBLEM 6.6

OR the contents of ports 1 and 2; put the result in external RAM location 0100h.

◆ **Thoughts on the Problem** The ports should be input ports for this problem to make any physical sense; otherwise, we would not know whether to use the pin data or the port SFR latch data.

The solution is as follows:

Mnemonic	Operation
MOV A,90h	Copy the pin data from port 1 to A
ORL A,0A0h	OR the contents of A with port 2; results in A
MOV DPTR,#0100h	Set the DPTR to point to external RAM address
MOVX @DPTR,A	Store the result

◆

◆ COMMENT ◆

Any time the port is the source of data, the pin levels are read; when the port is the destination, the latch is written. If the port is both source and destination (read–modify–write instructions), then the latch is used.

EXAMPLE PROBLEM 6.7

Find a number that, when XORed to the A register, results in the number 3Fh in A.

◆ **Thoughts on the Problem** Any number can be in A, so we will work backwards:

3Fh = A XOR N;   A XOR 3Fh = A XOR A XOR N = N

The solution is as follows:

Mnemonic	Operation
MOV R0,A	Save A in R0
XRL A,#3Fh	XOR A and 3Fh; forming N
XRL A,R0	XOR A and N yielding 3Fh

◆

◆ COMMENT ◆

Does this program work? Let's try several A's and see.

A = FFh	A XOR 3Fh = C0h	C0h XOR FFh = 3Fh
A = 00h	A XOR 3Fh = 3Fh	3Fh XOR 00h = 3Fh
A = 5Ah	A XOR 3Fh = 65h	65h XOR 5Ah = 3Fh

EXAMPLE PROBLEM 6.8

Swap every even-numbered bit of register R3 in bank 0 with the odd-numbered bit to its left. Swap bit 0 with bit 1, bit 2 with bit 3, and so on until bit 6 is swapped with bit 7.

◆ **Thoughts on the Problem** A rotate instruction can be used to swap the bits, and Boolean operators used to selectively save the rotated bits.

Mnemonic	Operation
MOV A,R3	Copy R3 to A
RL A	Even-numbered bits to odd-numbered positions
ANL A,#0AAh	Mask out (set to zero) the old odd-numbered bits
PUSH ACC	Save even-numbered bits
MOV A,R3	Copy R3 ro A
RR A	Odd-numbered bits to even-numbered positions
ANL A,#055h	Mask out the old odd-numbered bits
MOV R3,A	Save odd-numbered bits
POP ACC	Retrieve even-numbered bits
ORL 03h,A	OR the even and odd numbered bits into R3

◆

6.5 Summary

Boolean logic, rotate, and swap instructions are covered in this chapter. Byte-level operations involve each individual bit of a source byte operating on the same bit position in the destination byte; the results are put in the destination, while the source is not changed:

```
ANL destination,source
ORL destination,source
XRL destination,source
CLR A
CPL A
RR A
RL A
RRC A
RLC A
SWAP A
```

Bit-level operations involve individual bits found in one area of internal RAM and certain SFRs that may be addressed both by the assigned direct-byte address and eight individual bit addresses. The following Boolean logical operations may be done on each of these addressable bits:

ANL bit
ORL bit
CLR bit
CPL bit
SETB bit
MOV C, source bit
MOV destination bit, C

6.6 Problems

Write programs that perform the tasks listed using only opcodes that have been discussed in this and previous chapters. Write comments for each line of code and try to use as few lines as possible.

1. Set port 0, bits 1,3,5, and 7, to 1; set the rest to 0.
2. Clear bit 3 of RAM location 22h without affecting any other bit.
3. Invert the data on the port 0 pins and write the data to port 1.
4. Swap the nibbles of R0 and R1 so that the low nibble of R0 swaps with the high nibble of R1 and the high nibble of R0 swaps with the low nibble of R1.
5. Complement the lower nibble of RAM location 2Ah.
6. Make the low nibble of R5 the complement of the high nibble of R6.
7. Make the high nibble of R5 the complement of the low nibble of R6.
8. Move bit 6 of R0 to bit 3 of port 3.
9. Move bit 4 of RAM location 30h to bit 2 of A.
10. XOR a number with whatever is in A so that the result is FFh.
11. Store the most significant nibble of A in both nibbles of register R5; for example, if A = B6h, then R5 = BBh.
12. Store the least significant nibble of A in both nibbles of RAM address 3Ch; for example, if A = 36h, then 3Ch = 66h.
13. Set the Carry flag to 1 if the number in A is even; set the Carry flag to 0 if the number in A is odd.
14. Treat registers R0 and R1 as 16-bit registers, and rotate them one place to the left; bit 7 of R0 becomes bit 0 of R1, bit 7 of R1 becomes bit 0 of R0, and so on.
15. Repeat Problem 14 but rotate the registers one place to the right.

16. Rotate the DPTR one place to the left; bit 15 becomes bit 0.
17. Repeat Problem 16 but rotate the DPTR one place to the right.
18. Shift register B one place to the left; bit 0 becomes a 0, bit 6 becomes bit 7, and so on. Bit 7 is lost.
19. Find an additional single-instruction solution to Example Problem 6.1.
20. Write a program that will swap the bits of each nibble in register R5. Swap bits 0 and 1 with bits 2 and 3, and bits 4 and 5 with bits 6 and 7.

7

Arithmetic Operations

Chapter Outline

CHAPTER OBJECTIVES

On successful completion of this chapter you will be able to:

- Use instructions to increment and decrement the contents of registers and RAM.
- Do signed and unsigned addition and subtraction.
- Do unsigned multiplication and division.
- Do BCD addition.

7.0 Introduction

Applications of microcontrollers often involve performing mathematical calculations on data in order to alter program flow and modify program actions. A microcontroller is not designed to be a "number cruncher," as is a general-purpose computer. The domain of the microcontroller is that of controlling events as they change (real-time control). A sufficient number of mathematical opcodes must be provided, however, so that calculations associated with the control of simple processes can be done, in real time, as the controlled system operates. When faced with a control problem, the programmer must know whether the 8051 has sufficient capability to expeditiously handle the required data manipulation. If it does not, a higher performance model must be chosen.

The 24 arithmetic opcodes are grouped into the following types:

Mnemonic	Operation
INC destination	Increment destination by 1
DEC destination	Decrement destination by 1
ADD/ADDC destination, source	Add source to destination without/with Carry (C) flag
SUBB destination, source	Subtract, with carry, source from destination
MUL AB	Multiply the contents of registers A and B
DIV AB	Divide the contents of register A by the contents of register B
DA A	Decimal Adjust the A register

The addressing modes for the destination and source are the same as those discussed in Chapter 5: immediate, register, direct, and indirect.

7.1 Flags

A key part of performing arithmetic operations is the ability to store certain results of those operations that affect the way in which the program operates. For example, adding together two 1-byte numbers results in a 1-byte partial sum, because the 8051 is an 8-bit machine. But it is possible to get a 9-bit result when adding two 8-bit numbers. The ninth bit must be stored also, so the need for a 1-bit register, or Carry flag in this case, is identified. The program will then have to deal with the ninth bit, perhaps by adding it to a higher order byte in a multiple-byte addition scheme. Similar actions may have to be taken when a larger byte is subtracted from a smaller one. In this case, a borrow is necessary and must be dealt with by the program.

The 8051 has several dedicated latches, or flags, that store results of arithmetic operations. Opcodes covered in Chapter 8 are available to alter program flow based on the state of the flags. Not all instructions change the flags, but many a programming error has been made by a forgetful programmer who overlooked an instruction that does change a flag.

The 8051 has four arithmetic flags: the Carry (C), Auxiliary Carry (AC), Overflow (OV), and Parity (P).

Instructions Affecting Flags

The C, AC, and OV flags are arithmetic flags. They are set to 1 or cleared to 0 automatically, depending on the outcomes of the instructions shown in Table 7.1. The instruction set in Table 7.1 includes *all* instructions that modify the flags and is not confined to arithmetic instructions.

TABLE 7.1

Instruction Mnemonic	Flags Affected		
ADD	C	AC	OV
ADDC	C	AC	OV
ANL C,direct	C		
CJNE	C		
CLR C	C = 0		
CPL C	$C = \overline{C}$		
DA A	C		
DIV	C = 0	OV	
MOV C,direct	C		
MUL	C = 0	OV	
ORL C,direct	C		
RLC	C		
RRC	C		
SETB C	C = 1		
SUBB	C	AC	OV

One should remember, however, that the flags are all stored in the PSW. *Any* instruction that can modify a bit or a byte in that register (MOV, SETB, XCH, etc.) changes the flags. This type of change takes conscious effort on the part of the programmer.

A flag may be used for more than one type of result. For example, the C flag indicates a carry out of the lower byte position during addition and indicates a borrow during subtraction. The instruction that *last* affects a flag determines the use of that flag.

The Parity flag is affected by every instruction executed. The P flag will be set to a 1 if the number of 1's in the A register is odd and will be set to 0 if the number of 1's is even. All 0's in A yield a 1's count of 0, which is considered to be *even*. Parity check is an elementary error-checking method and is particularly valuable when checking data received via the serial port.

7.2 Incrementing and Decrementing

The simplest arithmetic operations involve adding or subtracting a binary 1 and a number. These simple operations become very powerful when coupled with the ability to repeat the operation—that is, to INCrement or DECrement—until a desired result is reached.[1] Register, direct, and indirect addresses may be INCremented or DECremented. No math flags (C, AC, OV) are affected.

The following list shows the increment and decrement mnemonics:

Mnemonic	Operation
INC A	Add a 1 to the A register
INC Rr	Add a 1 to register Rr
INC add	Add a 1 to contents of the direct memory address
INC @ Rp	Add a 1 to the contents of the memory address in Rp
INC DPTR	Add a 1 to the 16-bit DPTR
DEC A	Subtract a 1 from register A
DEC Rr	Subtract a 1 from register Rr
DEC add	Subtract a 1 from the contents of the direct memory address
DEC @ Rp	Subtract a 1 from the contents of the memory address in Rp

Note that increment and decrement instructions that operate on a port direct address alter the *latch* for that port.

The following list shows examples of increment and decrement arithmetic operations:

[1] This subject will be explored in Chapter 8.

Mnemonic	Operation
MOV A,#3Ah	A = 3Ah
DEC A	A = 39h
MOV R0,#15h	R0 = 15h
MOV 15h,#12h	Internal RAM address 15h = 12h
INC @R0	Internal RAM address 15h = 13h
DEC 15h	Internal RAM address 15h = 12h
INC R0	R0 = 16h
MOV 16h,A	Internal RAM address 16h = 39h
INC @R0	Internal RAM address 16h = 3Ah
MOV DPTR,#12FFh	DPTR = 12FFh
INC DPTR	DPTR = 1300h
DEC 83h	DPTR = 1200h (SFR 83h is the DPH byte)

◆ CAUTION ◆

- Remember: *No* math flags are affected.
- All 8-bit address contents overflow from FFh to 00h.
- DPTR is 16 bits; DPTR overflows from FFFFh to 0000h.
- The 8-bit address contents underflow from 00h to FFh.
- There is no DEC DPTR to match the INC DPTR.

7.3 Addition

All addition is done with the A register as the destination of the result. All addressing modes may be used for the source: an immediate number, a register, a direct address, and an indirect address. Some instructions include the Carry flag as an additional source of a single bit that is included in the operation at the *least* significant bit position.

The following list shows the addition mnemonics:

Mnemonic	Operation
ADD A,#n	Add A and the immediate number n; put the sum in A
ADD A,Rr	Add A and register Rr; put the sum in A
ADD A,add	Add A and the address contents; put the sum in A
ADD A,@Rp	Add A and the contents of the address in Rp; put the sum in A

Note that the C flag is set to 1 if there is a carry out of bit position 7; it is cleared to 0 otherwise. The AC flag is set to 1 if there is a carry out of bit position 3; it is cleared otherwise. The OV flag is set to 1 if there is a carry out of bit position 7, but not bit position 6 or if there is a carry out of bit position 6 but not bit position 7, which may be expressed as the logical operation

OV = C7 XOR C6

Unsigned and Signed Addition

The programmer may decide that the numbers used in the program are to be unsigned numbers—that is, numbers that are 8-bit positive binary numbers ranging from 00h to FFh. Alternatively, the programmer may need to use both positive and negative signed numbers.

Signed numbers use bit 7 as a *sign* bit in the most significant byte (MSB) of the *group* of bytes *chosen* by the programmer to represent the largest number to be needed by the program. Bits 0 to 6 of the MSB, and any other bytes, express the magnitude of the number. Signed numbers use a 1 in bit position 7 of the MSB as a negative sign and a 0 as a positive sign. Further, all negative numbers are not in true form, but are in two's complement form. When doing signed arithmetic, the programmer must *know* how large the largest number is to be—that is, how many bytes are needed for each number.

In signed form, a single byte number may range in size from 10000000b, which is −128d, to 01111111b, which is +127d. The number 00000000b is 000d and has a positive sign, so there are 128d negative numbers and 128d positive numbers. The C and OV flags have been included in the 8051 to enable the programmer to use either numbering scheme.

Adding or subtracting unsigned numbers may generate a Carry flag when the sum exceeds FFh or a Borrow flag when the minuend is less than the subtrahend. The OV flag is not used for unsigned addition and subtraction. Adding or subtracting signed numbers can lead to carries and borrows in a similar manner, and to overflow conditions as a result of the actions of the sign bits.

Unsigned Addition

Unsigned numbers make use of the Carry flag to detect when the result of an ADD operation is a number larger than FFh. If the carry is set to 1 after an ADD, then the carry can be added to a higher order byte so that the sum is not lost. For instance:

```
 95d  =    01011111b               =    5Fh
189d  =    10111101b               =    BDh
284d     1)00011100b = 284d           1)1Ch
```

where) indicates the state of the C flag. The C flag is set to 1 to account for the carry out from the sum. The program could add the Carry flag to another byte that forms the second byte of a larger number.

Signed Addition

Signed numbers may be added two ways: addition of like signed numbers and addition of unlike signed numbers. If unlike signed numbers are added, then it is not possible for the result to be larger than −128d or +127d, and the sign of the result will always be correct. For example:

```
 −001d  =   11111111b               =    FFh
 +027d  =   00011011b               =    1Bh
 +026d    1)00011010b  = +026d         1)1Ah
```

Here, there is a carry from bit 7 so the Carry flag is 1. There is also a carry from bit 6, and the OV flag is 0. For this condition, no action need be taken by the program to correct the sum.

If positive numbers are added, there is the possibility that the sum will exceed +127d, as demonstrated in the following example:

```
 +100d  =   01100100b               =    64h
 +050d  =   00110010b               =    32h
 +150d    0)10010110b  = −106d         0)96h
```

Ignoring the sign of the result, the magnitude is seen to be +22d, which would be correct if we had some way of accounting for the +128d, which, unfortunately, is larger than a single byte can hold. There is no carry from bit 7 and the Carry flag is 0; there is a carry from bit 6 so the OV flag is 1.

An example of adding two positive numbers that do not exceed the positive limit is this:

```
 +045d  =   00101101b               =    2Dh
 +075d  =   01001011b               =    4Bh
 +120d    0)01111000b  = 120d          0)78h
```

Note that there are no carries from bits 6 or 7 of the sum; the Carry and OV flags are both 0.

The result of adding two negative numbers together for a sum that does not exceed the negative limit is shown in this example:

```
 −030d  =   11100010b               =    E2h
 −050d  =   11001110b               =    CEh
 −080d    1)10110000b  = −080d         1)B0h
```

Here, there is a carry from bit 7 and the Carry flag is 1; there is a carry from bit 6 and the OV flag is 0. These are the same flags as the case for adding unlike numbers; no corrections are needed for the sum.

When adding two negative numbers whose sum does exceed −128d, we have:

−070d	=	10111010b		= BAh
−070d	=	10111010b		= BAh
−140d		1)01110100b	= +116d	1)74h

Or, the magnitude can be interpreted as −12d, which is the remainder after a carry out of −128d. In this example, there is a carry from bit position 7, and no carry from bit position 6, so the Carry and the OV flags are set to 1. The magnitude of the sum is correct; the sign bit must be changed to a 1.

From these examples the programming actions needed for the C and OV flags are as follows:

Flags		Action
C	**OV**	
0	0	None
0	1	Complement the sign
1	0	None
1	1	Complement the sign

A general rule is that *if the OV flag is set, then complement the sign.* The OV flag also signals that the sum exceeds the largest positive or negative numbers thought to be needed in the program.

Multiple-Byte Signed Arithmetic

The nature of multiple-byte arithmetic for signed and unsigned numbers is distinctly different from single-byte arithmetic. Using more than one byte in unsigned arithmetic means that carries or borrows are propagated from low-order to high-order bytes by the simple technique of adding the carry to the next highest byte for addition and subtracting the borrow from the next highest byte for subtraction.

Signed numbers appear to behave like unsigned numbers until the last byte is reached. For a signed number, the seventh bit of the highest byte is the sign; if the sign is negative, then the *entire* number is in two's complement form.

For example, using a 2-byte signed number, we have the following examples:

+32767d = 01111111 11111111b = 7FFFh
+00000d = 00000000 00000000b = 0000h
−00001d = 11111111 11111111b = FFFFh
−32768d = 10000000 00000000b = 8000h

Note that the lowest byte of the numbers 00000d and −32768d are exactly alike, as are the lowest bytes for +32767d and −00001d.

For multi-byte signed number arithmetic, then, the lower bytes are treated as unsigned numbers. All checks for overflow are done only for the highest order byte that contains the sign. An overflow at the highest order byte is not usually recoverable. The programmer has made a *mistake* and probably has made no provisions for a number larger than planned. Some error acknowledgment procedure, or user notification, should be included in the program if this type of mistake is a possibility.

The preceding examples show the need to add the Carry flag to higher order bytes in signed and unsigned addition operations. Opcodes that accomplish this task are similar to the ADD mnemonics: A C is appended to show that the carry bit is added to the sum in bit position 0.

The following list shows the add with carry mnemonics:

Mnemonic	Operation
ADDC A,#n	Add the contents of A, the immediate number n, and the C flag; put the sum in A
ADDC A,add	Add the contents of A, the direct address contents, and the C flag; put the sum in A
ADDC A,Rr	Add the contents of A, register Rr, and the C flag; put the sum in A
ADDC A,@Rp	Add the contents of A, the contents of the indirect address in Rp, and the C flag; put the sum in A

Note that the C, AC, and OV flags behave exactly as they do for the ADD commands.

The following list shows examples of ADD and ADDC multiple-byte signed arithmetic operations:

Mnemonic	Operation
MOV A,#1Ch	A = 1Ch
MOV R5,#0A1h	R5 = A1h
ADD A,R5	A = BDh; C = 0, OV = 0
ADD A,R5	A = 5Eh; C = 1, OV = 1
ADDC A,#10h	A = 6Fh; C = 0, OV = 0
ADDC A,#10h	A = 7Fh; C = 0, OV = 0

- ADDC is normally used to add a carry after the LSBY addition in a multi-byte process. ADD is normally used for the LSBY addition.

7.4 Subtraction

Subtraction can be done by taking the two's complement of the number to be subtracted, the subtrahend, and adding it to another number, the minuend. The 8051, however, has commands to perform direct subtraction of two signed or unsigned numbers. Register A is the destination address for subtraction. All four addressing modes may be used for source addresses. The commands treat the Carry flag as a borrow and always subtract the Carry flag as part of the operation.

The following list shows the subtract mnemonics:

Mnemonic	Operation
SUBB A,#n	Subtract immediate number n and the C flag from A; put the result in A
SUBB A,add	Subtract the contents of add and the C flag from A; put the result in A
SUBB A,Rr	Subtract Rr and the C flag from A; put the result in A
SUBB A,@Rp	Subtract the contents of the address in Rp and the C flag from A; put the result in A

Note that the C flag is set if a borrow is needed into bit 7 and reset otherwise. The AC flag is set if a borrow is needed into bit 3 and reset otherwise. The OV flag is set if there is a borrow into bit 7 and not bit 6 or if there is a borrow into bit 6 and not bit 7. As in the case for addition, the OV flag is the XOR of the borrows into bit positions 7 and 6.

Unsigned and Signed Subtraction

Again, depending on what is needed, the programmer may choose to use bytes as signed or unsigned numbers. The Carry flag is now thought of as a Borrow flag to account for situations when a larger number is subtracted from a smaller number. The OV flag indicates results that must be adjusted whenever two numbers of unlike signs are subtracted and the result exceeds the planned signed magnitudes.

Unsigned Subtraction

Because the C flag is always subtracted from A along with the source byte, it must be set to 0 if the programmer does not want the flag included in the subtraction. If a multi-byte subtraction is done, the C flag is cleared for the first byte and then included in subsequent higher byte operations.

The result will be in true form, with no borrow if the source number is smaller than A, or in two's complement form, with a borrow if the source is larger than A. These are *not* signed numbers, as all 8 bits are used for the magnitude. The range of numbers is from positive 255d (C = 0, A = FFh) to negative 255d (C = 1, A = 01h).

The following example demonstrates subtraction of a larger number from a smaller number:

```
         015d =    00001111b             =  0Fh
SUBB     100d =    01100100b             =  64h
        -085d    1)10101011b = 171d      = 1)ABh
```

The C flag is set to 1, and the OV flag is set to 0. The two's complement of the result is 085d.

The reverse of the example yields the following result:

	100d	=	01100100b		=	64h
SUBB	015d	=	00001111b		=	0Fh
	085d	=	0)01010101b	= 085d		0)55h

The C flag is set to 0, and the OV flag is set to 0. The magnitude of the result is in true form.

Signed Subtraction

As is the case for addition, two combinations of unsigned numbers are possible when subtracting: subtracting numbers of like and unlike signs. When numbers of like sign are subtracted, it is impossible for the result to exceed the positive or negative magnitude limits of +127d or −128d, so the magnitude and sign of the result do not need to be adjusted, as shown in the following example:

	+100d	=	01100100b	(Carry flag = 0 before SUBB)	=	64h
SUBB	+126d	=	01111110b		=	7Eh
	−026d		1)11100110b	= −026d		1)E6h

There is a borrow into bit positions 7 and 6; the Carry flag is set to 1, and the OV flag is cleared.

The following example demonstrates using two negative numbers:

	−061d	=	11000011b	(Carry flag = 0 before SUBB)	=	C3h
SUBB	−116d	=	10001100b		=	8Ch
	+055d		0)00110111b	= +55d		0)37h

There are no borrows into bit positions 6 or 7, so the OV and Carry flags are cleared to 0.

An overflow is possible when subtracting numbers of opposite sign because the situation becomes one of adding numbers of like signs, as can be demonstrated in the following example:

	−099d	=	10011101b	(Carry flag = 0 before SUBB)	=	9Dh
SUBB	+100d	=	01100100b		=	64h
	−199d		0)00111001b	= +057d		0)39h

Here, there is a borrow into bit position 6 but not into bit position 7; the OV flag is set to 1, and the Carry flag is cleared to 0. Because the OV flag is set to 1, the result must be adjusted. In this case, the magnitude can be interpreted as the two's complement of 71d, the remainder after a carry out of 128d from 199d. The magnitude is correct, and the sign needs to be corrected to a 1.

The following example shows a positive overflow:

	+087d =	01010111b (Carry flag = 0 before SUBB)	=	57h
SUBB	−052d =	11001100b	=	CCh
	+139d	1)10001011b = −117d		1)8Bh

There is a borrow from bit position 7, and no borrow from bit position 6; the OV flag and the Carry flag are both set to 1. Again the answer must be adjusted because the OV flag is set to 1. The magnitude can be interpreted as a +011d, the remainder from a carry out of 128d. The sign must be changed to a binary 0 and the OV condition dealt with.

The general rule is that *if the OV flag is set to 1, then complement the sign bit.* The OV flag also signals that the result is greater than −128d or +127d.

Again, it must be emphasized: When an overflow occurs in a program, an error has been made in the estimation of the largest number needed to successfully operate the program. Theoretically, the program could resize every number used, but this extreme procedure would tend to hinder the performance of the microcontroller.

Note that for all the examples in this section, it is *assumed* that the Carry flag = 0 before the SUBB. The Carry flag must be 0 before any SUBB operation that depends on C = 0 is done.

The following list shows examples of SUBB multiple-byte signed arithmetic operations:

Mnemonic	Operation
MOV 0D0h,#00h	Carry flag = 0
MOV A,#3Ah	A = 3Ah
MOV 45h,#13h	Address 45h = 13h
SUBB A,45h	A = 27h; C = 0, OV = 0
SUBB A,45h	A = 14h; C = 0, OV = 0
SUBB A,#80h	A = 94h; C = 1, OV = 1
SUBB A,#22h	A = 71h; C = 0, OV = 0
SUBB A,#0FFh	A = 72h; C = 1, OV = 0

◆ CAUTION ◆

- Remember to set the Carry flag to 0 if it is not to be included as part of the subtraction operation.

7.5 Multiplication and Division

The 8051 has the capability to perform 8-bit integer multiplication and division using the A and B registers. Register B is used solely for these operations and has no other use except as a location in the SFR space of RAM that could be used to hold data. The A register holds 1 byte of data before a multiply or divide operation, and 1 of the result bytes after a multiply or divide operation.

Multiplication and division treat the numbers in registers A and B as unsigned. The programmer must devise ways to handle signed numbers.

Multiplication

Multiplication operations use registers A and B as both source and destination addresses for the operation. The unsigned number in register A is multiplied by the unsigned number in register B, as follows:

Mnemonic	Operation
MUL AB	Multiply A by B; put the low-order byte of the product in A, put the high-order byte in B

The OV flag will be set if $A \times B > FFh$. Setting the OV flag does *not* mean that an error has occurred. Rather, it signals that the number is larger than 8 bits, and the programmer needs to inspect register B for the high-order byte of the multiplication operation. The Carry flag is always cleared to 0.

The largest possible product is FE01h when both A and B contain FFh. Register A contains 01h and register B contains FEh after multiplication of FFh by FFh. The OV flag is set to 1 to signal that register B contains the high-order byte of the product; the Carry flag is 0.

The following list gives examples of MUL multiple-byte arithmetic operations:

Mnemonic	Operation
MOV A,#7Bh	A = 7Bh
MOV 0F0h,#02h	B = 02h
MUL AB	A = F6h and B = 00h; OV Flag = 0
MOV B,#0FEh	B = FEh
MUL AB	A = 14h and B = F4h; OV Flag = 1

◆ CAUTION ◆

◆ Note there is no comma between A and B in the MUL mnemonic.

Division

Division operations use registers A and B as both source and destination addresses for the operation. The unsigned number in register A is divided by the unsigned number in register B, as follows:

Mnemonic	Operation
DIV AB	Divide A by B; put the integer part of quotient in register A and the integer part of the remainder in B

The OV flag is cleared to 0 unless B holds 00h before the DIV. Then the OV flag is set to 1 to show division by 0. The contents of A and B, when division by 0 is attempted, are undefined. The Carry flag is always reset.

Division always results in integer quotients and remainders, as shown in the following example:

$$\frac{\text{A1=213d}}{\text{B1=017d}} = 12 \text{ (quotient) and } 9 \text{ (remainder)}$$
$$[213\text{d} = (12 \times 17) + 9]$$

When done in hex:

$$\frac{\text{A=0D5h}}{\text{B=011h}} = \text{C (quotient) and 9 (remainder)}$$

The following list gives examples of DIV multiple-byte arithmetic operations:

Mnemonic	Operation
MOV A,#0FFh	A = FFh (255d)
MOV 0F0h,#2Ch	B = 2C (44d)
DIV AB	A = 05h and B = 23h [255d = (5 × 44) + 35]
DIV AB	A = 00h and B = 05h [05d = (0 × 35) + 5]
DIV AB	A = 00h and B = 00h [00d = (0 × 5) + 0]
DIV AB	A = ?? and B = ??; OV flag is set to 1

- The original contents of A and B are lost.
- Note there is no comma between A and B in the DIV mnemonic.

7.6 Decimal Arithmetic

Most 8051 applications involve adding intelligence to machines where the hexadecimal numbering system works naturally. There are instances, however, when the application involves interacting with humans, who insist on using the decimal number system. In such cases, it may be more convenient for the programmer to use the decimal number system to represent all numbers in the program.

Four bits are required to represent the decimal numbers from 0 to 9 (0000 to 1001) and the numbers are often called *binary coded decimal* (BCD) numbers. Two of these BCD numbers can then be packed into a single byte of data.

The 8051 does all arithmetic operations in pure binary. When BCD numbers are being used the result will often be a non-BCD number, as shown in the following example:

49BCD	= 01001001b		= 49h
+38BCD	= 00111000b		= 38h
87BCD	10000001b = 81BCD		81h

Note that to adjust the answer, an 06d needs to be added to the result.

The opcode that adjusts the result of BCD addition is the decimal adjust A for addition (DA A) command, as follows:

Mnemonic	**Operation**
DA A	Adjust the sum of two packed BCD numbers found in A register; leave the adjusted number in A.

The C flag is set to 1 if the adjusted number exceeds 99BCD and set to 0 otherwise. The DA A instruction makes use of the AC flag and the binary sums of the individual binary nibbles to adjust the answer to BCD. The AC flag has no other use to the programmer and no instructions—other than a MOV or a direct bit operation to the PSW—affect the AC flag.

It is important to remember that the DA A instruction assumes the added numbers were in BCD *before* the addition was done. Adding hexadecimal numbers and then using DA A will *not* convert the sum to BCD.

The DA A opcode only works when used with ADD or ADDC opcodes and does not give correct adjustments for SUBB, MUL, or DIV operations. The programmer might best consider the ADD or ADDC and DA A as a single instruction and use the pair automatically when doing BCD addition in the 8051.

The following list gives examples of BCD multiple-byte arithmetic operations:

Mnemonic	**Operation**
MOV A,#42h	A = 42BCD
ADD A,#13h	A = 55h; C = 0
DA A	A = 55h; C = 0
ADD A,#17h	A = 6Ch; C = 0
DA A	A = 72BCD; C = 0
ADDC A,#34h	A = A6h; C = 0
DA A	A = 06BCD; C = 1
ADDC A,#11h	A = 18BCD; C = 0
DA A	A = 18BCD; C = 0

◆ CAUTION ◆

- All numbers used must be in BCD form before addition.
- Only ADD and ADDC are adjusted to BCD by DA A.

7.7 Example Problems

The challenge of the programs presented in this section is writing them using only opcodes that have been covered to this point in the book. Experienced programmers may long for some of the opcodes to be covered in Chapter 8, but as we shall see, programs can be written without them.

EXAMPLE PROBLEM 7.1

Put a number into each internal RAM address from 20h to 27h, which equals the address number: put the number 20h into address 20h, the number 21h into address 21h, and so on.

◆ **Thoughts on the Problem** A series of direct address MOVes of an immediate number to each address is simpler, but this example will use INC and indirect MOVes instead. Register 1 of bank 0, which has the direct address 01h, is used to hold the indirect address in internal RAM.

Mnemonic	Operation
MOV R1,#20h	Put starting address in R1
MOV @R1,01h	Put the contents of R1 into the address pointed to by R1
INC R1	Point to the next address
MOV @R1,01h	And keep going until R1 = 27h
INC R1	Point to the next address
MOV @R1,01h	
INC R1	Point to the next address
MOV @R1,01h	
INC R1	Point to the next address
MOV @R1,01h	
INC R1	Point to the next address
MOV @R1,01h	
INC R1	Point to the next address
MOV @R1,01h	
INC R1	Point to the next address
MOV @R1,01h	Done

◆

EXAMPLE PROBLEM 7.2

Add a 1 to the A register using five different instructions.

◆ **Thoughts on the Problem** There are four ADD instructions and one INC instruction, which will add a 1 to the A register. Register bank 0 is assumed to be active.

Mnemonic	Operation
ADD A,#01h	Add an immediate number to A
MOV R5,#01h	Put the number in R5 and add it to A
ADD A,R5	
ADD A,05h	Or, add R5 using its direct address
MOV R0,#05h	Use R0 to point to R5
ADD A,@R0	
INC A	Which is the simplest way

◆

EXAMPLE PROBLEM 7.3

Subtract a 1 from the contents of the A register without using any address mode whose address contains the number one.

◆ **Thoughts on the Problem** SUBB also involves subtracting the C flag from A, so any of the four SUBB instructions may be used as long as C = 1. DECrementing A, or adding a −1 to it, as was done at the dawn of the computer age, also works.

Mnemonic	Operation
SETB C	Set the Carry flag to 1
SUBB A,#00h	Subtract the Carry flag
DEC A	Subtract a one from A
ADD A,#0FFh	FFh is a -1 signed number

◆

EXAMPLE PROBLEM 7.4

Add the unsigned numbers found in internal RAM locations 25h, 26h, and 27h together and put the result in RAM locations 31h (MSB) and 30h (LSB).

◆ **Thoughts on the Problem** The largest number possible is FFh + FFh = 01FEh + FFh = 02FDh, so that 2 bytes will hold the largest possible number. The MSB will be set to 0 and any carry bit added to it for each byte addition.

To solve this problem, use an ADD instruction for each addition and an ADDC to the MSB for each carry that might be generated. The first ADD will adjust any Carry flag that exists before the program starts.

The complete program is as follows:

Mnemonic	Operation
MOV 31h,#00h	Clear the MSB of the result to 0
MOV A,25h	Get the first byte to be added from location 25h
ADD A,26h	Add the second byte found in RAM location 26h
MOV R0,A	Save the sum of the first two bytes in R0
MOV A,#00h	Clear A to 00
ADDC A,31h	Add the carry to the MSB; carry = 0 after this operation
MOV 31h,A	Store MSB
MOV A,R0	Get partial sum back
ADD A,27h	Form final LSB sum
MOV 30h,A	Store LSB
MOV A,#00h	Clear A for MSB addition
ADDC A,31h	Form final MSB
MOV 31h,A	Store final MSB

◆

◆ COMMENT ◆

Notice how awkward it becomes to have to use the A register for all operations. Jump instructions, which will be covered in Chapter 8, require less use of A.

EXAMPLE PROBLEM 7.5

Repeat Example Problem 7.4 using BCD numbers.

◆ **Thoughts on the Problem** The numbers in the RAM locations *must* be in BCD before the problem begins. The largest number possible is 99d + 99d = 198d + 99d = 297d, so that up to two carries can be added to the MSB.

The solution to this problem is identical to that for unsigned numbers, except a DA A must be added after each ADD instruction. If more bytes were added so that the MSB could exceed 09d, then a DA A would also be necessary after the ADDC opcodes.

The complete program is as follows:

Mnemonic	Operation
MOV 31h,#00h	Clear the MSB of the result to 0
MOV A,25h	Get the first byte to be added from location 25h
ADD A,26h	Add the second byte found in RAM location 26h
DA A	Adjust the answer to BCD form
MOV R0,A	Save the sum of the first two bytes in R0
MOV A,#00h	Clear A to 00
ADDC A,31h	Add the carry to the MSB; carry = 0 after this operation
MOV 31h,A	Store MSB
MOV A,R0	Get partial sum back
ADD A,27h	Form final LSB sum
DA A	Adjust the final sum to BCD
MOV 30h,A	Store LSB
MOV A,#00h	Clear A for MSB addition
ADDC A,31h	Form final MSB
MOV 31h,A	Store final MSB

◆

◆ COMMENT ◆

When using BCD numbers, DA A can best be thought of as an integral part of the ADD instructions.

EXAMPLE PROBLEM 7.6

Multiply the unsigned number in register R3 by the unsigned number on port 2 and put the result in external RAM locations 10h (MSB) and 11h (LSB).

◆ **Thoughts on the Problem** The MUL instruction uses the A and B registers; the problem consists of MOVes to A and B followed by MOVes to the external RAM. The complete program is as follows:

Mnemonic	Operation
MOV A,0A0h	Move the port 2 pin data to A
MOV 0F0h,R3	Move the data in R3 to the B register
MUL AB	Multiply the data; A has the low order result byte
MOV R0,#11h	Set R0 to point to external RAM location 11h
MOV @R0,A	Store the LSB in external RAM
DEC R0	Decrement R0 to point to 10h
MOV A,0F0h	Move B to A
MOV @R0,A	Store the MSB in external RAM

◆

◆ COMMENT ◆

- Again we see the bottleneck created by having to use the A register for all external data transfers.
- More advanced programs that do signed math operations and multi-byte multiplication and division will have to wait for the development of Jump instructions in Chapter 8.

7.8 Summary

The 8051 can perform all four arithmetic operations: addition, subtraction, multiplication, and division. Signed and unsigned numbers may be used in addition and subtraction; an OV flag is provided to signal programmer errors in estimating signed number magnitudes needed and to adjust signed number results. Multiplication and division use unsigned numbers. BCD arithmetic may be done using the DA A and ADD or ADDC instructions.

The following list gives the arithmetic mnemonics:

Mnemonic	Operation
ADD A, source	Add the source byte to A; put the result in A and adjust the C and OV flags
ADDC A, source	Add the source byte and the carry to A; put the result in A and adjust the C and OV flags
DA A	Adjust the binary result of adding two BCD numbers in the A register to BCD and adjust the Carry flag
DEC source	Subtract a 1 from the source; roll from 00h to FFh
DIV AB	Divide the byte in A by the byte in B; put the quotient in A and the remainder in B; set the OV flag to 1 if B = 00h before the division
INC source	Add a 1 to the source; roll from FFh or FFFFh to 00h or 0000h
MUL AB	Multiply the bytes in A and B; put the high-order byte of the result in B, the low-order byte in A; set the OV flag to 1 if the result is > FFh
SUBB A, source	Subtract the source byte and the carry from A; put the result in A and adjust the C and OV flags

7.9 Problems

Write programs that perform the tasks listed using only opcodes that have been discussed in this and previous chapters. Use comments on each line of code and try to use as few lines as possible. All numbers may be considered to be unsigned numbers.

1. Add the bytes in RAM locations 34h and 35h; put the result in registers R5 (LSB) and R6 (MSB).

2. Add the bytes in registers R3 and R4; put the result in RAM locations 4Ah (LSB) and 4Bh (MSB).

3. Add the number 84h to the contents of RAM locations 17h (LSB) and 18h (MSB).

4. Add the byte in external RAM location 02CDh to internal RAM location 19h; put the result into external RAM locations 00C0h (LSB) and 00C1h (MSB).

5–8. Repeat Problems 1–4, assuming the numbers are in BCD format.

9. Subtract the contents of R2 from the number F3h; put the result in external RAM location 028Bh.

10. Subtract the contents of R1 from R0; put the result in R7.

11. Subtract the contents of RAM location 13h from RAM location 2Bh; put the result in RAM location 3Ch.

12. Subtract the contents of TH0 from TH1; put the result in TL0.

13. Increment the contents of RAM locations 13h, 14h, and 15h using indirect addressing only.

14. Increment TL1 by 05h.

15. Increment external RAM locations 0100h and 0200h.

16. Add a 1 to every external RAM address from 00h to 06h.

17. Add a 1 to every external RAM address from 0100h to 0106h.

18. Decrement TL0, TH0, TL1, and TH1.

19. Decrement external RAM locations 0123h and 01BDh.

20. Decrement external RAM locations 45h and 46h.

21. Multiply the data in RAM location 22h by the data in RAM location 15h; put the result in RAM locations 19h (low byte) and 1Ah (high byte).

22. Square the contents of R5; put the result in R0 (high byte) and R1 (low byte).

23. Divide the data in RAM location 3Eh by the number 12h; put the quotient in R4 and the remainder in R5.

24. Divide the number in RAM location 15h by the data in RAM location 16h; put the resulting quotient in external RAM location 7Ch.

25. Divide the data in RAM location 13h by the data in RAM location 14h, then restore the original data in 13h by multiplying the answer by the data in 14h.

8

Jump and Call Instructions

Chapter Outline

CHAPTER OBJECTIVES

On successful completion of this chapter you will be able to:

- Predict the range of jump opcodes.
- Use bit and byte conditional jump opcodes.
- Use unconditional jump opcodes.
- Write and call subroutines.
- Discuss how the CPU uses the stack to store call opcode return addresses.
- Place interrupt subroutines at predetermined program code addresses.
- Understand how hardware-generated interrupts operate.
- Enable interrupts using bits of the Interrupt Enable SFR.
- Assign priorities to interrupts using the Interrupt Priority SFR.
- Write small programs using calls and interrupts.

8.0 Introduction

The opcodes that have been examined and used in the preceding chapters may be thought of as action codes. Each instruction performs a single operation on bytes of data.

The jumps and calls discussed in this chapter are *decision* codes that alter the flow of the program by examining the results of the action codes and changing the contents of the program counter. A jump permanently changes the contents of the program counter if certain program conditions exist. A call temporarily changes the program counter to allow another part of the program to run. These decision codes make it possible for the programmer to let the program adapt itself, as it runs, to the conditions that exist at the time.

Although it is true that computers can't "think" (at least as of this writing), they can make decisions about events that the programmer can foresee, using the following types of decision opcodes:

Jump on bit conditions
Compare bytes and jump if *not* equal
Decrement byte and jump if zero
Jump unconditionally
Call a subroutine
Return from a subroutine

Jumps and calls may also be generically referred to as "branches," which emphasizes that two divergent paths are made possible by this type of instruction.

8.1 The Jump and Call Program Range

A jump or call instruction can replace the contents of the program counter with a new program address number that causes program execution to begin at the code located at the new address. The difference, in bytes, of this new address from the address in the program where the jump or call is located is called the *range* of the jump or call. For example, if a jump instruction is located at program address 0100h, and the jump causes the program counter to become 0120h, then the range of the jump is 20h bytes.

Jump or call instructions may have one of three ranges: a *relative* range of +127d, −128d bytes from the instruction *following* the jump or call instruction; an *absolute* range on the same 2K byte page as the instruction *following* the jump or call; or a *long* range of any address from 0000h to FFFFh, anywhere in program memory. Figure 8.1 shows the relative range of all the jump instructions.

Relative Range

Jumps that replace the program counter contents with a new address that is greater than the address of the instruction *following* the jump by 127d or less than the address of the instruction following the jump by 128d are called *relative* jumps. They are so named because the address that is placed in the program counter is relative to the address where the jump occurs. If the absolute address of the jump instruction changes, then the jump address changes also but remains the same distance away from the jump instruction. The address following the jump is used to calculate the relative jump because of the action of the PC. The PC is incremented to point to the *next* instruction *before* the current instruction is executed. Thus, the PC is set to the following address before the jump instruction is executed, or in the vernacular: "before the jump is taken."

Relative jumping has two advantages. First, only 1 byte of data need be specified, either in positive format for jumps ahead in the program or in two's complement negative format for jumps behind. The jump address displacement byte can then be added to the PC to get the absolute address. Specifying only 1 byte saves program bytes and speeds up program execution. Second, the program that is written using relative jumps can be located anywhere in the program address space without re-assembling the code to generate absolute addresses.

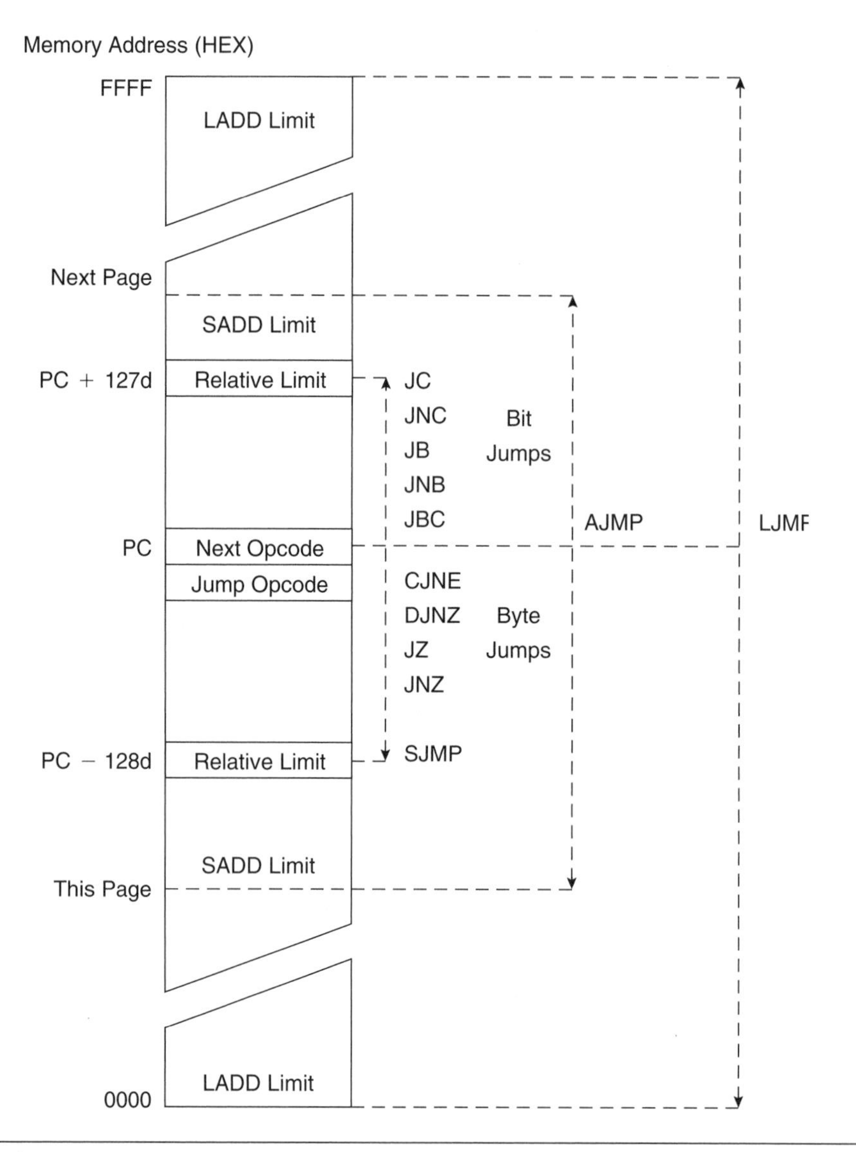

FIGURE 8.1 ◆ Jump Instruction Ranges

The disadvantage of using relative addressing is the requirement that all addresses jumped be within a range of +127d, −128d bytes (an 8-bit signed number range) of the jump instruction. This range is not a serious problem. Most jumps form program loops over short code ranges that are within the relative address capability. Jumps are the only branch instructions that can use the relative range.

If jumps beyond the relative range are needed, then a relative jump can be done to another relative jump until the desired address is reached. This need is better handled, however, by the jumps that are covered in the next sections.

Short Absolute Range

Absolute range makes use of the concept of dividing memory into logical divisions called *pages.* Program memory may be regarded as one continuous stretch of addresses from 0000h to FFFFh. Or, it may be divided into a series of pages of any convenient binary size, such as 256 bytes, 2K, 4K, and so on.

The 8051 program memory is arranged as 2K pages, giving a total of 32d (20h) pages. The hexadecimal address of each page is shown in Table 8.1.

Inspection of the page numbers shows that the upper 5 bits of the program counter hold the page *number,* and the lower 11 bits hold the *address* within each page. An absolute address is formed by taking the page number of the instruction *following* the branch and attaching the absolute page range address of 11 bits to it to form the 16-bit address.

Branches on page *boundaries* occur when the jump or call instruction finishes at X7FFh or XFFFh. The next instruction starts at X800h or X000h, which places the jump or call address on the same page as the *next* instruction after the jump or call. The page change presents no problem when branching ahead but could be troublesome if the branch is *backwards* in the program. The assembler will flag such problems as errors, so adjustments can be made by the programmer to use a different type of range.

Absolute range addressing has the same advantages as relative addressing; fewer bytes are needed and the code is relocatable as long as the relocated code remains on the following page. Absolute addressing has the advantage of

TABLE 8.1

Page	Address (hex)	Page	Address (hex)	Page	Address (hex)
00	0000 – 07FF	0B	5800 – 5FFF	16	B000 – B7FF
01	0800 – 0FFF	0C	6000 – 67FF	17	B800 – BFFF
02	1000 – 17FF	0D	6800 – 6FFF	18	C000 – C7FF
03	1800 – 1FFF	0E	7000 – 77FF	19	C800 – CFFF
04	2000 – 27FF	0F	7800 – 7FFF	1A	D000 – D7FF
05	2800 – 2FFF	10	8000 – 87FF	1B	D800 – DFFF
06	3000 – 37FF	11	8800 – 8FFF	1C	E000 – E7FF
07	3800 – 3FFF	12	9000 – 97FF	1D	E800 – EFFF
08	4000 – 47FF	13	9800 – 9FFF	1E	F000 – F7FF
09	4800 – 4FFF	14	A000 – A7FF	1F	F800 – FFFF
0A	5000 – 57FF	15	A800 – AFFF		

allowing jumps or calls over longer programming distances than does relative addressing.

Long Absolute Range

Addresses that can access the entire program space from 0000h to FFFFh use long-range addressing. Long-range addresses require more bytes of code to specify and are relocatable only at the beginning of 64K pages. Since we are limited to a nominal ROM address range of 64K, the program must be re-assembled every time a long-range address changes and these branches are not generally relocatable.

Long-range addressing has the advantage of using the entire program address space available to the 8051. It is most likely to be used in large programs.

8.2 Jumps

The ability of a program to respond quickly to changes in conditions depends largely on the number and types of jump instructions available to the programmer. The 8051 has a rich set of jumps that can operate at the bit and byte levels. These jump opcodes are one reason the 8051 is such a powerful microcontroller.

Jumps operate by testing for conditions that are specified in the jump mnemonic. If the condition is *true,* then the jump is taken—that is, the program counter is altered to the address that is part of the jump instruction. If the condition is *false,* then the instruction immediately following the jump instruction is executed because the program counter is not altered. Keep in mind that the condition of *true* does *not* mean a binary 1 and that *false* does *not* mean binary 0. The *condition* specified by the mnemonic is either true or false.

Bit Jumps

Bit jumps all operate according to the status of the Carry flag in the PSW or the status of any bit-addressable location. All bit jumps are relative to the program counter.

Jump instructions that test for bit conditions are shown in the following list:

Mnemonic	Operation
JC radd	Jump relative if the Carry flag is set to 1
JNC radd	Jump relative if the Carry flag is reset to 0
JB b,radd	Jump relative if addressable bit is set to 1
JNB b,radd	Jump relative if addressable bit is reset to 0
JBC b,radd	Jump relative if addressable bit is set, and clear the addressable bit to 0

Note that no flags are affected unless the bit in JBC is a flag bit in the PSW. When the bit used in a JBC instruction is a port bit, the SFR latch for that port is read, tested, and altered.

The following program example makes use of bit jumps:

Address	Mnemonic	Comment
LOOP:	MOV A,#10h	;A = 10h
	MOV R0,A	;R0 = 10h
ADDA:	ADD A,R0	;add R0 to A
	JNC ADDA	;if the Carry flag is 0, then no carry is
		;*true;* jump to address ADDA; jump until A
		;is F0h; the C flag is set to
		;1 on the next ADD and no carry is
		;*false;* do the next instruction
	MOV A,#10h	;A = 10h; do program again using JNB
ADDR:	ADD A,R0	;add R0 to A (R0 already equals 10h)
	JNB 0D7h,ADDR	;D7h is the bit address of the Carry flag
	JBC 0D7h,LOOP	;the carry bit is 1; the jump to LOOP
		;is taken, and the Carry flag is cleared
		;to 0
END		

- All jump addresses, such as ADDA and ADDR, must be within +127d, −128d of the instruction following the jump opcode.
- If the addressable bit is a flag bit and JBC is used, the flag bit will be cleared.
- *Do not* use any label names that are also the names of registers in the 8051. These are called *reserved* words and will cause great agitation in the assembler. See Appendix B for a listing of reserved names (also called *symbols.*)

Byte Jumps

Byte jumps—jump instructions that test bytes of data—behave as bit jumps. If the condition that is tested is *true,* the jump is taken; if the condition is *false,* the instruction after the jump is executed. All byte jumps are relative to the program counter.

The following list shows examples of byte jumps:

Mnemonic	Operation
CJNE A,add,radd	Compare the contents of the A register with the contents of the direct address; if they are *not* equal, then jump to the relative address; set the Carry flag to 1 if A is less than the contents of the direct address; otherwise, set the Carry flag to 0
CJNE A,#n,radd	Compare the contents of the A register with the immediate number n; if they are *not* equal, then jump to the relative address; set the Carry flag to 1 if A is less than the number; otherwise, set the Carry flag to 0
CJNE Rr,#n,radd	Compare the contents of register Rr with the immediate number n; if they are *not* equal, then jump to the relative address; set the Carry flag to 1 if Rn is less than the number; otherwise, set the Carry flag to 0
CJNE @Rp,#n,radd	Compare the contents of the address contained in register Rp to the number n; if they are *not* equal, then jump to the relative address; set the Carry flag to 1 if the contents of the address in Rp are less than the number; otherwise, set the Carry flag to 0
DJNZ Rr,radd	Decrement register Rr by 1 and jump to the relative address if the result is *not* 0; no flags are affected
DJNZ add,radd	Decrement the direct address by 1 and jump to the relative address if the result is *not* 0; no flags are affected unless the direct address is the PSW
JZ radd	Jump to the relative address if A is 0; the flags and the A register are not changed
JNZ radd	Jump to the relative address if A is *not* 0; the flags and the A register are not changed

Note that if the direct address used in a DJNZ is a port, the port SFR is decremented and tested for 0.

Unconditional Jumps

Unconditional jumps do not test any bit or byte to determine whether the jump should be taken. The jump is *always* taken. All jump ranges are found in this group of jumps, and these are the only jumps that can jump to any location in memory.

The following list shows examples of unconditional jumps:

Mnemonic	Operation
JMP @A+DPTR	Jump to the address formed by adding A to the DPTR; this is an unconditional jump and will always be done; the address can be anywhere in program memory; A, the DPTR, and the flags are unchanged
AJMP sadd	Jump to absolute short range address *sadd;* this is an unconditional jump and is always taken; no flags are affected
LJMP ladd	Jump to absolute long range address *ladd;* this is an unconditional jump and is always taken; no flags are affected
SJMP radd	Jump to relative address *radd;* this is an unconditional jump and is always taken; no flags are affected
NOP	Do nothing and go to the next instruction; NOP (no operation) is used to waste time in a software timing loop, or to leave room in a program for later additions; no flags are affected

The following program example uses byte and unconditional jumps:

Address	Mnemonic	Comments
	ORG 0000h	;code begins at 0000h. NOTE PC = 0000h
	MOV A, #30h	;compare A to contents of 50h, (50h)
	MOV 50h, #00h	;(50h) is set to 00h
AGN:	CJNE A, 50h, AEQ	;jump will be taken at first
	SJMP NXT	;relative jump taken when (50h) = 30h
AEQ:	DJNZ 50h, AGN	;count down location 50h until = 30h
NXT:	MOV R0, #0FFh	;count down R0 until = 0
DWN:	DJNZ R0, DWN	;jump here until R0 counts down
	MOV A, R0	;A = R0 = 00h
	JNZ ABIG	;will not take this jump
	JZ AZRO	;will take this jump
ABIG:	NOP	
AZRO:	LJMP WHERE	;not necessary, just to demo
	ORG 500h	;code assembled at 500h. NOTE PC = 500h
WHERE:	MOV A, #02h	;now jump to code at 0702h
	MOV DPTR, #700h	;still on the same 2K page
	JMP@A+DPTR	;jump to code at 702h
	ORG 700h	;NOTE PC = 702h
	NOP	;do nothing
	NOP	
HERE:	JMP LOOP	;absolute jump back to code at 0000h
END		

◆ **CAUTION** ◆

- DJNZ decrements *first, then* checks for 0. A location set to 00h and then decremented goes to FFh, then FEh, and so on, down to 00h.
- CJNE does not change the contents of any register or RAM location. It can change the Carry flag to 1 if the destination byte is less than the source byte.
- There is no zero flag; the JZ and JNZ instructions check the contents of the A register for 0.
- JMP @A+DPTR does not change A, DPTR, or any flags.

8.3 Calls and Subroutines

The life of a microcontroller would be very tranquil if all programs could run with no thought as to what is going on in the real world outside. However, a microcontroller is specifically intended to interact with the real world and to react, very quickly, to events that require program attention to correct or control.

A program that does not have to deal unexpectedly with the world outside of the microcontroller could be written using jumps to alter program flow as external conditions require. This sort of program can determine external conditions by moving data from the port pins to a location and jumping on the conditions of the port pin data. This technique is called *polling* and requires that the program does not have to respond to external conditions quickly. (Quickly means in microseconds; slowly means in milliseconds.)

Another method of changing program execution is using *interrupt* signals on certain external pins or internal registers to automatically cause a branch to a smaller program that deals with the specific situation. When the event that caused the interruption has been dealt with, the program resumes at the point in the program where the interruption took place. Interrupt action can also be generated using software instructions named *calls.*

Call instructions may be included explicitly in the program as mnemonics or implicitly included using hardware interrupts. In both cases, the call is used to execute a smaller, stand-alone program, which is termed a *routine* or, more often, a *subroutine.*

Subroutines

A *subroutine* is a program that may be used many times in the execution of a larger program. The subroutine could be written into the body of the main program everywhere it is needed, resulting in the fastest possible code execution. Using a subroutine in this manner has several serious drawbacks.

Common practice when writing a large program is to divide the total task among many programmers in order to speed completion. The entire program can be broken into smaller parts and each programmer given a part to write and debug. The main program can then call each of the parts, or subroutines, that have been developed and tested by each individual of the team.

Even if the program is written by one individual, it is more efficient to write an often-used routine once and then call it many times as needed. Also, when writing a program, the programmer does the main part first. Calls to subroutines, which will be written later, enable the larger task to be defined before the programmer becomes bogged down in the details of the application.

Finally, it is quite common to buy *libraries* of common subroutines that can be called by a main program. Again, buying libraries leads to faster program development.

Calls and the Stack

A call, whether hardware or software initiated, causes a jump to the address where the called subroutine is located. At the end of the subroutine the program resumes operation at the opcode address immediately *following* the call. As calls can be located anywhere in the program address space and used many times, there must be an automatic means of storing the address of the instruction following the call so that program execution can continue after the subroutine has executed.

The *stack* area of internal RAM is used to automatically store the address, called the return address, of the instruction found immediately after the call. The Stack Pointer register holds the address of the *last* space used on the stack. It stores the return address above this space, adjusting itself upward as the return address is stored. The terms *stack* and *stack pointer* are often used interchangeably to designate the *top* of the stack area in RAM that is pointed to by the stack pointer.

Figure 8.2 diagrams the following sequence of events:

1. A call opcode occurs in the program software, or an interrupt is generated in the hardware circuitry.
2. The return address of the next instruction after the call instruction or interrupt is found in the program counter.
3. The return address bytes are pushed on the stack, *low* byte *first*.
4. The stack pointer is incremented for each push on the stack.
5. The subroutine address is placed in the program counter.
6. The subroutine is executed.
7. A RET (return) opcode is encountered at the end of the subroutine.
8. Two pop operations restore the return address to the PC from the stack area in internal RAM.
9. The stack pointer is decremented for each address byte pop.

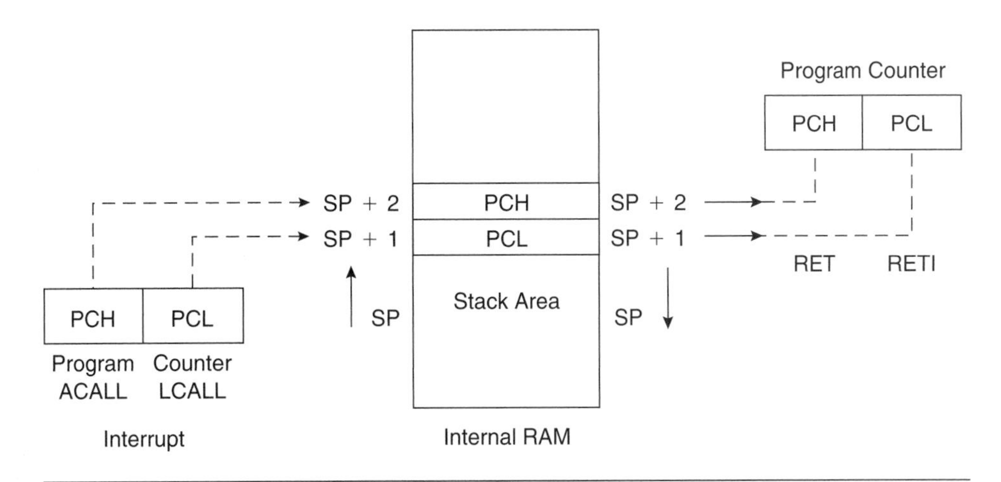

FIGURE 8.2 ◆ Storing and Retrieving the Return Address

- All of these steps are automatically handled by the 8051 hardware. It is the *responsibility* of the programmer to ensure that the subroutine ends in a RET instruction *and* that the stack does not grow up into data areas that are used by the program.
- Remember to set the SP above your data area.

Calls and Returns

Calls use short- or long-range addressing; returns have no addressing mode specified but are always long range. The following list shows examples of call opcodes:

Mnemonic	Operation
ACALL sadd	Call the subroutine located on the same page as the address of the opcode immediately following the ACALL instruction; push the address of the instruction immediately after the call on the stack
LCALL ladd	Call the subroutine located anywhere in program memory space; push the address of the instruction immediately following the call on the stack
RET	Pop 2 bytes from the stack into the program counter

8.4 Interrupts and Returns

As mentioned previously, an *interrupt* is a hardware-generated call. Just as a call opcode can be located within a program to automatically access a subroutine, certain pins on the 8051 can cause a call when external electrical signals on them go to a low state. Internal operations of the timers and the serial port can also cause an interrupt call to take place.

The subroutines called by an interrupt are located at fixed hardware addresses discussed in Chapter 3. Table 8.2 shows the interrupt subroutine addresses.

TABLE 8.2

Interrupt	Address (hex) Called
IE0	0003
TF0	000B
IE1	0013
TF1	001B
SERIAL	0023

When an interrupt call takes place, hardware interrupt disable flip-flops are set to prevent another interrupt of the same priority level from taking place until an interrupt return instruction has been executed in the interrupt subroutine. The action of the interrupt return instruction is shown below:

Mnemonic	Operation
RETI	Pop 2 bytes from the stack into the program counter and reset the interrupt enable flip-flops

Note that the only difference between the RET and RETI instructions is the enabling of the interrupt logic when RETI is used. RET is used at the ends of subroutines called by an opcode. RETI is used by subroutines called by an interrupt.

The following program example uses a call to a subroutine:

Address	Mnemonic	Comment
MAIN:	MOV 81h,#30h	;set the stack pointer to 30h in RAM
	LCALL SUB	;push address of NOP; PC = #SUB; SP = 32h
	NOP	;return from SUB to this opcode
	SJMP MAIN	;loop main program
SUB:	MOV A,#45h	;SUB loads A with 45h and returns
	RET	;pop return address to PC; SP = 30h

- Set the stack pointer above any area of RAM used for additional register banks or data memory.
- The stack may only be 128 bytes *maximum,* which limits the number of successive calls with no returns to 64.
- Using RETI at the end of a *software-called* subroutine may enable the interrupt logic erroneously.
- To jump out of a subroutine (not recommended), adjust the stack for the two return address bytes by POPing it twice or by moving data to the stack pointer to reset it to its original value.
- Use the LCALL instruction if your subroutines are normally placed at the end of your program.

In the following example of an interrupt call to a routine, timer 0 is used in mode 0 to overflow and set the timer 0 Interrupt flag. When the interrupt is generated, the program vectors to the interrupt routine, resets the timer 0 Interrupt flag, stops the timer, and returns.

```
Address   Mnemonic            Comment
          ORG 0000h           ;begin program at 0000
          AJMP OVER           ;jump over interrupt subroutine
          ORG 000Bh           ;put timer 0 interrupt subroutine here
          CLR 8Ch             ;stop timer 0; set TR0 = 0
          RETI                ;return and enable interrupt structure
          .
          .
OVER:     MOV 0A8h,#82h       ;enable the timer 0 interrupt in the IE
          MOV 89h,#00h        ;set timer operation, mode 0
          MOV 8Ah,#00h        ;clear TL0
          MOV 8Ch,#00h        ;clear TH0
          SET 8Ch             ;start timer 0; set TR0 = 1
HERE:     SJMP HERE           ;loop until interrupt
END
```

◆ CAUTION ◆

The programmer must enable any interrupt by setting the appropriate enabling bits in the IE register.

8◆5 More Detail on Interrupts

The preceding section on interrupts reviews material first introduced in Chapter 3. Those readers who have no immediate interest in interrupts may skip this section and inspect the simple call and interrupt examples. Programmers who are currently caught up (or bogged down) in writing *interrupt-driven* programs should find this section useful.

Interrupt concepts are basically simple, and involve the following *sequence* of events:

1. The programmer enables interrupt circuit action by setting interrupt enable flag bit(s) to 1. The 8051 has a total of five interrupt *sources,* each of which may generate an interrupt *signal.*
2. External or internal *circuit action* causes one of the interrupt signals to be generated. Typically, the signal involves setting an external flip-flop (provided by the system designer) or an internal flag bit (provided by the 8051).
3. The CPU *finishes the current instruction,* pushes the PC on the stack, and replaces the original PC contents with the address of the first instruction of the program code for the particular source that caused the interrupt. All of the other interrupting source enable bits are *temporarily* disabled.
4. The interrupt program executes. While executing, the interrupt program *must* reset the flip-flop (the system designer must incorporate this in the

external flip-flop circuit) or internal flag (means to do this are provided in the 8051) that generated the interrupt signal.

5. At the end of the interrupt program, a RETI instruction resets all the interrupt-enable circuitry and pops the original PC contents from the stack back into the PC. The CPU resumes executing the interrupted program.

Note that if the interrupting signal is not reset before a RETI instruction, the *same* interrupt will occur again. This process will never stop, and the program will *loop forever* (or until reset) at the interrupt program location.

An interrupt-driven program is one that makes intensive use of hardware interrupts to call subroutines that deal with events that are not a direct result of software instructions. Interrupt-driven programs are also called *real-time* programs to call attention to the fact that hardware events cause the CPU to service, or *handle,* those events quickly without waiting for *software polling* action. Subroutines that are written to specifically deal with interrupts are often called *interrupt handlers.*

Perhaps the most important difference between a call *instruction* and an interrupt *action* is that the place in the program where a call instruction executes is completely under the control of the programmer. Interrupt execution, however, may occur *at any place* in the program, with little programmer control beyond turning the interrupts on or off. Call instructions are placed in the program by the programmer. *When* the call is executed is determined by *where* it is placed in the program. The call can access its associated subroutine only when it is fetched and executed.

An interrupt, however, once it has been enabled by the programmer, may call its associated subroutine at any place and at *any time* the program is executing. The "any time" is the scary part: How do we know an interrupt won't come along just when the program is doing something really crucial and really mess things up? The answer to that question is that we have to be careful that interrupts can not do any damage to the program. The 8051 provides the programmer with two function registers by which rogue interrupts are kept under control: the Interrupt Enable (IE) and the Interrupt Priority (IP) registers.

The Interrupt Enable (IE) Special Function Register

The IE SFR is discussed in Chapter 3 and is also shown in Appendix F. The main purpose of the IE register is to allow the programmer to control *all* interrupt action with one master (*global*) register bit, and then to control *each* individual interrupt with an associated IE register enable bit.

The Enable All Interrupts (EA) Bit

The EA bit (bit 7 of IE) is the global interrupt enable bit. No interrupts will be acknowledged unless the EA bit is set to 1. Note, however, that setting EA does

TABLE 8.3

Interrupting Source	IE Name	IE Bit
Reserved (for 8052)	ET2	5
Serial Port	ES	4
Timer 1 Overflow	ET1	3
External Interrupt 1 (port 3, bit 3)	EX1	2
Timer 0 Overflow	ET0	1
External Interrupt 0 (port 3, bit 2)	EX0	0

not mean that interrupts will be serviced; setting EA means they *can* be serviced. Each interrupt must also be *individually* enabled.

Individual Interrupt Enable Bits

Each interrupting source has an associated enabling bit in IE. The individual interrupt sources are shown in Table 8.3.

To summarize: For any interrupt to be enabled, the EA bit and the enabling bit for that interrupt must *both be set to 1.*

The Interrupt Priority (IP) Special Function Register

The IP SFR may be used by the programmer to determine which interrupts are the most important. Importance means which of two interrupts will be serviced *first* should both occur at the same time. The more important will be acknowledged. The less important interrupt must wait, *and keep its interrupt signal active,* to be acknowledged by the CPU.

As was the case with the IE register, each interrupt source is assigned a priority bit in the IP register as shown in Table 8.4.

The programmer can assign a high priority to an interrupt by setting its priority bit to 1, or a low priority by setting the priority bit to 0.

TABLE 8.4

Interrupting Source	IP Name	IE Bit
Reserved (for 8052)	PT2	5
Serial Port	PS	4
Timer 1 Overflow	PT1	3
External Interrupt 1 (port 3, bit 3)	PX1	2
Timer 0 Overflow	PT0	1
External Interrupt 0 (port 3, bit 2)	PX0	0

Ties

If two interrupts happen at the same time, and they *both* have the same priority, there must be a way to choose one over the other. The dilemma of ties has been solved for us by the 8051 designers; the following order of priority is established in the 8051 circuitry:

1. External Interrupt 0 (port 3, bit 2)
2. Timer 0 Overflow
3. External Interrupt 1 (port 3, bit 3)
4. Timer 1 Overflow
5. Serial Port

Note that the 8051 does not remember any interrupt source but the one that has been acknowledged; interrupting sources must keep their interrupting signal active until the interrupt is handled. Internal interrupts (timers and serial) can keep their flag flip-flops active until the interrupt is serviced. External interrupts, however, are the *responsibility of the system designer/programmer.*

Interrupts and Interrupt Handler Subroutines

The designers of the 8051 have established *fixed* addresses in the code ROM where the subroutine handlers for each interrupt type *must* be placed. As discussed in Chapter 3, the fixed addresses that must contain the beginning instructions for each interrupt type are as shown in Table 8.5.

Note that there is not a lot of code space between the various interrupt handler addresses in code ROM. Clearly, if the handler subroutine is lengthy, a jump to the rest of the handler can be the first instruction used at an interrupt handler address. Note also that the main program, which must begin at address 0000h, must jump over any interrupt(s) addresses that are in use. For example, a program in which all of the interrupts are used should be arranged as shown next:

TABLE 8.5

Interrupt Source	Handler Subroutine Must Begin at Address
External Interrupt 0	0003h
Timer 0 Overflow	000Bh
External Interrupt 1	0013h
Timer 1 Overflow	001Bh
Serial Port	0023h

```
Address   Mnemonic          Comment
          sjmp over         ;over is a label above the
                            ;serial interrupt code
          org 0003h         ;ORG code for EI0 handler
          ljmp x0h          ;jump to EI0 handler routine
          org 000bh         ;ORG code for T0 handler
          ljmp t0h          ;jump to T0 handler routine
          org 0013h         ;ORG code for EI1 handler
          ljmp x1h          ;jump to EI1 handler routine
          org 001bh         ;ORG code for T1 handler
          ljmp t1h          ;jump to T1 handler routine
          org 0023h         ;ORG code for serial handler
          ljmp serh         ;jump to serial data routine
over:
          move sp,#40h      ;the rest of the program
          —
          —
          —
x0h:                        ;EI0 handler subroutine
          reti
t0h:                        ;T0 handler subroutine
          reti
x1h:                        ;EI1 handler subroutine
          reti
t1h:                        ;T1 handler subroutine
          reti
serh:                       ;Serial handler subroutine
          reti
end
```

The Fine Print

There are several "gotchas" the programmer should keep in mind when dealing with interrupt-driven programs:

- It is *crucial* that all of the registers and SFRs used by the interrupted program be unchanged when the main program resumes after the handler RETI instruction.
- If the interrupt handler is going to make use of the PSW register (to do math involving the Carry flag, or to change to another register bank, for instance) then *the old PSW must be saved,* usually on the stack. If the interrupt handler changes any register or SFR, then the old registers and SFRs must also be saved.
- Then, just before the RETI, the handler must restore (pop) all of the saved registers from the stack so that the CPU state is *exactly* as it was before the interrupt.

In conjunction with the points just listed, remember that the stack is not infinite. Be sure the stack is placed above any data areas you are using in your program, and do not push more on the stack than it can hold. To save stack space, *it is highly recommended* that the handler switch to a register bank that is never used by any other program and conduct its business there. In fact, that is exactly why the 8051 is equipped with four register banks: so that handlers can quickly switch from the main program bank (usually bank 0) to another. Bank switching saves stack space, and push and pop time, so that interrupts are really handled in real time.

In Conclusion

Interrupt-driven programs are considered to be the bread-and-butter of the microcontroller society. The programmer must take precautions so that the CPU is not overwhelmed or does not get lost. Bugs that arise from inattention to detail, such as forgetting to save the main program registers or letting the stack overflow, can be extremely difficult to find because of their seemingly *random* nature. But, when done correctly, interrupt-driven programs are the only way many industrial problems can be solved. For some examples of interrupt-driven programs, please refer to Chapters 9, 10, and 11.

8.6 Example Problems

We now have all of the tools needed to write powerful, compact programs. The addition of the decision jump and call opcodes permits the program to alter its operation as it runs.

EXAMPLE PROBLEM 8.1

Repeat Example 7.1 using a jump instruction from this chapter.

◆ **Thoughts on the Problem** A DJNZ instruction may be used to loop through the program until the memory is filled. NOTE: Simply changing the number placed in R0 enables large areas of memory to be filled.

Address	Mnemonic	Comment
LOOP:		;loop as many times as desired for debugging
	MOV R0,#08h	;20 to 27 takes 8 program actions
	MOV R1,#20h	;begin at RAM address 20h
MORE:		;program loops back to this label
	MOV @R1,01h	;move contents of R1 to RAM address in R1
	INC R1	;count R1 to point to next address
	DJNZ R0,MORE	;count R0 down to 0; jump to MORE until R0 = 0
	SJMP LOOP	;for debugger use to watch again
END		

◆

EXAMPLE PROBLEM 8.2

Write a program that uses all of the unconditional jumps.

◆ **Thoughts on the Problem** There are four truly unconditional jumps. To demonstrate them, ORG directives must be used to place some jump labels ("targets") hundreds or thousands of bytes from the jump instruction.

Address	Mnemonic	Comment
	ORG 000h	;direct the assembler to place code at xxxxh
		;NOTE: 0000h is the *default* starting place
BEGIN:		;a label for code address 0000h
	SJMP THERE	;short-range jump to the THERE label
	ORG 0050h	;direct the assembler to put code at 0050h
THERE:		;label for code address 0050h
	AJMP YONDER	;a page jump to the YONDER label
	ORG 0400h	;direct the assembler to put code at 0400h
YONDER:		;label for code address 0400h
	MOV A,#80h	;put an *offset* into A, 80h for example
	MOV DPTR,#0FF00h	;put a *base* number into DPTR, FF00h for
		;example
	JMP @A+DPTR	;jump to address FF80h. NOTE: no label
		;needed
	ORG 0FF80h	;direct assembler to put next code byte at
		;FF80h
	LJMP BEGIN	;long jump back to code address 0000h
END		

◆

EXAMPLE PROBLEM 8.3

Write a program that demonstrates all of the CALL instructions.

◆ **Thoughts on the Problem** It is crucial that the "main" program not reach a CALLed subroutine except by CALLing it. For this reason, programmers place all of the CALLed subroutines after the end of the main program. The main program loops back into itself to avoid "running into" any subroutine.

Address	Mnemonic	Comment
	MOV A,#34h	;assembler starts at default code address 0000h
	ACALL DOUBLE	;call subroutine assembled at the DOUBLE label
HERE:		;assembler keeps track of exactly where HERE is
	SJMP HERE	;main program loops forever at the HERE address
DOUBLE;		;code address for the DOUBLE subroutine
	ADD A,0E0h	;add A to itself
	LCALL INVERT	;now call the INVERT subroutine

```
THERE:                          ;not needed; used only to demonstrate a point
            RET                 ;return to the HERE label in the main program
            ORG 5000h           ;INVERT subroutine can be at any suitable code address
INVERT:                         ;INVERT starts at code address 5000h
            XRL, A,#0FFh        ;invert all the bits in A
            RET                 ;return to the THERE label in DOUBLE subroutine
END
```

◆

EXAMPLE PROBLEM 8.4

Turn on a low-current LED connected port 0, bit 0, after 10 cycles of an external pulse connected to port 3, bit 5.

◆ **Thoughts on the Problem** Port pin 3.5 is the counting mode input pin to Timer T1. For this example, the T1 Overflow flag, TF1, will be polled until it is set. The LED will be turned on by bringing pin P0.0 low. When using the debugger, RST the program after it loops back to the beginning.

```
Address     Mnemonic                    Comment
            SETMODE EQU 01010000b       ;program T1 to act as a 16-bit counter
            LIGHTITUP EQU 0FEh          ;latch pin P0.0 low with this bit pattern
LOOP:                                   ;for debugger use
            MOV TMOD,#SETMODE           ;program T1
            MOV TH1,#0FFh               ;start T1 at a count 10 less than 0000h
            MOV TL1,#0F6h               ;T1 counts to FFFFh after 9 input
                                        ;pulses
                                        ;and then overflows to 0000h on pulse 10
            SETB TR1                    ;enable T1 to count
HERE:                                   ;loop here until TF1 sets after 10 counts
            JNB TF1,HERE                ;if TR1 not set then loop back
            CLR TF1                     ;NOTE: flag will not automatically reset
            MOV P0, LIGHTITUP           ;turn on the LED
            SJMP LOOP                   ;loop back to debug again
END
```

◆

EXAMPLE PROBLEM 8.5

Repeat the previous problem using TF1 to initiate an interrupt to the main program.

◆ **Thoughts on the Problem** Two additions to the previous program must be made: An interrupt program for TF1 *must* be located in program code at the mandatory TF1 interrupt address, and the interrupt register *must* be programmed to enable a TF1 interrupt. ("These additions are noted by asterisks in the comment line.) NOTE: TF1 is automatically reset after causing an interrupt.

Address	Mnemonic	Comment
	SETMODE EQU 01010000b	;program T1 to act as a 16-bit counter
	LIGHTITUP EQU 0FEH	;latch pin P0.0 low with this bit pattern
	SJMP OVER	;*jump over the TF1 interrupt program
ORG	001Bh	;*TF1 interrupt program *must* begin here
	MOV P0,#LIGHTITUP	;turn on LED
	RETI	;*return from interrupt program to main program
OVER:		;main program resumes here
	MOV TMOD,#SETMODE	;program T1
	MOV TH1,#0FFh	;start T1 at a count 10 less than 0000h
	MOV TL1,#0F6h	;T1 counts to FFFFh after 9 input pulses
		;and then overflows to 0000h on pulse 10
	SETB TR1	;enable T1 to count
	SETB EA	;*enable interrupts, in general
	SETB ET1	;*enable TF1 interrupt, in particular
HERE:		;main program loops here until 10 pulses
		;then resumes looping after the RETI
	SJMP HERE	
END		

◆

EXAMPLE PROBLEM 8.6

Place any number in internal RAM location 3Ch and increment it until the number equals 2Ah.

◆ **Thoughts on the Problem** The number can be incremented and then tested to see whether it equals 2Ah. If it does, then the program is over; if not, then loop back and increment the number again.

Three methods can be used to accomplish this task.

◆ **Method 1:**

Address	Mnemonic	Comment
ONE:	CLR C	;this program will use SUBB to detect equality
	MOV A,#2Ah	;put the target number in A
	SUBB A,3Ch	;subtract the contents of 3Ch; C is cleared
	JZ DONE	;if A = 00h, then the contents of 3Ch = 2Ah
	INC 3Ch	;if A is not 0, then loop until it is
	SJMP ONE	;loop to try again
DONE:	NOP	;when finished, jump here and continue
END		

◆

◆ COMMENT ◆

As there is no compare instruction for the 8051, the SUBB instruction is used to compare A against a number. The SUBB instruction subtracts the C flag also, so the C flag has to be cleared before the SUBB instruction is used.

◆ **Method 2:**

Address	Mnemonic	Comment
TWO:	INC 3Ch	;incrementing 3Ch first saves a jump later
	MOV A,#2Ah	;this program will use XOR to detect equality
	XRL A,3Ch	;XOR with the contents of 3Ch; if equal, A = 00h
	JNZ TWO	;this jump is the reverse of Program 1
	NOP	;finished when the jump is *false*
END		

◆

◆ **COMMENT** ◆

Many times if the loop is begun with the action that is to be repeated until the loop is satisfied, only one jump, which repeats the loop, is needed.

◆ **Method 3:**

Address	Mnemonic	Comment
THREE:	INC 3Ch	;begin by incrementing the direct address
	MOV A,#2Ah	;this program uses the very efficient CJNE
	CJNE A,3Ch,THREE	;jump if A and (3Ch) are not equal
	NOP	;all done
END		

◆

◆ **COMMENT** ◆

CJNE combines a compare and a jump into one compact *instruction.*

EXAMPLE PROBLEM 8.7

The number A6h is placed somewhere in external RAM between locations 0100h and 0200h. Find the address of that location and put that address in R6 (LSB) and R7 (MSB).

◆ **Thoughts on the Problem** The DPTR is used to point to the bytes in external memory, and CJNE is used to compare and jump until a match is found.

Address	Mnemonic	Comment
	MOV 20h,#0A6h	;load 20h with the number to be found
	MOV DPTR, #00FFh	;start the DPTR below the first address
MOR:	INC DPTR	;increment first and save a jump
	MOVX A,@DPTR	;get a number from external memory to A
	CJNE A,20h,MOR	;compare the number against (20h) and
		;loop to MOR if not equal
	MOV R7,83h	;move DPH byte to R7
	MOV R6,82h	;move DPL byte to R6; finished
END		

◆

◆ COMMENT ◆

This program might loop forever unless we know the number will be found; a check to see whether the DPTR has exceeded 0200h can be included to leave the loop if the number is not found before DPTR = 0201h.

EXAMPLE PROBLEM 8.8

Find the address of the first two internal RAM locations between 20h and 60h which contain consecutive numbers. If so, set the Carry flag to 1, else clear the flag.

◆ **Thoughts on the Problem** A check for end of memory will be included as a Called routine, and CJNE and a pointing register will be used to search memory.

Address	Mnemonic	Comment
	MOV 81h,#65h	;set the stack above memory area
	MOV R0,#20h	;load R0 with address of memory start
NXT:	MOV A,@R0	;get first number
	INC A	;increment and compare to next number
	MOV 1Fh,A	;store incremented number at 1Fh
	INC R0	;point to next number
	ACALL DUN	;see if R0 greater than 60h
	JNC THRU	;DUN returns C = 0 if over 60h
	MOV A,@R0	;get next number
	CJNE A,1Fh,NXT	;if not equal then look at next pair
	SETB 0D7h	;set the carry to 1; finished
THRU:	SJMP THRU	;jump here if beyond 60h
DUN:	CLR C	;clear the carry
	MOV A,#61h	;use XOR as a compare
	XRL A,R0	;A will be 0 if equal
	JNZ BCK	;if not 0 then continue
	RET	;A 0, signal calling routine
BCK:	CPL C	;A not 0, set C to indicate not done
	RET	
END		

◆

◆ COMMENT ◆

Set the stack pointer to put the stack out of the memory area in use.

8.7 Summary

Jumps

Jumps alter program flow by replacing the PC counter contents with the address of the jump address. Jumps have the following ranges:

Relative: up to PC +127 bytes, PC −128 bytes away from the PC
Absolute short: anywhere on a 2K-byte page
Absolute long: anywhere in program memory

Jump opcodes can test an individual bit, or a byte, to check for conditions that make the program jump to a new program address. The bit jumps are as follows:

Instruction Type	Result
JC radd	Jump relative if Carry flag set to 1
JNC radd	Jump relative if Carry flag cleared to 0
JB b,radd	Jump relative if addressable bit set to 1
JNB b,radd	Jump relative if addressable bit cleared to 0
JBC b,radd	Jump relative if addressable bit set to 1 and clear bit to 0

Byte jumps are as follows:

Instruction Type	Result
CJNE destination,source, address	Compare destination and source; jump to address if *not* equal
DJNZ destination,address	Decrement destination by 1; jump to address if the result is *not* zero
JZ radd	Jump A = 00h to relative address
JNZ radd	Jump A > 00h to relative address

Unconditional jumps make no test and are always made. They are shown in the following list:

Instruction Type	Result
JMP @A+DPTR	Jump to 16-bit address formed by adding A to the DPTR
AJMP sadd	Jump to absolute short address
LJMP ladd	Jump to absolute long address
SJMP radd	Jump to relative address
NOP	Do nothing and go to next opcode

Call and Return

Software calls may use short- and long-range addressing; returns are to any long-range address in memory. Interrupts are calls forced by hardware action and call subroutines located at predefined addresses in program memory. The following list shows calls and returns:

Instruction Type	Result
ACALL sadd	Call the routine located at absolute short address
LCALL ladd	Call the routine located at absolute long address
RET	Return to anywhere in the program at the address found on the top 2 bytes of the stack
RETI	Return from a routine called by a hardware interrupt and reset the interrupt logic

8.8 Problems

Write programs for each of the following problems using as few lines of code as you can. Place comments on each line of code.

1. Put a random number in R3 and increment it until it equals E1h.
2. Put a random number in address 20h and increment it until it equals a random number put in R5.
3. Put a random number in R3 and decrement it until it equals E1h.
4. Put a random number in address 20h (LSB) and 21h (MSB) and decrement them as if they were a single 16-bit counter until they equal random numbers in R2 (LSB) and R3 (MSB).
5. Random unsigned numbers are placed in registers R0 to R4. Find the largest number and put it in R6.
6. Repeat Problem 5, but find the smallest number.
7. If the lower nibble of any number placed in A is larger than the upper nibble, set the C flag to 1; otherwise clear it.
8. Count the number of 1s in any number in register B and put the count in R5.
9. Count the number of 0s in any number in register R3 and put the count in R5.
10. If the signed number placed in R7 is negative, set the Carry flag to 1; otherwise clear it.
11. Increment the DPTR from any initialized value to ABCDh.
12. Decrement the DPTR from any initialized value to 0033h, as a 16-bit register.
13. Use R4 (LSB) and R5 (MSB) as a single 16-bit counter, and decrement the pair until they equal 0000h.
14. Get the contents of the PC to the DPTR.
15. Get the contents of the DPTR to the PC.
16. Get any 2 bytes you wish to the PC.
17. Write a simple subroutine, call it, and jump back to the calling program after adjusting the stack pointer.
18. Put one random number in R2 and another in R5. Increment R2 and decrement R5 until they are equal.
19. Fill external memory locations 100h to 200h with the number AAh.
20. Transfer the data in internal RAM locations 10h to 20h to internal RAM locations 30h to 40h.

21. Set every third byte in internal RAM from address 20h to 7Fh, to FFh.
22. Count the number of bytes in external RAM locations 100h to 200h that are greater than the random unsigned number in R3 *and* less than the random unsigned number in R4. Use registers R6 (LSB) and R7 (MSB) to hold the count.
23. Assuming the crystal frequency is 10 megahertz, write a program that will use timer 1 to interrupt the program after a delay of 2 ms.
24. Put the address of every internal RAM byte from 50h to 70h in the address; for instance, internal RAM location 6Dh would contain 6Dh.
25. Put the byte AAh in all internal RAM locations from 20h to 40h, then read them back and set the Carry flag to 1 if any byte read back is not AAh.
26. Explain why ending an interrupt routine with an RET instruction will only work once.
27. Explain why the main program cannot be allowed to get to a subroutine except by CALLing it.

9

An 8051 Microcontroller Design

Chapter Outline

CHAPTER OBJECTIVES

On successful completion of this chapter you will be able to:

- Discuss the design of a small 8051(8031)-based system.
- Calculate external memory access times.
- List the steps used to test the design.
- Use timing subroutines.
- Construct lookup tables.
- Discuss serial data transmission/reception fundamentals.

9.0 Introduction

In this chapter a hardware configuration for an 8051 microcontroller, which will be used for all of the example applications in Chapters 10 and 11, is defined. Programs that check the initial prototype of the design (debugging programs) are given in this chapter, followed by several common subroutines that can be used by programs in succeeding chapters.

The design of the microcontroller system begins with an identified need and a blank piece of paper or computer screen. The evolution of the microcontroller-based system follows these steps:

1. Define a specification.
2. Design a microcontroller system to this specification.
3. Write programs that will assist in checking the design.
4. Write several common subroutines and test them.

The most important step is the first one. If the application is for high-volume production (greater than 10,000 units), then the task must be very carefully analyzed. A precise or "tight" specification is evolved for what will become a major investment in factory-programmed parts. As the volume goes down for any particular application, the specifications become more general as the designers attempt to write a specification that might fit a wider range of applications.

The list leaves out a few real-world steps, most notably the redesign of the microcontroller after it is discovered that the application has grown beyond the original specification or, as is more common, the application was not well understood in the beginning. Experienced designers learn to add a little "fat" to the specification in anticipation of the inexorable need for "one more bit of I/O and one more kilobyte of memory."

9.1 A Microcontroller Specification

A typical outline for a microcontroller design might read as follows:

"A requirement exists for an intelligent controller for real-time control and data monitoring applications. The controller is part of a networked system of identical units that are connected to a host computer through a serial data link. The controller is to be produced in low volumes, typically less than one thousand units for any particular application, and it must be low cost."

The 8051 family is chosen for the following reasons:

- Low part cost
- Multiple vendors
- Available in NMOS and CMOS technologies
- Software tools available and inexpensive
- High-level language compilers available

The first three items are very important from a production cost standpoint. The software aids available reduce first costs and enable projects to be completed in a timely manner.

The low-volume production requirement and the need for changing the program to fit particular applications establish the necessity of using external EPROM to hold the application program. In turn, ports 0 (AD0 – AD7) and 2 (A8 – A15) must be used for interfacing to the external ROM and will not be available for I/O.

Because one possible use of the controller will be to gather data, RAM beyond that available internally may be needed. External RAM is added for this eventuality. The immediate consequence of this decision is that port 3 bits 6 ($\overline{WR}$) and 7 ($\overline{RD}$) are needed for the external RAM and are not available for I/O. External memory uses the 28-pin standard configuration, which enables memories as large as 64K to be inserted in the memory sockets.

Commercially available EPROM parts that double in size beginning at 2K can be purchased. The minimum EPROM size selected is 8K and the maximum size is 64K. These choices reflect the part sizes that are most readily available from vendors.

Static RAM parts are commonly available in 2K, 8K, 32K, and 64K sizes; the RAM sizes are chosen to be 8K or 32K to reflect commercial realities. The various memory sizes can be incorporated by including jumpers for the additional address lines needed by larger memories and pullup resistors to enable alternate pin uses on smaller memories.

The serial data needs can be handled by the internal serial port circuitry. Once again, two more I/O pins of port 3 are used: bits 3.0 (RXD) and 3.1 (TXD). We are left with all of port 1 for general-purpose I/O and port 3 pins 2 – 5 for general-purpose I/O or for external interrupts and timing inputs.

Note that rapid loss of I/O capability occurs as the alternate port functions are used and should be expected unless volumes are high enough to justify factory-programmed parts. The handicap is not as great as it appears, however; two methods exist that are commonly used to expand the I/O capability of any computer application: port I/O and memory-mapped I/O. I/O expansion is discussed in the next section.

Finally, we select a 12 megahertz crystal for ease of timing and UART baud rates, and the specification is complete. To summarize, we have

- 80C31-12 (ROMless) microcontroller
- 64K of external EPROM
- 32K of external RAM
- 8 general-purpose I/O lines
- 4 general-purpose or programmable I/O lines
- 1 full-duplex serial port
- 12 megahertz crystal clock

Now that the specification is complete, the design can be done.

9.2 A Microcontroller Design

The final design, shown in Figure 9.1, is based on the external memory circuit found in Chapter 3. Any I/O circuitry needed for a particular application will be added to the basic design as required. A design may be done in several ways; the choices made for this design are constrained by cost and the desire for flexibility. Remember that an 8051 may be used if ROM and RAM requirements do not exceed 4K and 128 bytes, respectively.

External Memory and Memory Space Decoding

External memory is added by using port 0 as a data and low-order address bus and port 2 as a high-order address bus. The data and low addresses are time multiplexed on port 0. An external 373 type address latch is connected to port 0 to store the low address byte whenever external memory is accessed. The low-order address is gated into the transparent latch by the ALE pulse from the 8031. Port 0 then becomes a bidirectional data bus during the read or write phase of a machine cycle.

RAM and ROM are addressed by entirely different control lines from the 8031: $\overline{PSEN}$ for the ROM and $\overline{WR}$ or $\overline{RD}$ for the RAM. The result is that each occupies one of two parallel 64K address spaces. The decoding problem becomes one of simply adding suitable jumpers and pullup resistors so that the user can insert the memory capacity needed. Jumpers are inserted so that the correct address line reaches the memory pin or the pin is pulled high as required by the

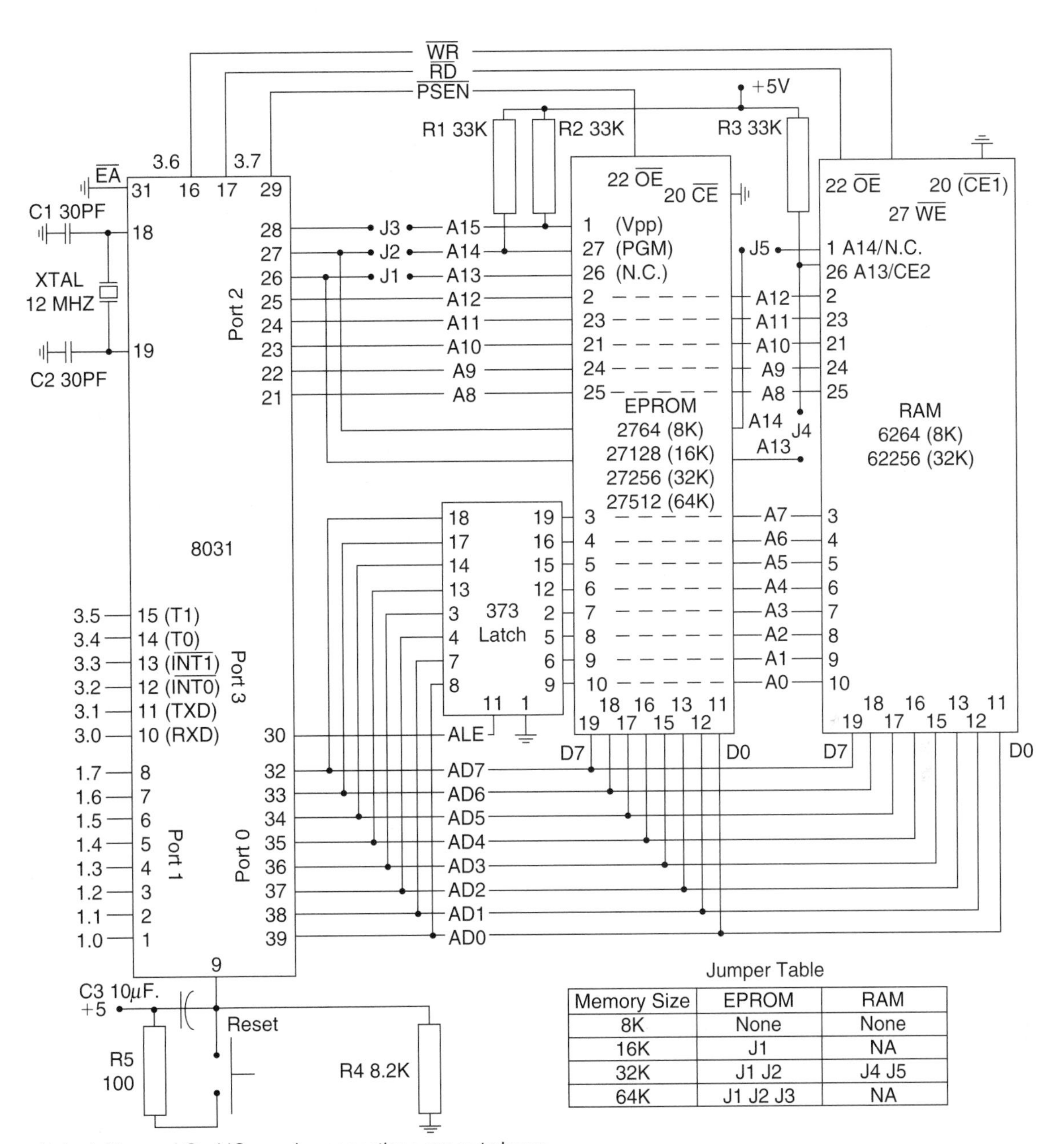

Jumper Table

Memory Size	EPROM	RAM
8K	None	None
16K	J1	NA
32K	J1 J2	J4 J5
64K	J1 J2 J3	NA

Note 1: Vcc and Gnd IC supply connections are not shown.
Note 2: Lines shown dashed (— — —) indicate connections from the EPROM address pins to the RAM address pins.

FIGURE 9.1 ◆ 8031 Microcontroller with External ROM and RAM

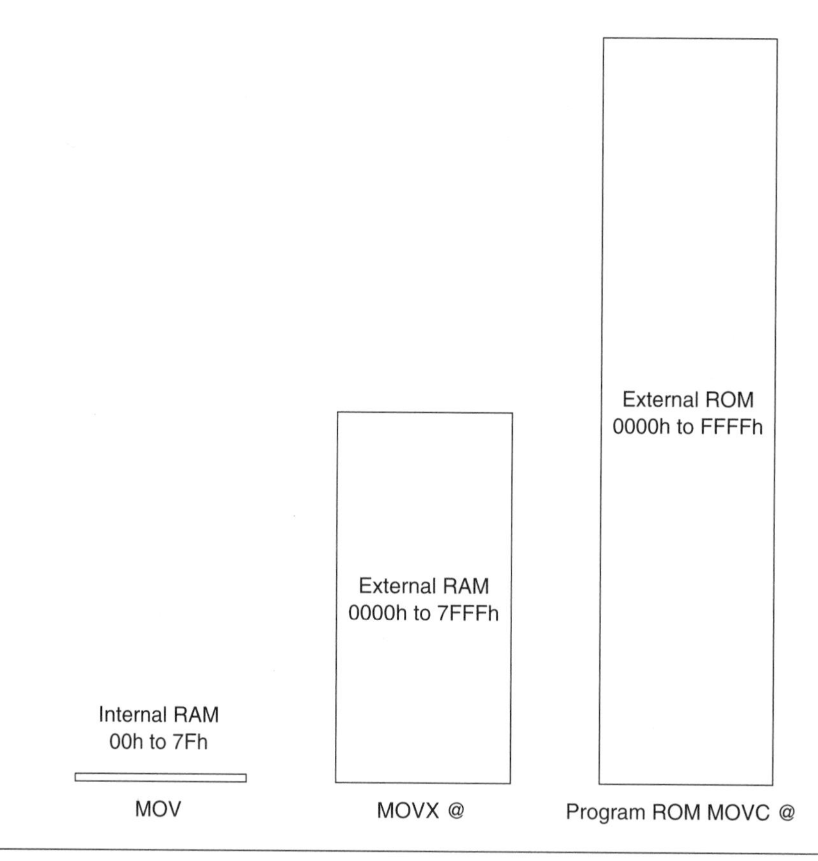

FIGURE 9.2 ◆ 8031 Memory Sizes

memory used. The jumper table in Figure 9.1 for the EPROM and RAM memories that can be inserted in the memory sockets shows the jumper configuration. Figure 9.2 graphically demonstrates the relative sizes of the internal and external memories available to the programmer.

Reset and Clock Circuits

The 8051 uses an active high reset pin. The reset input must go high for two machine cycles when power is first applied and then sink low. The simple RC circuit used here will supply system voltage (Vcc) to the reset pin until the capacitor begins to charge. At a threshold of about 2.5 V, the reset input reaches a low level and the system begins to run. Internal reset circuitry has hysteresis necessitated by the slow fall time of the RC circuit. The addition of a reset button enables the user to reset the system without having to turn power off and on.

The clock circuit of Chapter 3 is added, and the design is finished.

Expanding I/O

Ports 1 and 3 can be used to form small control and bidirectional data busses. The data busses can interface with additional external circuits to expand I/O up to any practical number of lines.

There are many popular families of programmable port chips. The one chosen here is the popular 8255 programmable interface adaptor, which is available from a number of vendors. Details on the full capabilities of the 8255 are given in Appendix D. The 8255 has an internal mode register to which control words are written by the host computer. These control words determine the actions of the 8255 ports, named A, B, and C, enabling them to act as input ports, output ports, or some combination of both.

Figure 9.3 shows a circuit that adds an 8255 port expansion chip to the design. The number of ports is now three 8-bit ports for the system. The penalty paid for expanding I/O in this manner is a reduction in speed that occurs due

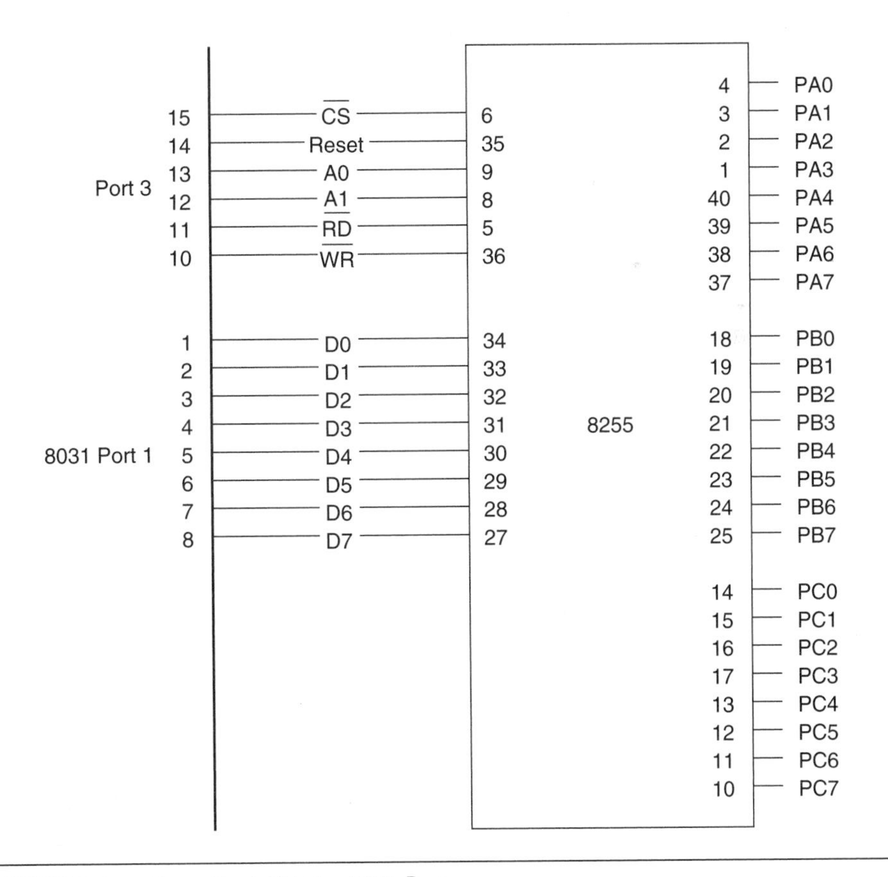

FIGURE 9.3 ◆ Expanding I/O Using 8031 Ports

to the overhead time needed to write control bits to ports 1 and 3 before the resulting I/O lines selected can be accessed. The advantage of using I/O port expansion is that the entire range of 8051 instructions can be used to access the added ports via ports 1 and 3.

Memory-Mapped I/O

The same programmable chip used for port expansion can also be added to the RAM memory space of the design, as shown in Figure 9.4. The present design uses only 32K of the permitted 64K of RAM address space; the upper 32K is vacant. The port chip can be addressed any time A15 is high (8000h or above), and

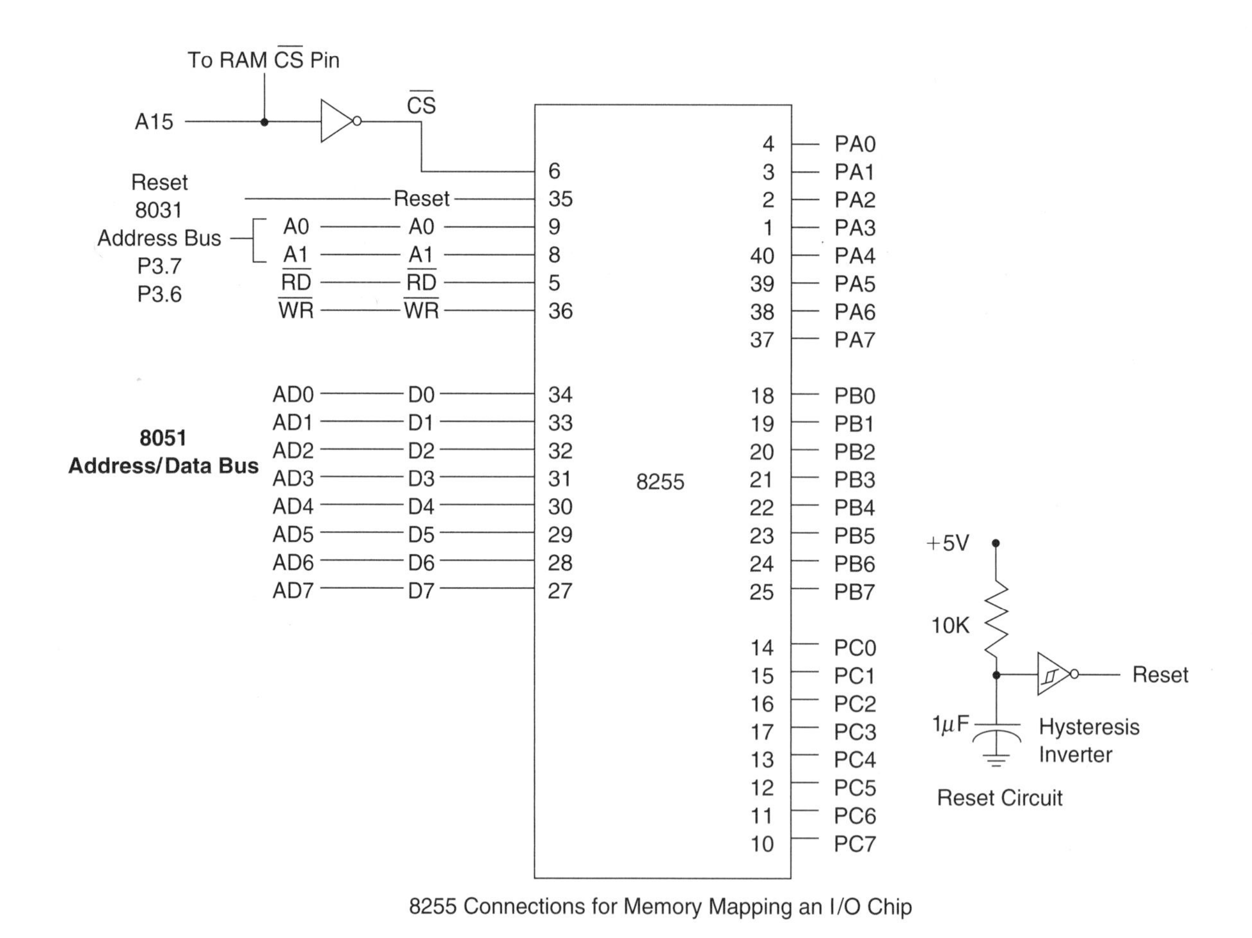

FIGURE 9.4 ◆ Expanding I/O Using Memory Mapping

the 32K RAM can be addressed whenever A15 is low (7FFFh and below). This decoding scheme requires only the addition of an inverter to decode the memory space for RAM and I/O.

Memory Address Decoding

If additional external RAM above 32K is to be added to the design, then the simple inverter scheme of Figure 9.4 must be replaced with a memory *decoder* circuit. A decoder is the familiar combinational logic circuit that has n binary inputs, and up to 2^n outputs. Only one of the outputs is active at any time.

The task facing the decoder designer is to determine which memory addresses to use to enable each I/O chip. The decoding scheme of Figure 9.4 uses all of the addresses from 8000h to FFFFh simply to enable one I/O chip! Clearly, if RAM above address 7FFFh is added to the system, the decoder design must waste less of the RAM memory address range (memory *space*). A sample decoder design, one that divides the upper 32K RAM memory addresses into 4, 20h spaces, is shown in Figure 9.5.

The decoder circuit shown has 11 binary inputs and 5 outputs. Each output is active low, a commonly used active level for enabling many integrated circuits. The circuit may be realized using programmable logic. The decoder is programmed to operate as shown in Table 9.1.

The decoding scheme shown in Table 9.1 yields the memory address ranges given in Table 9.2.

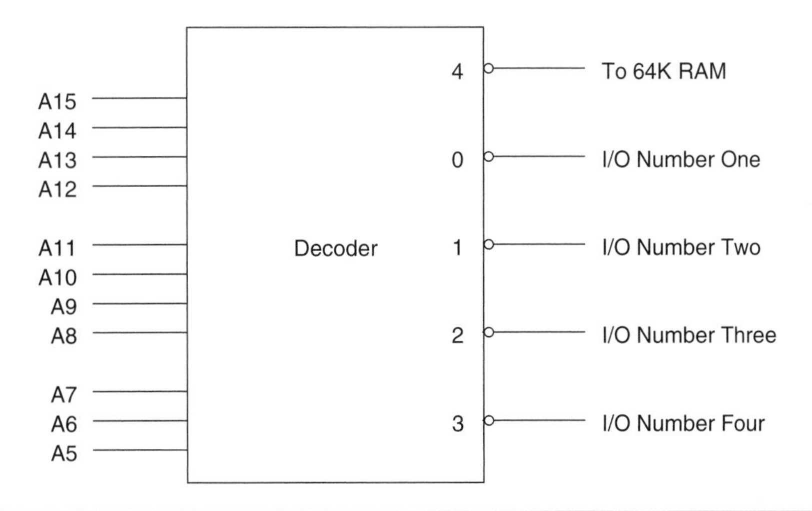

FIGURE 9.5 ◆ Memory Space Decoder

TABLE 9.1

Input											Output				
A15	A14	A13	A12	A11	A10	A9	A8	A7	A6	A5	0	1	2	3	4
1	1	1	1	1	1	1	1	1	0	0	L	H	H	H	H
1	1	1	1	1	1	1	1	1	0	1	H	L	H	H	H
1	1	1	1	1	1	1	1	1	1	0	H	H	L	H	H
1	1	1	1	1	1	1	1	1	1	1	H	H	H	L	H
All other address combinations											H	H	H	H	L

Note: Output 4 = NOT Outputs (0 AND 1 AND 2 AND 3)

TABLE 9.2

RAM Memory Address Range (hex)	Active Output
0000 – FF7F	4
FF80 – FF9F	0
FFA0 – FFBF	1
FFC0 – FFDF	2
FFE0 – FFFF	3

A 64K RAM memory may be enabled in the system by output 4. The last 80h addresses of the RAM are "wasted" so that up to four I/O chips may be selected. Each I/O chip may have 20h individually addressed internal locations.

Memory-mapped I/O has the advantage of not using any of the 8051 ports. Disadvantages include the loss of memory space for RAM that is used by the I/O address space, or the addition of memory decoding chips in order to limit the RAM address space loss. Programming overhead is about the same as for port I/O because only the cumbersome MOVX instruction may be used to access the memory-mapped I/O.

For both types of I/O expansion, the cost of the system begins to mount. At some point, a conventional microprocessor, with a rich set of I/O and memory instructions, may become a more economical choice.

Part Speed

One consideration, that does not appear on the design drawings, is the selection of memory parts that will work at the system speeds determined by the crystal frequency. All memory parts are priced according to the nanosecond of access time. The longer the access time (the time it takes for a byte of data to be read or written from or to the device after the address is valid), the cheaper the part. We shall now discuss how memory speeds are specified and calculated.

Memory Access Times

Memory access times are divided into two types: read access times and write access times. The read times are generally the most important because the memory must be addressed and enabled, and then data read *from* the chip. Furthermore, EPROMs generally are slower than SRAMs for any given generation of memory technology.

The designer also must take into account any circuit delays caused by the lower address latch and the decoder. All circuit delays serve to reduce the required memory chip access times. Memory chips generally have three read access times, defined as follows:

Taccess = Time from addresses valid to data out of the memory
Tenable = Time from chip enable to data out of the memory
Tread = Time from read signal active to data out of the memory

A review of Figure 3.9 shows that memory addresses are valid at the beginning of a machine cycle (actually, at the beginning of State 2.) The $\overline{\text{PSEN}}$ pulse goes active low (at the start of State 3), and returns high after 3 clock cycles (in the middle of State 4.) Remember there are two clock cycles per state, and that there are two $\overline{\text{PSEN}}$ pulses per machine cycle. Assuming that a 12 MHz crystal is used in the design, a machine cycle lasts 1 microsecond, and each clock period is 83.33 nanoseconds.

Some address delay is encountered as the low-order addresses pass through the '373-type data latch. Assuming the delay is 20 nanoseconds, the 8051 design allows time from addresses valid at the EPROM address pins until data must be at the EPROM pins of:

Taccess = State 2 to State 4.5 − 20 = 5 × 83.33 − 20 = 396.66 ns

The decoder has a delay (for example let us assume 40 ns), so that the time to enable the memory is later than the time the addresses go valid by 40 ns. The time available to the memory from addresses valid until data must be at the EPROM data pins becomes:

Tenable = State 2 to State 4.5 − 40 = 376.66 ns

Finally, the PSEN pulse determines the time from read active until data is available at the memory data pins:

Tread = Width of $\overline{\text{PSEN}}$ = 3 × 83.33 = 250 ns

The least expensive 8K EPROMs available today have Taccess and Tenable times of 250 ns, which leaves a generous safety margin for our design. (Tread times are generally less than half of Taccess times.)

Production Concerns

The design omits many features that would be incorporated by a design-manufacturing team. Chief among these are the inclusion of test-points, LED indicators, and other items that should be added to enhance manufacturing and field service of the microcontroller. These concerns are well beyond the scope of this book, but the wise designer always ensures that the legitimate concerns of the technical, manufacturing, and service departments are addressed.

9.3 Testing the Design

Once the hardware has been assembled, it is necessary to verify that the design is correct and that the prototype is built to the design drawing. This verification of the design is done by running several small programs, beginning with the most basic program and building on the demonstrated success of each.

Crystal Test

The initial test is to ensure that both the crystal and the reset circuit are working. The 8051 is inserted in the circuit, and the ALE pulse is checked with an oscilloscope to verify that the ALE frequency is 1/6 of the crystal frequency. Next, the reset button is pushed, and all ports are checked to see that they are in the high (input) state.

ROM Test

The most fundamental program test is to ensure that the microcontroller can fetch and execute programs from the EPROM. Code byte fetching can be tested by verifying that each address line of the ROM is properly wired by using a series of repeated jump instructions that exercise all of the address lines. The test used here will jump to addresses that are a power of 2. Only one address line will be high, and all of the rest will be low. The address pattern tests for proper wiring of each address line and also checks for shorts between any two lines.

If the test is successful, the program stops at the highest possible ROM address. The address bus can then be checked with a logic probe to verify that the highest address has been reached. Correct operation is indicated by the highest order address bus bit, which will appear constant. If not, the probe will blink indicating random program fetches.

The test is run by inserting the '373 latch, the programmed 64K EPROM, and jumpers 1–3, and resetting the 8051. The test can be stopped at any address by jumping to that address, as is done in the last statement in the following ROM test program:

```
Address    Mnemonic      Comment
           org 0000h     ;start at the bottom of ROM
begin:     ljmp add2     ;test address lines A0 and A1
           org 0004h     ;next jump at address 0004h (A2)
add2:      ljmp add3     ;test address line A2
           org 0008h     ;next jump at address 0008h (A3)
add3:      ljmp add4     ;test address line A3
           org 0010h     ;next jump at address 0010h (A4)
add4:      ljmp add5     ;test address line A4
           org 0020h     ;next jump at address 0020h (A5)
add5:      ljmp add6     ;test address line A5
           org 0040h     ;next jump at address 0040h (A6)
add6:      ljmp add7     ;test address line A6
           org 0080h     ;next jump at address 0080h (A7)
add7:      ljmp add8     ;test address line A7
           org 0100h     ;next jump at address 0100h (A8)
add8:      ljmp add9     ;test address line A8
           org 0200h     ;next jump at address 0200h (A9)
add9:      ljmp add10    ;test address line A9
           org 0400h     ;next jump at address 0400h (A10)
add10:     ljmp add11    ;test address line A10
           org 0800h     ;next jump at address 0800h (A11)
add11:     ljmp add12    ;test address line A11
           org 1000h     ;next jump at address 1000h (A12)
add12:     ljmp add13    ;test address line A12
           org 2000h     ;next jump at address 2000h (A13)
add13:     ljmp add14    ;test address line A13
           org 4000h     ;last jump at address 4000h (A14)
add14:     ljmp add15    ;test address line A14
           org 8000h     ;test address line A15 and remain here
add15:     ljmp add15    ;jump here in a loop
           end           ;assembler use
;
;This address, A15, will remain latched while A2 – A14 will
;remain low. A0 and A1 will vary as the bytes of the jump
;instruction are fetched.
```

Inspection of the listing for this program in Figure 9.6 shows that all the address lines are exercised.

RAM Test

Once sure of the ability of the microcontroller to execute code, the RAM can be checked. A common test is to write a so-called checkerboard pattern to RAM—that is, an alternating pattern of 1 and 0 in memory. Writing bytes of 55h or AAh will generate such a pattern.

```
0000                          org 0000h       ;start at the bottom of ROM
0000    020004   begin:       ljmp add2       ;test address lines A0 and A1
0004                          org 0004h       ;next jump at address 0004h (A2)
0004    020008   add2:        ljmp add3       ;test address line A2
0008                          org 0008h       ;next jump at address 0008h (A3)
0008    020010   add3:        ljmp add4       ;test address line A3
0010                          org 0010h       ;next jump at address 0010h (A4)
0010    020020   add4:        ljmp add5       ;test address line A4
0020                          org 0020h       ;next jump at address 0020h (A5)
0020    020040   add5:        ljmp add6       ;test address line A5
0040                          org 0040h       ;next jump at address 0040h (A6)
0040    020080   add6:        ljmp add7       ;test address line A6
0080                          org 0080h       ;next jump at address 0080h (A7)
0080    020100   add7:        ljmp add8       ;test address line A7
0100                          org 0100h       ;next jump at address 0100h (A8)
0100    020200   add8:        ljmp add9       ;test address line A8
0200                          org 0200h       ;next jump at address 0200h (A9)
0200    020400   add9:        ljmp add10      ;test address line A9
0400                          org 0400h       ;next jump at address 0400h (A10)
0400    020800   add10:       ljmp add11      ;test address line A10
0800                          org 0800h       ;next jump at address 0800h (A11)
0800    021000   add11:       ljmp add12      ;test address line A11
1000                          org. 1000h      ;next jump at address 1000h (A12)
1000    022000   add12:       ljmp add13      ;test address line A12
2000                          org 2000h       ;next jump at address 2000h (A13)
2000    024000   add13:       ljmp add14      ;test address line A13
4000                          org 4000h       ;last jump at address 4000h (A14)
4000    028000   add14:       ljmp add15      ;test address line A14 and remain
8000                          org 8000h       ;test address line A15 and remain
                                              ;here
8000    028000   add15:       ljmp add15      ;jump here in a loop
8003                          end             ;assembler use
```

FIGURE 9.6 ◆ Assembled ROM Check Program

The next program writes this pattern to external RAM, then reads the pattern back and checks each byte read back against the byte that was written. If a check fails, then the address where the failure occurred is in the DPTR register. Port 1 and the free bits of port 3 can then be used to indicate the contents of DPTR.

There are 14 bits available using these ports (the serial port is not in use now, so bits 3.0 and 3.1 are free), and 15 are needed to express a 32K address range. The program will test a range of 8K at a time, using 13 bits to hold the 8K address on failure. Four versions have to be run to cover the entire RAM address space. If the test is passed, then bit 14 (port 3.5) is a 1. If the test fails, then bit 14 is a 0, and the other 13 bits hold the address (in the 8K page) at which the failure occurred.

Interestingly, this test does not check for correct wiring of the RAM address lines. As long as all address lines end on some valid address, the test will work. A wiring check requires that a ROM be programmed with some unique pattern at each address that is a power of 2 and read using a check program that inspects each unique address location for a unique pattern.

The RAM test program is listed next:

```
ramstart equ 0000h              ;set 8K RAM test start address
rmstphi equ 20h                 ;set 8K RAM test stop address
pattern equ 55h                 ;bit test pattern = 0101 0101
;P3.5 = 1 if RAM passes test, P3.5 = 0 if RAM fails test
;and P1 = LSB of fail address, P3.0-4 = MSB of fail address (in 8K page)
                                ;
        setb p3.5               ;set Port 3.5 pin high (assume pass)
        mov dptr,#ramstart      ;point to RAM start address
test:   mov a,#pattern          ;test bits in A
        movx@dptr,a             ;write test bits to address in RAM
        inc dptr                ;point to next address in RAM
        mov a,#rmstphi          ;see if beyond address 1fffh in RAM
        cjne a,dph,test         ;if not done then write next address
        mov dptr,#ramstart      ;done writing, now read back
check:  movx a,@dptr            ;read back RAM address
        cjne a,#pattern, fail   ;read data should = write data
        inc dptr                ;point to next address
        mov a,#rmstphi          ;check to see if past last address
        cjne a,dph,check        ;if not continue
                                ;pass - P3.5 = 1 (is not reset)
here:   sjmp here               ;loop at this address until reset
fail:   mov p3, dph             ;P3.0 to P3.4 = MSB of fail address (00 to 1fh)
        mov pl, dpl             ;P1 = LSB of fail address
        clr p3.5                ;fail - P3.5 = 0
        sjmp here               ;loop at halt address
end
```

◆ **COMMENT** ◆

- Change the ramstart and rmstphi equ hex numbers to check pages 2000h to 3FFFh, 4000h to 5FFFh, and 6000h to 7FFFh.
- Note that a full 16-bit check for end of memory does not have to be done because of page boundaries of (20)00, (40)00, (60)00, and (80)00h.
- There is no *halt* command for the 8051; jumps in place serve to perform the halt function.

We have now tested all the external circuitry that has been added to the 8031. The remainder of the chapter is devoted to several subroutines that can be used by the application programs in Chapters 10 and 11.

9.4 Timing Subroutines

Subroutines are used by call programs in what is known as a *transparent* manner—that is, the calling program can use the subroutines without being bothered by the details of what is actually going on in the subroutine. Usually, the call program preloads certain locations with data, calls the subroutine, then gets the results back in the preload locations.

The subroutine must take great care to save the values of all memory locations in the system that the subroutine uses to perform internal functions and restore these values before returning to the call program. Failure to save values results in occasional bugs in the main program. The main program assumes that everything is the same both before and after a subroutine is called.

Finally, good documentation is essential so that the user of the subroutine knows precisely how to use it.

Time Delays

Perhaps the most-used subroutine is one that generates a programmable time delay. Time delays may be done by using software loops that essentially do nothing for some period, or by using hardware timers that count internal clock pulses.

The hardware timers may be operated in either a software or a hardware mode. In the software mode, the program inspects the timer Overflow flag and jumps when it is set. The hardware mode uses the interrupt structure of the 8051 to generate an interrupt to the program when the timer overflows.

The interrupt method is preferred whenever processor time is scarce. The interrupt mode allows the processor to continue to execute useful code while the time delay is taking place. Both the pure software and timer-software modes tie up the processor while the delay is taking place.

If the interrupt mode is used, then the program *must* have an interrupt handling routine at the dedicated interrupt program vector location specified in Chapter 3. The program must also have programmed the various interrupt control registers. This degree of *nontransparency* generally means that interrupt-driven subroutines are normally written by the user as needed and not used from a purchased library of subroutines.

Pure Software Time Delay

The subroutine named *softime* generates delays ranging from 1 to 65,535 milliseconds by using register R7 to generate the basic 1 millisecond delay. The call program loads the desired delay into registers A (LSB) and B (MSB) before calling softime.

The key to writing this program is to calculate the exact time each instruction will take at the clock frequency in use. For a crystal of 12 megahertz, each machine cycle (12 clock pulses) is 1 microsecond. Should the crystal frequency be changed, the subroutine would have to have the internal timing loop number "delay" changed.

Softime

Softime will delay the number of milliseconds expressed by the binary number, from 1 to 65,535d, found in registers A (LSB) and B (MSB). The call program loads the desired delay into registers A and B and calls softime. Loading 0s into A and B results in an immediate return.

Comments in the program show how the basic delay loop takes 1,000 1-microsecond cycles to complete, or a delay of 1 millisecond. Additional instructions that test for A rollovers at FFh, and A counting down to 0, add an additional 4 microseconds for a total "normal" delay of 1,004 microseconds each time A is decremented. When A does rollover at FFh, then B must be decremented, and when A is 0 B must be tested for 0 also. There are B decrements at rollovers (1 cycle) and B + 1 tests (2 cycles) for B equal to 0, which add 3B + 2 cycles to the total delay.

Any routine must have additional instructions, or *overhead,* for entering and leaving the routine that add to the total time delay of the loop. Overhead time includes saving and restoring registers, testing for A and B set to 00h, and returning to the calling program. The total delay is then: (1,004 × Desired delay) + 3B + 15 microseconds. The error due to the extra instructions included in the software timing loop is as follows:

$$\text{Error} = \frac{\text{Actual delay} - \text{Desired delay}}{\text{Desired delay}}$$

$$\text{Error} = \frac{1.004 \times \text{Desired delay} + .003\text{B} + .015 - \text{Desired delay}}{\text{Desired delay (milliseconds)}}$$

$$\%\text{Error} = .4 + \frac{.3\text{B} + 1.5}{\text{Desired delay}}$$

The worst case error occurs when the desired delay is 1 millisecond (A = 01h, B = 00h), or an error of 1.9%. The error for a desired delay of 256 milliseconds (A = 00h, B = 01h) is .41%, and the error for a desired delay of 65,535 milliseconds (A = B = FFh) is .40%. Note that any real crystal is rarely exactly 12 MHz, so additional errors will be introduced by the crystal tolerance.

The reader may wish to "fix up" Softime by deleting some of the NOPs in the basic loop and changing the definition of delay to bring the total delay closer to that passed to the routine.

◆ COMMENT ◆

- Note that register A, when used in a defined mnemonic, is used as "A." When used as a direct address in a mnemonic (where any add could be used), the equate name ACC is used. The equate usage is also seen for R7, where the name of the register may be used in those mnemonics for which it is specifically defined. For mnemonics that use any add, the actual address must be used.
- The restriction on A = B = 00 is due to the fact that the program would initially count A from 00 . . . FFh . . . 00 then exit. If it were desired to be able to use this initial condition for A and B, then an all = zero condition could be handled by the test for 0000 used, set a flag for the condition, decrement B from 00 to FFh the first time B is decremented, then reset the flag for the remainder of the program.

```
Address   Mnemonic            Comment
;Softime, using a 12MHz crystal for a loop delay of 1 ms.

          delay equ 0a6h      ;166 × 6 cycles = 996 loop cycles
          dlylsb equ 0f4h     ;try 500 (1f4h) milliseconds; LSB in A
          dlymsb equ 01h      ;MSB in B
          org 0000h
blink:
          setb p1.0           ;demonstrate a 1-second LED blink rate
          mov a,#dlylsb       ;pass the desired delay time to Softime
          mov b,#dlymsb
          acall softime       ;call for 500 ms delay @ 12MHz
          clr p1.0            ;turn off LED
          mov a,#dlylsb       ;delay another 500 ms
          mov b,#dlymsb
          acall softime
          sjmp blink          ;loop to start
;
softime:                      ;time delay routine named "softime"
          push 07h            ;overhead of:  2 cycles
          push acc            ;              2     Test for AB = 0000
          orl a,b             ;              1
          cjne a,#00h,ok      ;              2
          pop acc             ;                2  only if A = B = 00h
          sjmp done           ;                2
ok:                           ;
```

```
            pop acc             ;               2
                                ;               9  overhead cycles to enter
timer:                          ;
mov r7,#delay                   ;1 for MOV R7, total loop: 1004 cycles
onemil:                         ;
            nop                 ;1 for each NOP
            nop                 ;
            nop                 ;this loop takes 6 cycles; total 996 cycles
            nop                 ;
            djnz r7,onemil      ;2 for the DJNZ
            nop                 ;1 for the NOP
            nop                 ;1 for the NOP
            dec a               ;1 for DEC, 1,000 cycles at this point
            cjne a,#0ffh,noroll ;2 cycles to see if rollover
            dec b               ;overhead of 1 cycle each B decrement
noroll:                         ;adds B cycles to total delay
            cjne a,#00h,timer   ;2 cycles to see if done
            cjne a,b,timer      ;overhead of 2 cycles each B + 1 decrement
done:                           ;adds 2B + 2 cycles to total delay
            pop 07h             ;overhead of    2  cycles to return
            ret                 ;overhead of    2  cycles to return
            end                 ;
                                ;               4  cycles to return
```

The restriction on A = B = 00 is due to the fact that the program would count A down from 00 to FFh after a 1 millisecond delay. B would then be decremented to FFh, and the subroutine would return after a delay of 10000h milliseconds.

Software-Polled Timer

A delay that uses the timers to generate the delay and a continuous software flag test (the flag is "polled" to see whether it is set) to determine when the timers have finished the delay is given in this section. The user program signals the total delay desired by passing delay variables in the A and B registers in a manner similar to the pure software delay subroutine. A basic interval of 1 millisecond is again chosen so that the delay may range from 1 to 65,535 ms.

The clock frequency for the timer is the crystal frequency divided by 12, or 1 machine cycle, which makes each count of the timer 1 microsecond for a 12 megahertz crystal. Twelve megahertz is an excellent choice for generating accurate time delays, such as for use in systems that maintain a time-of-day clock.

Timer 0 will be used to count 1,000 (03E8h) internal clock pulses to generate the basic 1 millisecond delay; registers A and B will be counted down as T0 overflows. The timer counts up, so it will be necessary to put the two's complement of the desired number in the timer and count up until it overflows.

Timer

The time delay routine named *timer* uses timer 0 and registers A and B to generate delays from 1 to 65,535d milliseconds. The calling program loads registers A (LSB) and B (MSB) with the desired delay in milliseconds. Loading a delay of 0000h results in an immediate return.

Note that the error due to overhead instructions for the timer program is slightly worse, over the entire range of delays, than a pure software delay program. The total timing error for the program is as shown next:

$$\text{Error} = \frac{\text{Actual delay} - \text{Desired delay}}{\text{Desired delay}} = \frac{\text{Overhead} + \text{Desired} - \text{Desired}}{\text{Desired Delay}}$$

$$\text{Error} = \frac{\text{Desired delay} \times 15 + 3\text{B} + 10(\text{microseconds})}{\text{desired delay (milliseconds)}}$$

$$\%\text{Error} = 1.5 + \frac{.3\text{B} + 1}{\text{Desired delay}}$$

The error for a desired 1-millisecond delay is 2.5%; for a desired delay of 256 milliseconds is 1.5%; and for a desired delay of 65,535 milliseconds is also 1.5%. As an exercise, the reader may decide how to adjust the number loaded into T0 so that the 15 overhead cycles used to decrement A may be eliminated and the error made quit small.

```
Address   Mnemonic                 Comment
;Timer, a program that uses T0 to generate the basic delay. No check
;for A = B = 00 is done.
          onemshi equ 0fch         ;two's complement of 03E8h = FC18h
          onemslo equ 18h          ;used to set number in T0
          dlylsb equ 0f4h          ;set delay LSB in register A
          dlymsb equ 01h           ;set delay MSB in register B
          org 0000h
blink:
          setb p1.0                ;Blink P1.0 LED; LED on
          mov a,#dlylsb            ;pass desired 500 ms delay to routine
          mov b,#dlymsb
          acall timer              ;call the Timer subroutine
          clr p1.0                 ;Blink P1.0 LED; LED off
          mov a,#dlylsb            ;pass desired 500 ms delay to routine
          mov b,#dlymsb
          acall timer
          sjmp blink
timer:
```

```
          anl tcon,#0cfh       ;clear T0 overflow & run flags only          2
          anl tmod,#0f0h       ;T0 enabled; T1 control unchanged            2
          orl tmod,#01h        ;set T0 to mode 1 (16 bit up counter)        2
                               ;                                        ______
                               ;overhead cycles to begin timing             6
onems:                         ;total cycles delay per decrement of A:
          mov tl0,#onemslo     ;set T0 to count up from FC18h               2
          move th0,#onemshi    ;                                            2
          orl tcon,#10h        ;start T0 timer                              2
wait:                          ;                                         1000
          jbc tf0,dwnab        ;poll T0 Overflow flag until set             2
          sjmp wait            ;loop until Overflow flag set
dwnab:
          anl tcon,#0efh       ;stop T0; other TCON bits unchanged          2
          dec a                ;count A down                                1
          cjne a,#0ffh,noroll  ;2 cycles to test for rollover               2
                               ;
          dec b                ;1 cycle to count B down
noroll:                        ;B extra cycles per delay
          cjne a,#00h,onems    ;2 overhead cycles every A decrement         2
                               ;
                               ; normal decrement A total =              1015
done:                          ;2B + 2 extra cycles per delay
          ret                  ;2 overhead cycles; total overhead =        10
                               ;
                               ;decrement B total = 3B + 2
          end                  ;delay = (desired delay) × 1015 + (3B) + 10
                               ;microseconds
```

◆ **COMMENT** ◆

- T0 cannot be used accurately for other timing or counting functions in the user program; thus, there is no need to save the TCON and TMOD bits for T0. T0 itself could be used to store data if it is saved.
- This program has no inherent advantage over the pure software delay program; both take up all processor time. The software polled timer has a slight advantage in flexibility in that the number loaded into T0 can be easily changed in the program to shorten or lengthen the basic timing loop. Thus, the call program could also pass the basic timing delay (in other memory locations) and get delays that could be programmed in microseconds or hours.
- One way for the program to continue to run while the timer times out is to have the program loop back on itself periodically, checking the timer Overflow flag. This looping is the normal operating mode for most programs; if the program execution time is small compared with the desired delay, then the error in the total time delay will be small.

Pure Hardware Delay

If delays must be done in a program, but the program can not wait while the delay is computed, then the delay will have to be done using a timer enabled to interrupt the main program. A timer interrupt allows a timer to run *and generate a time delay at the same time as the main program loop is executing.*

The program given in this section, *hardtime,* operates in the following manner:

1. Timer T1 is initialized to count a delay of 1,000 microseconds. Timer T1 will interrupt the main program loop, named *wait,* whenever the T1 Overflow flag is set, or every millisecond. The wait loop continuously blinks port bit P1.7 at a rate of 3.7 Hertz to demonstrate that it is running during the desired delay.
2. An interrupt routine must be located in the program at the T1 interrupt address 001Bh. In the example program, the time delay subroutine hardtime is called from the interrupt address subroutine named *intT1* located at 001Bh.
3. Hardtime determines if the total desired delay is finished and sets a flag named *doneflag,* if the desired time delay is finished, before returning to intT1.
4. intT1 jumps to the user routine named *blink* if the total desired time is up. If the total desired time is not up, then intT1 returns to the main wait loop. Port bit P1.0 is blinked at a rate of .5 Hertz by blink. Blink returns to the main wait loop.

The program begins by jumping over the interrupt at 001Bh, and proceeds to initialize the various registers needed to enable T1 to time and a T1 interrupt to be done. Registers R0 and R1 of bank 3 are used to hold the desired total delay of 1,000 milliseconds.

Note that the time delay is, at most, 1,008 microseconds every time T1 overflows. The delay is due to the latency time before an interrupt may be acknowledged by the CPU, the time to call hardtime, and the time to re-initialize TL1. Any overhead time used by the subroutine hardtime does not change the basic delay after TL1 is set because T1 is running and timing out while hardtime is determining if the total desired time is up. The *first* delay to blink will be late by the overhead time needed to find if the desired time is up. After the first delay, subsequent delays to blink will be the desired delay multiplied by 1.008 milliseconds plus an extra 5 microseconds needed by hardtime to reload the desired time delay and set doneflag if the desired time is up.

```
;hardtime, Uses R0 and R1 of bank 3 to hold the desired time delay

        onemshi equ 0fch        ;two's complement of 03E8h = FC18h
        onemslo equ 18h
```

```
        dlylsb equ 0f4h          ;set delay LSB in register R0 of bank 3
        dlymsb equ 01h           ;set delay MSB in register R1 of bank 3
        doneflag equ 00h         ;addressable bit 00 = 1 when delay done
        org 0000h
        mov sp, #30h             ;set SP above bit addressable area
        sjmp overint             ;jump over T1 interrupt address in program
;
        org 001bh                ;T1 interrupt subroutine at 001Bh
;the interrupt subroutine for T1 must be located here in the program
;the time to interrupt to address 001Bh is a maximum of 3 cycles
intT1:
        acall hardtime           ;call "hardtime" time delay subroutine
;2 cycles of overhead for the acall hardtime instruction
        jbc doneflag,blink       ;"hardtime" returns flag set when delay up
        reti                     ;else direct return to user's main program
;
;begin by initializing the T1 and Interrupt registers
overint:
        orl ie,#88h              ;enable interrupts and T1 interrupt
        mov 18h,#dlylsb          ;pass desired delay to timer subroutine
        mov 19h,#dlymsb          ;in bank 3 R0(lsb) and R1(msb)
        move tl1,#onemslo        ;set T1 to count up from FC18h
        mov th1,#onemshi         ;
        anl tmod,#0fh            ;set T1 to time
        orl tmod,#10h            ;set T1 to time as a mode 1 timer
        clr doneflag             ;start with delay not finished
        orl tcon,#40h            ;start T1
;
;place main program loop here to run continuously
wait:
        setb p1.7                ;main program continues on, shown here
        acall softdelay          ;as a software delay loop that blinks an
        clr p1.7                 ;LED connected to P1.7
        acall softdelay
        sjmp wait                ;loop
softdelay:                       ;softdelay = .131 seconds
        mov r0, 0ffh
outside:
        mov r1, 0ffh
inside:
        djnz r1,inside
        djnz r0,outside
        ret
;
;hardtime is called approximately every millisecond by T1 interrupt
hardtime:                        ;time delay called from T1 interrupt
        mov tl1,#onemslo         ;reset TL1 to 18h, overhead = 3 cycles
        mov th1,#onemshi         ;interrupt 1,000 cycles from here
```

```
;total time between interrupts is actually 1,008 cycles
      push psw              ;save PSW contents
      push acc              ;save A register
      orl psw,#18h          ;switch to register bank 3
      dec r0                ;count down LSB in R0, 00 becomes FF
      cjne r0,#0ffh,noroll  ;check for R0 = FF
      dec r1                ;decrement R1 if so, else check for done
noroll:
      mov a,r1              ;test for R0 = R1 = 00
      orl a,r0              ;if so, then delay is up
      jnz notdone           ;                          2 NOT DONE CYCLES
done:                       ;                          8 DONE CYCLES
      mov r0,#dlylsb        ;restore desired delay time       3
      mov r1,#dlymsb        ;in bank 3 R0(lsb) and R1(msb)    3
      setb doneflag         ;show delay is up                 2
notdone:                    ;                          DIFFERENCE = 6
      pop acc               ;restore original A
      pop psw               ;restore original PSW
      ret                   ;return to T1 interrupt call
;
;the user routine that uses the T1 time delay placed here
;
blink:                      ;jump here whenever delay is up
      xrl p1,#01h           ;XOR P1.0 alternately 1 or 0
      reti                  ;return to main program loop
      .end
```

◆ COMMENT ◆

- The minimum usable delay is 1 millisecond because a 1 millisecond delay is done to begin the delay interrupt cycle.
- All timing routines can be assembled at interrupt location 001Bh if stack space is limited.
- The RETI instruction is used when returning to the main program, after each interrupt.
- There is no check for an initial delay of 0000h.

9.5 Lookup Tables for the 8051

There are many instances in computing when one number must be converted into another number, or a group of numbers, on a one-to-one basis. A common example is to change an ASCII character for the decimal numbers 0 to 9 into the binary equivalent (BCD) of those numbers. ASCII 30h is used to represent 00d, 31h is 01d, and so on until ASCII 39h is used for 09d.

Clearly, one way to convert from ASCII to BCD is to subtract a 30h from the ASCII character. Another approach uses a table in ROM that contains the BCD numbers 00 to 09. The table is stored in ROM at addresses that are related to the ASCII character that is to be converted to BCD. The ASCII character is used to form part of the address where its equivalent BCD number is stored. The contents of the address "pointed" to by the ASCII character are then moved to a register in the 8051 for further use. The ASCII character is then said to have "looked up" its equivalent BCD number.

For example, using ASCII characters 30h to 9h we can construct the following program, at the addresses indicated, using db commands:

Address	Mnemonic	Comment
	org 1030h	;start table at ROM location 1030h
	db 00h	;location 1030h contains 00 BCD
	db 01h	;location 1031h contains 01 BCD
	db 02h	;location 1032h contains 02 BCD
	db 03h	;location 1033h contains 03 BCD
	db 04h	;location 1034h contains 04 BCD
	db 05h	;location 1035h contains 05 BCD
	db 06h	;location 1036h contains 06 BCD
	db 07h	;location 1037h contains 07 BCD
	db 08h	;location 1038h contains 08 BCD
	db 09h	;location 1039h contains 09 BCD

Each address whose low byte is the ASCII byte contains the BCD equivalent of that ASCII byte. If the DPTR is loaded with 1000h and A is loaded with the desired ASCII byte, then a MOVC A,@A+DPTR will move the equivalent BCD byte for the ASCII byte in A to A.

Lookup tables may be used to perform very complicated data translation feats, including trigonometric and exponential conversions. Although lookup tables require space in ROM, they enable conversions to be done very quickly, far faster than using computational methods.

The 8051 is equipped with a set of instructions that facilitate the construction and use of lookup tables: the MOVC A,@A+DPTR and the MOVC A,@A+PC. In both cases A holds the pointer, or some number calculated from the pointer, which is also called an *offset*. DPTR or PC holds a *base* address that allows the data table to be placed at any convenient location in ROM. In the ASCII example just illustrated, the base address is 1000h, and A holds an offset number ranging from 30h to 39h.

Typically, PC is used for small *local* tables of data that may be included in the body of the program. DPTR might be used to point to large tables that are normally assembled at the end of program code.

In both cases, the desired byte of data is found at the address in ROM that is equal to base + offset. Figure 9.7 demonstrates how the final address in the lookup table is calculated using the two base registers.

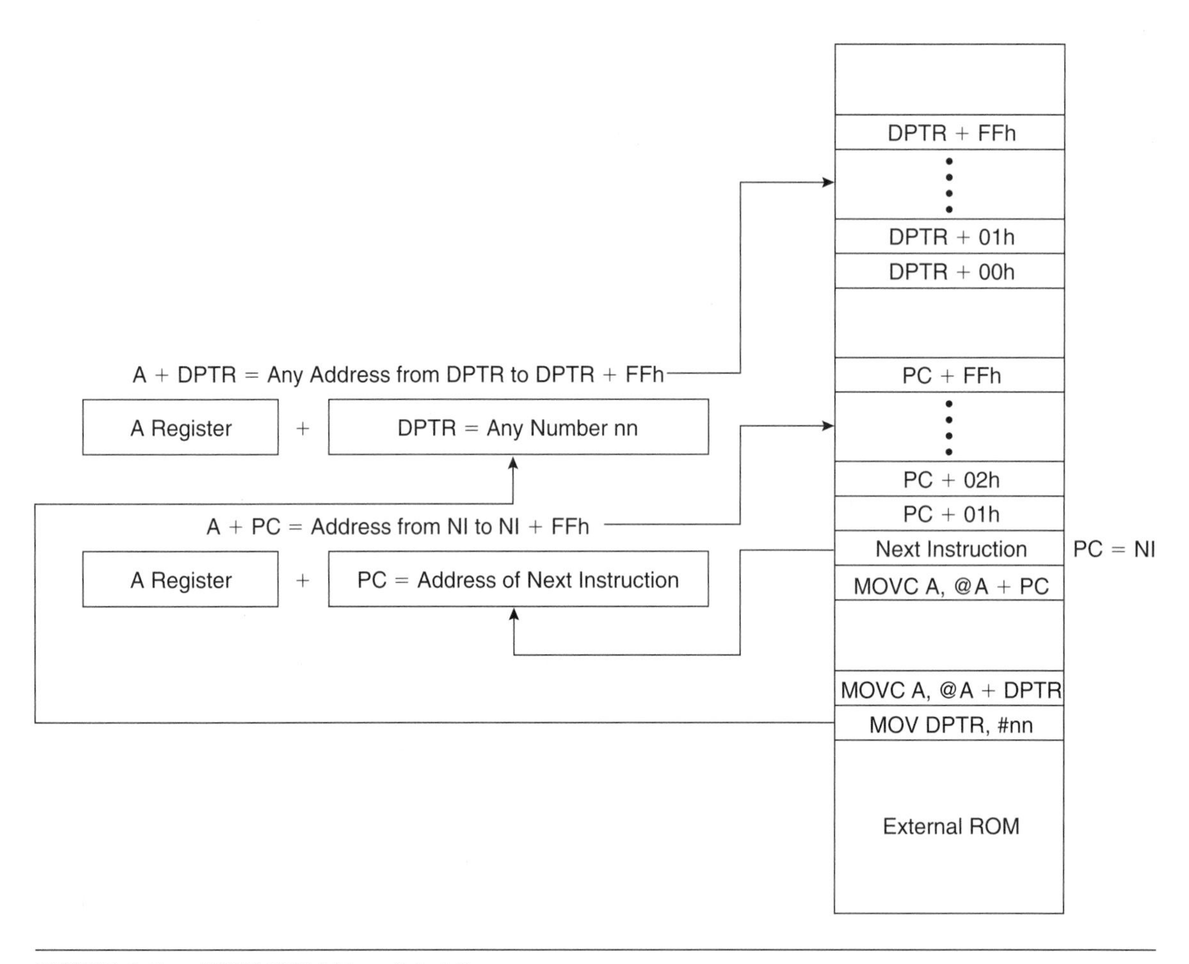

FIGURE 9.7 ◆ MOVC ROM Address Calculations

One limitation of lookup tables might be the appearance that only 256 different values—corresponding to the 256 different values that A might hold—may be put in a table. This limitation can be overcome by using techniques to alter the DPTR such that the base address is changed in increments of 256 bytes. The same offset in A can point to any number of data bytes in tables that differ only by the beginning address of the base. For example, by changing the number loaded in DPTR from 1000h to 1100h in the ASCII-to-BCD table given previously, the ASCII byte in A can now point to an entirely new set of conversion bytes.

Both PC and DPTR base address programs are given in the examples that follow.

PC as a Base Address

Suppose that the number in A is known to be between 00h and 0Fh and that the number in A is to be squared. A could be loaded into B and a MUL AB done or a local lookup table constructed.

The table cannot be placed directly after the MOVC instruction. A jump instruction must be placed between the MOVC and the table, or the program soon fetches the first data byte of the table and executes it as code. Remember also that the PC contains the address of the jump instruction (the next instruction after the MOVC command) when the table address is computed.

Pclook

The program *pclook* looks up data in a table that has a base address in the PC and the offset in A. After the MOVC instruction, A contains the number that is the square of the original number in A.

```
Address   Mnemonic          Comment
pclook:   mov a,#0ah        ;find the square of 0Ah (64h)
          add a,#02h        ;adjust for 2-byte sjmp over
          movc a,@a+pc      ;get equivalent data from table to A
          sjmp over         ;jump over the lookup table
;
;the lookup table is inserted here, at PC + 2. (PC = 0005h)
;
          db 00h            ;begin table here, 00^2 = 00
          db 01h            ;01^2 = 01d
          db 04h            ;02^2 = 04d
          db 09h            ;03^2 = 09d
          db 10h            ;04^2 = 16d
          db 19h            ;05^2 = 25d
          db 24h            ;06^2 = 36d
          db 31h            ;07^2 = 49d
          db 40h            ;08^2 = 64d
          db 51h            ;09^2 = 81d
          db 64h            ;0A^2 = 100d
          db 79h            ;0B^2 = 121d
          db 90h            ;0C^2 = 144d
          db 0a9h           ;0D^2 = 169d
          db 0c4h           ;0E^2 = 196d
          db 0e1h           ;0F^2 = 225d
over:     sjmp over         ;simulate rest of user program
          end
```

Figure 9.8 shows the assembled listing of this program and the resulting address of the table relative to the MOVC instruction.

```
0000
0000 740A  pclook:   mov a,#0ah        ;find the square of 0Ah (64h)
0002 2402            add a,#02h        ;adjust for 2-byte sjmp over
0004 83              movc a,@a+pc      ;get equivalent data from table
                                       ;to A
0005 8010            sjmp over         ;jump over the lookup table
0007            ;the lookup table is inserted here, at PC + 2 (PC = 0005h)
0007 00              db 00h            ;begin table here, 00^2 = 00
0008 01              db 01h            ;01^2 = 01d
0009 04              db 04h            ;02^2 = 04d
000A 09              db 09h            ;03^2 = 09d
000B 10              db 10h            ;04^2 = 16d
000C 19              db 19h            ;05^2 = 25d
000D 24              db 24h            ;06^2 = 36d
000E 31              db 31h            ;07^2 = 49d
000F 40              db 40h            ;08^2 = 64d
0010 51              db 51h            ;09^2 = 81d
0011 64              db 64h            ;0A^2 = 100d
0012 79              db 79h            ;0B^2 = 121d
0013 90              db 90h            ;0C^2 = 144d
0014 A9              db 0A9h           ;0D^2 = 169d
0015 C4              db 0C4h           ;0E^2 = 196d
0016 E1              db 0E1h           ;0F^2 = 225d
0017 80F  over:      sjmp over         ;simulate rest of user program
0019                 end
```

FIGURE 9.8 ◆ Lookup Table Using the PC

◆ COMMENT ◆

The number added to A reflects the number of bytes in the SJMP instruction. If more code is inserted between the MOVC and the table, a similar number of bytes must be added. Adding bytes can result in overflowing A when the sum of these adjusting bytes and the contents of A exceed 255d. If this happens, the lookup data must be limited to the number of bytes found by subtracting the number of adjustment bytes from 255d.

DPTR as a Base Address

The DPTR is used to construct a lookup table in the next example. Remove the restriction that the number in A must be less than 10h and let A hold any number from 00h to FFh. The square of any number larger than 0Fh results in a 4-byte result; store the result in registers R0 (LSB) and R1 (MSB).

Two tables are constructed in this section: one for the LSB and the second for the MSB. A points to both bytes in the two tables, and the DPTR is used to hold two base addresses for the two tables. The entire set of two tables, each with 256 entries, will not be constructed for this example. The beginning and example values are shown as a skeleton of the entire table.

Dplook

The lookup table program *dplook* holds the square of any number found in the A register. The result is placed in R0 (LSB) and R1 (MSB). A is stored temporarily in R1 in order to point to the MSB byte.

```
Address   Mnemonic               Comment
          lowbyte equ 0200h      ;base address of LSB table
          hibyte equ 0300h       ;base address of MSB table
          org 0000h
dplook:   mov a,#5ah             ;find the square of 5Ah (1FA4h)
          mov r1,a               ;store A for later use
          mov dptr,#lowbyte      ;set DPTR to base address of LSB
          movc a,@a+dptr         ;get LSB
          mov r0,a               ;store LSB in R0
          mov a,r1               ;recover A for pointing to MSB
          mov dptr,#hibyte       ;set DPTR to base address of MSB
          movc a,@a+dptr         ;get MSB
          mov r1,a               ;store MSB in R1
here:     sjmp here              ;simulate rest of user program
          org lowbyte            ;place LSB table starting here
          db 00h                 ;00^2 = 0000
          db 01h                 ;01^2 = 0001
;place rest of table up to the LSB of 59^2 here
          org lowbyte + 5ah      ;put LSB of 5A^2 here
          db 0a4h                ;LSB is A4h
;place rest of LSB table here
          org hibyte             ;place MSB table starting here
          db 00h                 ;00^2 = 0000
          db 00h                 ;01^2 = 0001
;place rest of table up to the MSB of 59^2 here
          org hibyte + 5ah       ;put MSB of 5A^2 here
          db 1fh                 ;MSB is 1Fh
;place rest of MSB table here
          end
```

◆ COMMENT ◆

- Note that there are no jumps to "get over" the tables; the tables are normally placed at the end of the program code.
- A does not require adjustment; DPTR is a constant.

Figure 9.9 shows the assembled code; location 025Ah holds the LSB of 5A^2, and location 035Ah holds the MSB.

```
0200                    lowbyte equ 0200h      ;base address of LSB table
0300                    hibyte equ 0300h       ;base address of MSB table
0000                    org 0000h
0000 745A  dplook:      mov a,#5ah             ;find the square of 5Ah
                                               ;(1FA4h)
0002 F9                 mov r1,a               ;store A for later use
0003 900200             mov dptr,#lowbyte      ;set DPTR to base address
                                               ;of LSB
0006 93                 movc a,@a+dptr         ;get LSB
0007 F8                 mov r0,a               ;store LSB in R0
0008 E9                 mov a,r1               ;recover A for pointing
                                               ;to MSB
0009 900300             mov dptr,#hibyte       ;set DPTR to base address
                                               ;of MSB
000C 93                 movc a,@a+dptr         ;get MSB
000D F9                 mov r1,a               ;store MSB in R1
000E 80FE  here:        sjmp here              ;simulate rest of user
                                               ;program
0200                    org lowbyte            ;place LSB table starting
                                               ;here
0200 00                 db 00h                 ;00^2 = 0000
0201 01                 db 01h                 ;01^2 = 0001
0202            ;place rest of table up to the LSB of 59^2 here
025A                    org lowbyte + 5ah      ;put LSB of 5A^2 here
025A A4                 db 0a4h                ;LSB is A4h
025B            ;place rest of LSB table here
0300                    org hibyte             ;place MSB table starting
                                               ;here
0300 00                 db 00h                 ;00^2 = 0000
0301 00                 db 00h                 ;01^2 = 0001
0302            ;place rest of table up to the MSB of 59^2 here
035A                    org hibyte + 5ah       ;put MSB of 5A^2 here
035A 1F                 db 1fh                 ;MSB is 1Fh
035B            ;place rest of MSB table here
035B                    end
```

FIGURE 9.9 ◆ Lookup Table Using the DPTR

9.6 Serial Data Transmission

The hallmark of contemporary industrial computing is the linking together of multiple processors to form a *local area network* or LAN. The degree of complexity of the LAN may be as simple as a microcontroller interchanging data with an I/O device, as complicated as linking multiple processors in an automated robotic manufacturing cell, or as truly complex as the linking of many computers in a very high speed, distributed system with shared disk and I/O resources.

All of these levels of increasing sophistication have one feature in common: the need to send and receive data from one location to another. The most cost-effective way to meet this need is to send the data as a serial stream of bits in order to reduce the cost (and bulk) of multiple conductor cable. Optical fiber bundles, which are physically small, can be used for parallel data transmission. However, the cost incurred for the fibers, the terminations, and the optical interface to the computer currently prohibit optical fiber use, except in those cases where speed is more important than economics.

So pervasive is serial data transmission that special integrated circuits dedicated solely to serial data transmission and reception appeared commercially in the early 1970s. These chips, commonly called *universal asynchronous receiver transmitters* or UARTS, perform all the serial data transmission and reception timing tasks of the most popular data communication scheme still in use today: serial 8-bit ASCII-coded characters at predefined bit rates of 300 to 19,200 bits per second.

Asynchronous transmission utilizes a start bit and 1 or more stop bits, as shown in Figure 9.10, to alert the receiving unit that a character is about to arrive and to signal the end of a character. This "overhead" of extra bits, with the attendant slowing of data byte rates, has encouraged the development of synchronous data transmission schemes. Synchronous data transmission involves alerting the receiving unit to the arrival of data by a unique pattern that starts data transmission, followed by a long string of characters. The end of transmission is signaled by another unique pattern, usually containing error-checking characters.

Each scheme has its advantages. For relatively short or infrequent messages, the asynchronous mode is best; for long messages or constant data transmission, the synchronous mode is superior.

The 8051 contains serial data transmission/reception circuitry that can be programmed to use four asynchronous data communication modes numbered from 0 to 3. One of these, mode 1, is the standard UART mode, and three simple asynchronous communication programs using this mode will be developed

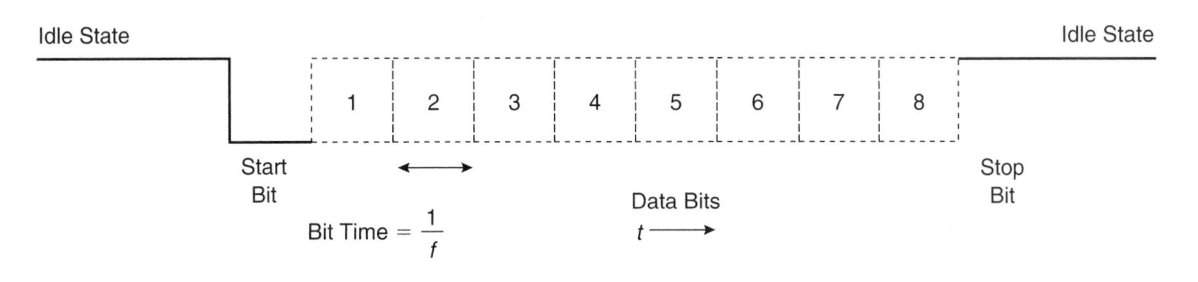

FIGURE 9.10 ◆ Asynchronous 8-Bit Character

here. More complicated asynchronous programs that use all of the communication modes will be written in Chapter 11.

Character Transmission Using a Time Delay

Often data transmission is unidirectional from the microcontroller to an output device, such as a display or a printer. Each character sent to the output device takes from 33.3 to .5 milliseconds to transmit, depending on the baud rate chosen. The program must wait until one character is sent before loading the next, or data will be lost. A simple way to prevent data loss is to use a time delay that delays the known transmission time of one character before the next is sent.

Sendchar

A program called *Sendchar* takes the character in the A register, transmits it, delays for the transmission time, and then returns to the calling program. Timer 1 must be used to set the baud rate, which is 2400 baud in this example. The delay for one 10-bit character is 1,000/240 or 4.16 milliseconds. The software delay softime is used for a 5 millisecond delay. Timer 1 needs to generate a final baud rate of 2400 at SBUF. Using a 12 megahertz crystal, the reload number is $256 - 12E6/(32 \times 12 \times 2400)$, which is 242.98 or integer 243. This yields an actual rate of 2404.

```
;Sendchar, using a 12MHz crystal for UART timing. 2400 nominal baud
;rate for an actual rate of 2403.8 (2404). Delay between characters
;for 5 milliseconds.
;
    baudnum equ 0f3h        ;number loaded into TH1 for 2403.8 baud
    delay equ 0a6h          ;166 × 6 cycles = 996 loop cycles
    dlylsb equ 05h          ;try 5 (0005) millisecond; LSB in A
    dlymsb equ 00h          ;MSB in B
    org 0000h
    anl pcon,#7fh           ;set SMOD bit to 0 for Baud × 32 rate
    anl tmod,#30h           ;alter timer T1 configuration only
    orl tmod,#20h           ;set timer T1 as an 8-bit autoload
    mov th1,#baudnum        ;TH1 set for divide clock by 13d
    setb tr1                ;T1 running at 1E6/13 = 76923 Hz.
    mov scon,#40h           ;UART mode 1; Baud = 76923/32 = 2403.8
xmit:
    mov sbuf,#'A'           ;transmit ASCII "A" continuously
    acall xmitime           ;wait 5 ms before next transmission
    sjmp xmit               ;loop again
    ;
xmitime:
    mov a,#dlylsb           ;pass the desired delay time to softime
    mov b,#dlymsb
    acall softime           ;call for 5 ms delay @ 12MHz
```

```
        ret
;
softime:                        ;time delay routine named "softime"
        push 07h                ;see previous program
        push acc
        orl a,b
        cjne a,#00h,ok
        pop acc
        sjmp done
ok:
        pop acc
timer:
        mov r7,#delay
onemil:
        nop
        nop
        nop
        nop
        djnz r7,onemil
        nop
        nop
        dec a
        cjne a,#0ffh,noroll
        dec b
noroll:
        cjne a,#00h,timer
        cjne a,b,timer
done:
        pop 07h
        ret
        end
```

◆ COMMENT ◆

- If timer 1 and the serial port have different uses in the user program, then push and pop affected control registers. But remember, T1 and SBUF can only be used for one function at any given time.
- The 12 megahertz crystal yields convenient standard baud rates of 300, 1200, 2400, 4800, 9600, or 19,200. The errors using this crystal for these rates are given in the following table:

Divisor	Rate	Error (%)
32	300	.16
32	1,200	.16
32	2,400	.15
16	4,800	.16
16	9,600	8.50
16	19,200	6.3

The error grows for higher baud rates as ever smaller reload numbers are rounded to the nearest integer. Using an 11.059 megahertz crystal reduces the errors to less than .002% at the slight cost of speed of program execution.

Character Transmission by Polling

An alternative to waiting a set time for transmission is to monitor the TI flag in the SCON register until it is set by the transmission of the last character written to SBUF. The polling routine must reset TI before returning to the call program. Failure to reset TI will inhibit all calls after the first, stopping all data transmission except the first character.

This technique has the advantage of simplicity; less code is used, and the routine *xmit,* does not care what the actual baud rate is.

```
;Xmit, using a 12MHz crystal for UART timing. 2400 nominal baud rate
;for a 12MHz crystal (2404). Poll TI flag to know when to send next
;character.
        baudnum equ 0f3h        ;number loaded into TH1 for 2403.8 baud
        org 0000h
        anl pcon,#7fh           ;set SMOD bit to 0 for Baud × 32 rate
        anl tmod,#30h           ;alter timer T1 configuration only
        orl tmod,#20h           ;set timer T1 as an 8-bit autoload
        mov th1,#baudnum        ;TH1 set for divide clock by 13d
        setb tr1                ;T1 running at 1E6/13 = 76923 Hz
        mov scon,#40h           ;set UART to mode 1
xmit:
        mov sbuf,#'U'           ;transmit ASCII "U" (55h) continuously
wait:                           ;NOTE: character pattern = 0101010101
                                ;continuously – frequency = 1201.9 Hz
        jbc ti,xmit             ;wait for TI set before next xmission
        sjmp wait               ;else poll flag again
        end
```

◆ COMMENT ◆

TI remains a 0 until SBUF is empty; when the 8051 is reset, or on power up, TI is set to 0.

Interrupt-Driven Character Transmission

The third method of determining when transmission is finished is to use the interrupt structure of the 8051. One interrupt vector address in program code, location 0023h, is assigned to both the transmit interrupt, TI, and the receive interrupt, RI. When a serial interrupt occurs, a hardware call to location 0023h accesses the interrupt handling routine placed there by the programmer.

The program begins by loading the first character to be sent into SBUF and enabling the serial interrupt bit in the EI register. The user program can then continue executing. When SBUF becomes empty, TI will be set, resulting in an immediate vector to 0023h and the subroutine placed there executed. The subroutine at 0023h, called *serial,* will reset TI, re-load SBUF, and return to the user program at the place where it was interrupted.

SBUFR

An interrupt-driven data transmission routine for continuous character transmission, which is assembled at the interrupt vector location 0023h. A portion of the user program that activates the interrupt routine is shown.

```
;Sbufr, using a 12MHz crystal for UART timing. 2400 nominal baud
;rate for a 12MHz crystal (2404). Use TI interrupt to program
;located at address 0023h to send next character.
;
        baudnum equ 0f3h        ;number loaded into TH1 for 2403.8 baud
        org 0000h
        sjmp over               ;jump over serial interrupt location
;
serial:                         ;place TI interrupt routine here
        org 0023h               ;place serial interrupt program here
        clr ti                  ;reset TI flag for next interrupt
        mov sbuf,#'U'           ;repeat character transmission forever
        reti                    ;return from interrupt to "wait" loop
;
over:
        anl pcon,#7fh           ;set SMOD bit to 0 for Baud × 32 rate
        anl tmod,#30h           ;alter timer T1 configuration only
        orl tmod,#20h           ;set timer T1 as an 8-bit autoload
        mov th1,#baudnum        ;TH1 set for divide clock by 13d
        setb tr1                ;T1 running at 1E6/13 = 76923 Hz
        mov scon,#40h           ;set UART to mode 1
        orl ie,#90h             ;enable global and serial interrupts
        mov sbuf,#'U'           ;send first character
wait:
        sjmp wait               ;simulate the rest of the program here
        end
```

◆ COMMENT ◆

- If TI is not cleared before the RETI instruction is used, there will be an immediate interrupt and vector back to 0023h.
- RETI is used to reset the entire interrupt structure, not to clear any interrupt bits.

Receiving Serial Data

Transmissions from outside sources to the 8051 are not predictable unless an elaborate time-of-day clock is maintained at the sender and receiver. Messages can then be sent at predefined times. A time-of-day clock generally ties up timers at both ends to generate the required "wake-up" calls.

Two methods are normally used to alert the receiving program that serial data has arrived: software polling or interrupt driven. The sending entity, or *talker,* transmits data at random times, but uses an agreed-upon baud rate and data transmission mode. The receiving unit, commonly dubbed the *listener,* configures the serial port to the mode and baud rate to be used and then proceeds with its program.

If one programmer were responsible for the talker and another for the listener, lively discussions would ensue when the units were connected and data interchange did not take place. One common method used to test communication programs is for each programmer to use a terminal to simulate the other unit. When the units are connected for the final test, a CRT terminal in a transparent mode, which shows all data transmitted in both directions, is connected between the two systems to show what is taking place in the communication link.

Polling for Received Data

Polling involves periodically testing the Received Data flag RI and calling the data receiving subroutine when it is set. Care must be taken to remember to reset RI, or the same character will be read again. Reading SBUF does *not* clear the data in SBUF or the RI flag.

The program can sit in a loop, constantly testing the flag until data is received, or run through the entire program in a circular manner, testing the flag on each circuit of the program. The loop approach guarantees that the data will be read as soon as it is received; however, very little else will be accomplished by the program while waiting for the data. The circular approach lets the program run while awaiting the data.

In order not to miss any data, the circular approach requires that the program be able to run a complete circuit in the time it takes to receive 1 data character. The time restraint on the program is not as stringent a requirement as it may first appear. The receiver is double buffered, which lets the reception of a second character begin while a previous character remains unread in SBUF. If the first character is read before the last bit of the second is complete, then no data will be lost. This means that, after a 2-character burst, the program still must finish in 1-character time to catch a third.

The character time is the number of bits per character divided by the baud rate. For serial data transmission mode 1, a character uses 10 bits: start, 8 code bits, and stop. A 1200 baud rate, which might be typical for a system where the talker and listener do not interchange volumes of data, results in a character rate of 120 characters per second, or a character time of 8.33 milliseconds. Using an average of 18 oscillator periods per instruction, each instruction will require 1.5 microseconds to execute, enabling a program length of 5,555 instructions. Such a large machine language program will suffice for many simple control and monitoring applications where data transmission rates are low. If more time is needed, the baud rate could be reduced to as low as 300 baud, yielding a program size of over 20K, which approaches half the maximum size of the ROM in our example 8051 design.

Recv

A polling-type program named *Recv* follows. *Recv* is a circular-type program that tests RI at the label here: and jumps to the receive subroutine if RI is set. If RI is not set the program continues with the section represented by the NOP instruction. At the end of the program a jump back to here: re-tests RI.

```
;Recv, a program that polls for a received character. 12MHz crystal
;for a nominal rate of 2400 (actual of 2404) baud. The transmitter
;used to test this program transmits at 2400 baud with no loss of
;characters.
     baudnum equ 0f3h        ;number loaded into TH1 for 2403.8 baud
     org 0000h
     anl pcon,#7fh           ;set SMOD bit to 0 for Baud × 32 rate
     anl tmod,#30h           ;alter timer T1 configuration only
     orl tmod,#20h           ;set timer T1 as an 8-bit autoload
     mov th1,#baudnum        ;TH1 set for divide clock by 13d
     setb tr1                ;T1 running at 1E6/13 = 76923 Hz
     mov scon,#40h           ;UART mode 1; Baud = 76923/32 = 2403.8
     setb ren                ;for emphasis – remember to enable REN
here:
     jbc ri,receive          ;loop here until character received
     nop                     ;rest of program could be placed here
     sjmp here               ;poll RI again
receive:
     clr ri                  ;clear bit to prevent a false read
     mov p1,sbuf             ;display character on port 1
     sjmp here               ;
     end
```

Interrupt-Driven Data Reception

When large volumes of data must be received, the data rate will overwhelm the polling approach unless the user program is extremely short, a feature not usually found in systems in which large amounts of data are interchanged. Interrupt-driven systems allow the program to run with brief pauses to read the received data. In Chapter 11, a program is developed that allows for the reception of long strings of data in a manner completely transparent to the user program.

Intdat

A short interrupt-driven program named *Intdat* is listed next. Intdat demonstrates how RI may be enabled to automatically cause an interrupt call to location 0023h in the program. Once RI is enabled to interrupt the main program, data is read as it is received by the subroutine at 0023h. Received data could be then stored by the subroutine in the RAM locations (a data *buffer*) for later reading by the main program.

Intdat enables RI to interrupt and then enters an endless loop at label there: to represent the main program. As each character is received, an interrupt to the subroutine named receive: at 0023h is done. Received data is displayed on P1 and then a RETI is done back to the main program.

Note that a program that both transmits and receives would have to test both the TI and the RI flags to determine the proper response to an interrupt to 0023h.

```
;Intdat interrupts to 0023h for each received character. 12MHz
;crystal for a nominal baud rate of 2400, actual of 2404.
    baudnum equ 0f3h        ;number loaded into TH1 for 2403.8 baud
    org 0000h
    sjmp over               ;jump over serial interrupt at 0023h
;
receive:
    org 0023h
    clr ri                  ;be sure to reset RI
    mov p1,sbuf             ;get character and copy to P1
    reti                    ;return to looping program
;
over:
    anl pcon,#7fh           ;set SMOD bit to 0 for Baud × 32 rate
    anl tmod,#0fh           ;alter timer T1 configuration only
    orl tmod,#20h           ;set timer T1 as an 8-bit autoload
    mov th1,#baudnum        ;TH1 set for divide clock by 13d
    setb tr1                ;T1 running at 1E6/13 = 76923 Hz
    mov scon,#40h           ;UART mode 1; Baud = 76923/32 = 2403.8
    setb ren                ;for emphasis only – must enable REN
    orl ie,#90h             ;enable global and serial interrupts
there:
    sjmp there              ;represents rest of a looping main program
    end
```

◆ COMMENT ◆

If the receive subroutine that is called takes longer to execute than the character time, then data will be lost. Long subroutine times would be highly unusual; however, it is possible to overload any system by constant data reception.

9.7 Summary

An 8051-based microprocessor system has been designed that incorporates many features found in commercial designs. The design can be easily duplicated by the reader and uses external EPROM and RAM so that test programs may be exercised. Various size memories may be used by the impecunious to reduce system cost.

The design features are:

- External RAM: 8K to 32K
- External ROM: 8K to 64K
- I/O ports: 1–8 bit, port 1
- Other ports: port 3.0 (RXD)
3.1 (TXD)
3.2 ($\overline{\text{INT0}}$)
3.3 ($\overline{\text{INT1}}$)
3.4 (T0)
3.5 (T1)
- Crystal: 12 megahertz

Other crystal frequencies may be used to generate convenient timing frequencies. The design can be modified to include a single-step capability (see Problem 2).

Methods of adding additional ports to the basic design are discussed and several example circuits that indicate the expansion possibilities of the 8051 are presented.

Programs written to test the design can be used to verify any prototypes that are built by the reader. These tests involve verifying the proper operation of the ROM and RAM connections.

Several programs and subroutines are developed that let the user begin to exercise the 8051 instruction code and hardware capabilities. This code can be run on the simulator or on an actual prototype. These programs cover the most common types found in most applications:

- Time delays: software; timer, software-polled; timer, interrupt-driven
- Lookup tables: PC base; DPTR base
- Serial data communications transmission: time delay; software-polled; interrupt-driven
- Serial data communications reception: software-polled; interrupt-driven

The foundations laid in this chapter will be built upon by example application programs and hardware configurations found in Chapters 10 and 11.

9.8 Questions and Problems

1. Determine whether the 8051 can be made to execute a single program instruction (single-stepped) using external circuitry (no software) only.
2. Outline a scheme for single-stepping the 8051 using a combination of hardware and software. (Hint: Use an $\overline{\text{INTX}}$.)
3. While running the EPROM test, it is found that the program cannot jump from 2000h to 4000h successfully. Determine what address line(s) is(are) faulty.
4. Calculate the error for the delay program softime when values of 2d, 10d, and 1000d milliseconds are passed in A and B.
5. Find the shortest and longest delays possible using softime by changing only the equate value of the variable delay.
6. Give a general description of how you would test any time delay program. (Hint: Use a port pin.)
7. Calculate the shortest and longest delays possible using the program named timer by changing the initial value of T0.
8. Why is there no check for an initial timing value of 0000h in the program named hardtime?
9. Write a lookup table program, using the PC as the base, that finds a 1-byte square root (to the nearest whole integer) of any number placed in A. For example, the square roots of 01 and 02 are both 01, while the roots of 03 and 04 are 02. Calculate the first four and last four table values.
10. Write a lookup table, using the DPTR as the base, that finds a 2-byte square root of the number in A. The first byte is the integer value of the root, and the second byte is the fractional value. For example, the square root of 02 is 01.6Ah. Calculate four first and last table values.
11. Write a lookup table program that converts the hex number in A (0 – F) to its ASCII equivalent.
12. A PC-based lookup table, which contains 256d values, is mistakenly placed 50h bytes after the MOVC instruction that accesses it. Construct the table, showing where the byte associated with A = 00h is located. Find the largest number that can be placed in A to access the table.
13. Construct a lookup table program that converts the hex number in A to an equivalent BCD number in registers R4 (MSB) and R5 (LSB).
14. Reverse Problem 13 and write a lookup table program that takes the BCD number in R4 (MSB) and R5 (LSB) and converts it to a hex number in A.
15. Does asynchronous communication between two microprocessors have to be done at standard baud rates? Name one reason why you might wish to use standard rates.

16. Write a test program that will "loop test" the serial port. The output of the serial port (TXD) is connected to the input (RXD), and the test program is run. Success is indicated by port 1 pin 1 going high.
17. What is the significance of the Transmit flag, TI, when it is cleared to 0? When set to 1?
18. Using the programmable port of Figure 9.3, write a program that will configure all ports as outputs, and write a 55h to each.
19. Repeat problem 18 using the memory-mapped programmable port of Figure 9.4.

10

Applications

Chapter Outline

CHAPTER OBJECTIVES

On successful completion of this chapter you will be able to:

- Interface keyboards to the 8031-based microcontroller system.
- Interface LED and LCD displays to the microcontroller system.
- Use the microcontroller system to determine the frequency of external pulses.
- Interface the microcontroller system to A/D and D/A converters.
- Expand the interrupt capability of the microcontroller system.
- Analyze a system configuration that includes a keyboard, an LCD display, and serial data transmission.

10.0 Introduction

Microcontrollers tend to be underutilized in many applications. There is one main reason for this anomaly. Principally, the devices are so inexpensive that it makes little economic sense to try to select an optimal device for each new application. A new microcontroller involves the expense of new development software and training for the designers and programmers that could easily cost more than the part savings.

The result of this application bias is that microcontrollers tend to become obsolete at a slower rate than their CPU cousins. The 8051 is a good example of microcontroller design staying power. The 8051 architecture and assembly code are well over 20 years old, but the 8051 remains popular today. During the same time span, the IBM PC processor changed from the 8088 to the 80286 to the 80386 to the 80486 to the Pentium 1, 2, 3, and 4.

Application examples in a textbook present a picture of use that supports the previously made claim of underutilization. Limitations on space, time, and the patience of the reader preclude the inclusion of involved, multi-thousand-line, real-time examples. We will, instead, look at pieces of larger problems, each piece representing a task commonly found in most applications.

One of the best ways to get a "feel" for a new processor is to examine circuits and programs that address easily visualized applications and then to write variations. To assist in this process, we will study in detail the following typical hardware configurations and their accompanying programs:

- Keyboards
- Displays
- Pulse measurements
- A/D and D/A conversions
- Multi-source interrupts

The hardware and software are inexorably linked in the examples in this chapter. The choice of the first leads to the programming techniques of the second. The circuit designer should have a good understanding of the software limitations faced by the programmer. The programmer should avoid the temptation of having all the tricky problems handled by the hardware.

10.1 Keyboards

The predominant interface between humans and computers is the keyboard. Keyboards range in complexity from the "up-down" buttons used for elevators to the personal computer QWERTY layout, with the addition of function keys and numeric keypads. One of the first mass uses for the microcontroller was to interface between the keyboard and the main processor in personal computers. Industrial and commercial applications fall somewhere in between these extremes, using layouts that might feature from six to twenty keys.

The one constant in all keyboard applications is the need to accommodate the human user. Human beings can be irritable. They have little tolerance for machine failure; watch what happens when the product isn't ejected from the vending machine. Sometimes they are bored by routine, or even hostile towards the machine. The hardware designer has to select keys that will survive in the intended environment. The programmer must write code that will anticipate and defeat inadvertent and also deliberate attempts by the human to confuse the program. It is very important to give instant feedback to the user that the key hit has been acknowledged by the program. By light of a light, beep of a buzzer, display of the key hit, or whatever, the human user must know that the key has been recognized. Even feedback sometimes is not enough; note the behavior of people at an elevator. Even if the "up" light is lit when we arrive, we will push it again to let the machine know that "I'm here too."

Human Factors

The keyboard application program must guard against the following possibilities:

- More than one key pressed (simultaneously or released in any sequence)
- Key pressed and held
- Rapid key press and release

All of these situations can be addressed by hardware or software means; software, which is the most cost-effective, is emphasized here.

Key Switch Factors

The universal key characteristic is the ability to bounce: The key contacts vibrate open and closed for a number of milliseconds when the key is hit and often when it is released. These rapid pulses are not discernible to the human, but they last a relative eternity in the microsecond-dominated life of the microcontroller. Keys may be purchased that do not bounce, or keys may be debounced with RS flip-flops or debounced in software with time delays.

Keyboard Configurations

Keyboards are commercially produced in one of the three general hypothetical wiring configurations for a 16-key layout shown in Figure 10.1. The lead-per-key

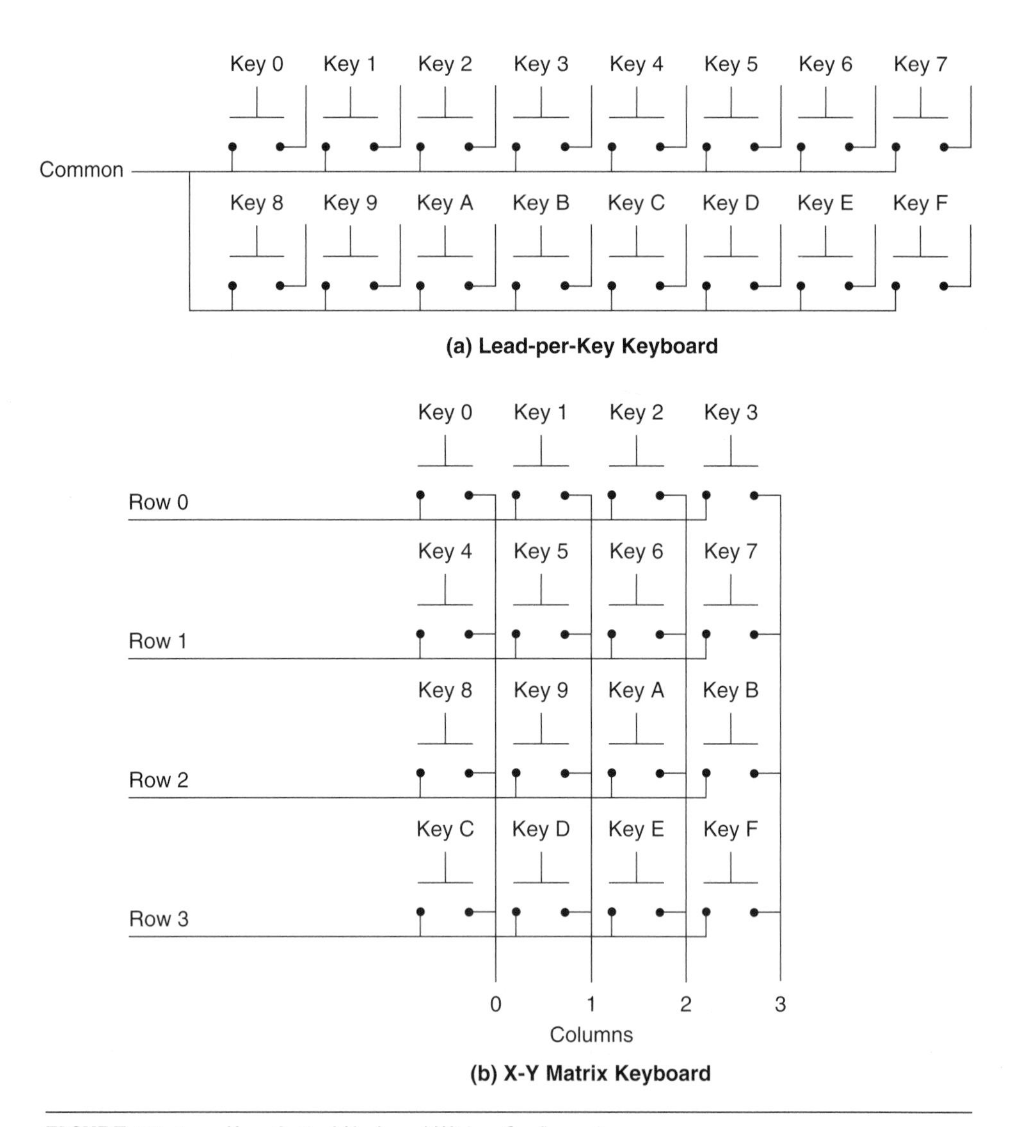

FIGURE 10.1 ◆ Hypothetical Keyboard Wiring Configurations

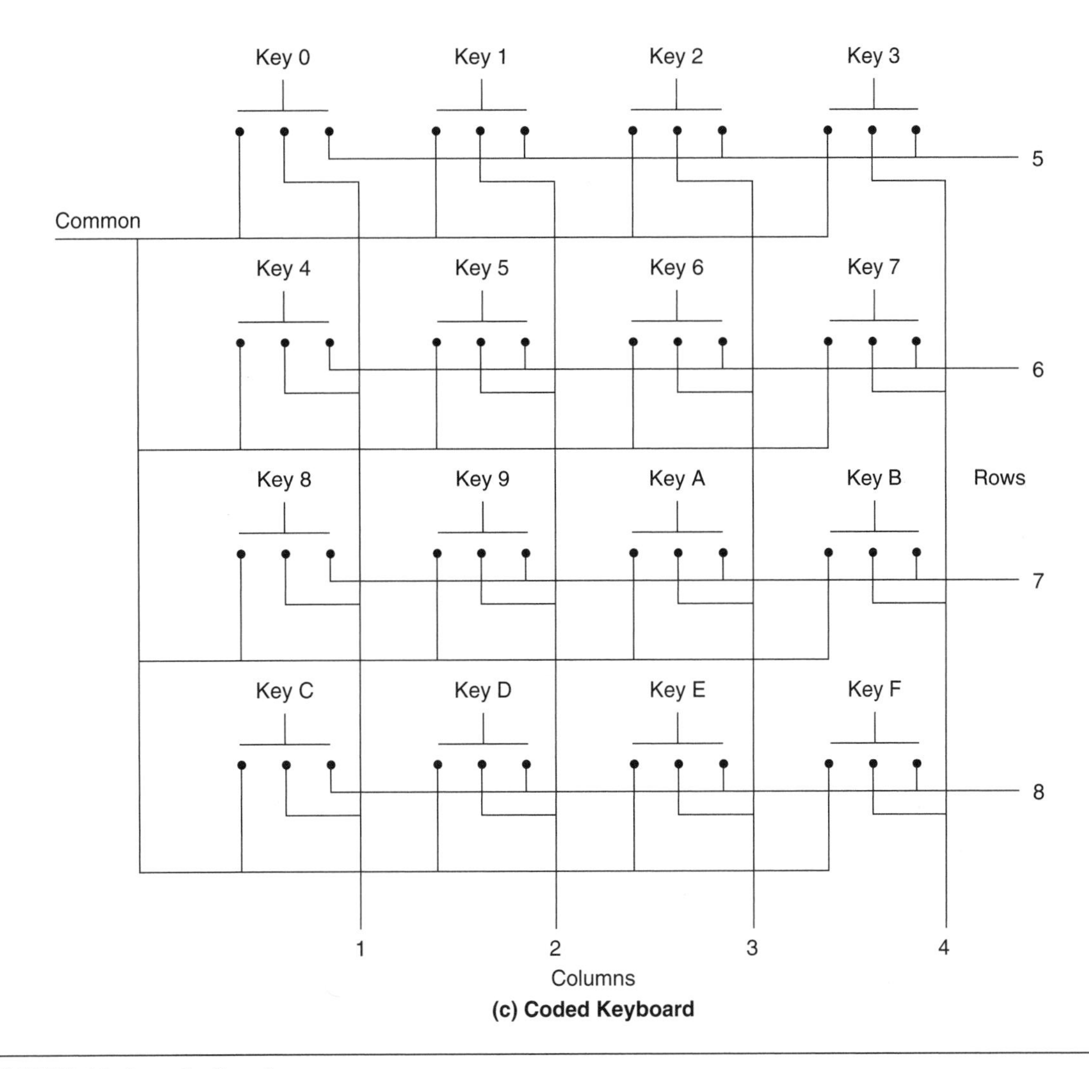

FIGURE 10.1 ◆ *Continued*

configuration is typically used when there are very few keys to be sensed. Since each key could tie up a port pin, it is suggested that the number be kept to 16 or fewer for this keyboard type. This configuration is the most cost-effective for a small number of keys.

The X-Y matrix connections shown in Figure 10.1 are very popular when the number of keys exceeds 10. The matrix is most efficient when arranged as a square so that N leads for X and N leads for Y can be used to sense as many as N^2 keys. Matrices are the most cost-effective for a large number of keys.

Coded keyboards were evolved originally for telephonic applications involving touch-tone signaling. The coding permits multiple key presses to be easily detected. The quality and durability of these keypads are excellent as a result of the high production volumes and intended use. They are generally limited to 16 keys or fewer, and tend to be the most expensive of all keyboard types.

Programs for Keyboards

Programs that deal with humans via keyboards approach the human and key-switch factors identified in the following manner:

- *Bounce:* A time delay that is known to exceed the manufacturer's specification is used to wait out the bounce period in both directions.
- *Multiple keys:* Only patterns that are generated by a valid key pressed are accepted—all others are ignored—and the first valid pattern is accepted.
- *Key held:* Valid key pattern accepted after valid debounce delay; no additional keys accepted until all keys are seen to be up for a certain period of time.
- *Rapid key hit:* The design is such that the keys are scanned at a rate faster than any human reaction time.

The last item brings up an important point: Should the keyboard be read as the program loops (software-polled) or read only when a key has been hit (interrupt-driven)?

In general, the smaller keyboards (lead-per-key and coded) can be handled either way. The common lead can be grounded and the key pattern read periodically. Or, the lows from each can be active-low ORed, as shown in Figure 10.2, and connected to one of the external $\overline{\text{INTX}}$ pins.

Matrix keyboards are scanned by bringing each X row low in sequence and detecting a Y column low to identify each key in the matrix. X-Y scanning can be done by using dedicated keyboard scanning circuitry or by using the microcontroller ports under program control. The scanning circuitry adds cost to the system. The programming approach takes processor time, and the possibility exists that response to the user may be sluggish if the program is busy elsewhere when a key is hit. Note how long your personal computer takes to respond to a break key when it is executing a print command, for instance. The choice between adding scanning hardware or program software is decided by how busy the processor is and the volume of entries by the user.

A Scanning Program for Small Keyboards

Assume that a lead-per-key keyboard is to be interfaced to the microcontroller. The keyboard has 10 keys (0 – 9), and the debounce time, when a key is pressed

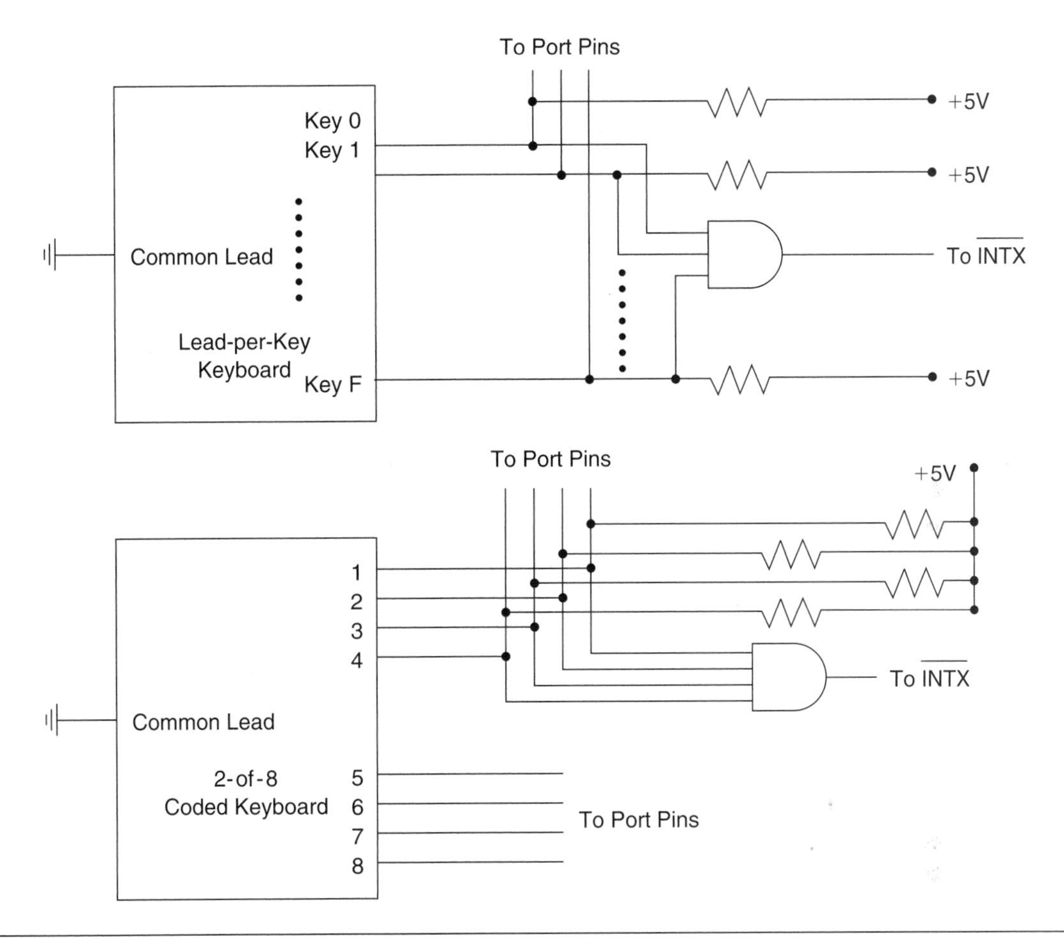

FIGURE 10.2 ◆ Lead-per-Key and Coded Keyboard Interrupt Circuits

or released, is 50 milliseconds. The keyboard is used to select snacks from a vending machine, so the processor is only occupied when a selection is made. The program constantly scans the keyboard waiting for a key to be pressed before calling the vending machine actuator subroutine. The keys are connected to port 1 (0 – 7) and ports 3.2 and 3.3 (8 – 9), as shown in Figure 10.3.

The 8031 works best when handling data in byte-sized packages. To save internal space, the 10-bit word representing the port pin configuration is converted to a single-byte number.

Because the processor has nothing to do until the key has been detected, the time delay softime (see Chapter 9) is used to debounce the keys.

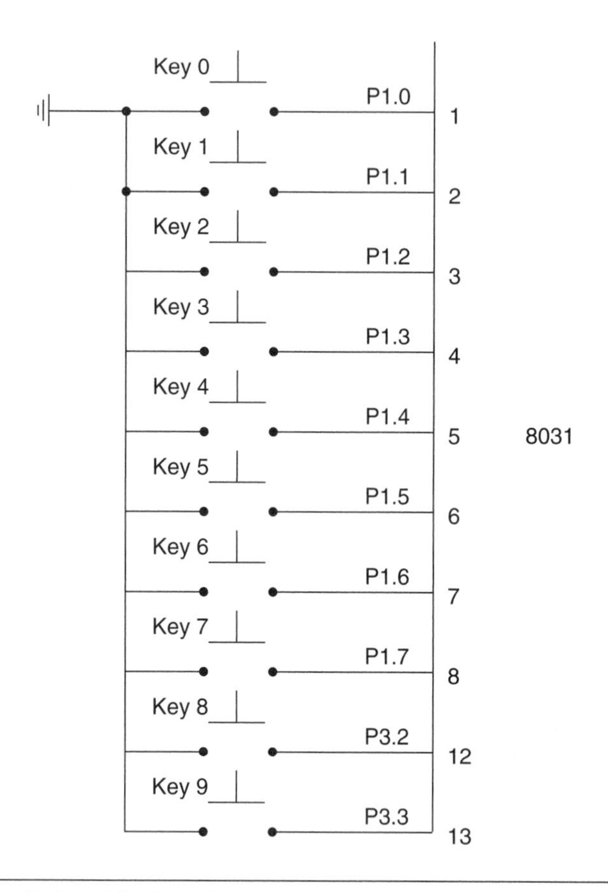

FIGURE 10.3 ◆ Keyboard Configuration for Getkey and Inkey Programs

Getkey

The routine *getkey* constantly scans a 10-key pad via ports 0 and 3. The keys are debounced in both directions and an "all-up" period of 100 milliseconds must be seen before a new key will be accepted. Invalid key patterns (more than one port pin low) are rejected.

```
;Getkey, uses "softime" as a software time delay to debounce the
;keys.

        bounce equ 32h          ;bounce time delay is 50d milliseconds
        next equ 64h            ;set time interval between keys to 100 ms
        newkey equ 70h          ;RAM variable stores valid new key
        flag equ 00h            ;bit addressable flag
```

```
        delay equ 0a6h              ;delay constant for "softime" routine
        org 0000h
getkey:
        mov p1,#0ffh                ;set P1 and P3 as input ports
scan:
        mov p3,#0ffh                ;if no key then P3 MSN = Fh
        acall keydown               ;look for key(s) grounded (down)
        jz scan                     ;loop until key(s) pressed: A=00 if none
        acall convert               ;convert; set flag bit if not valid key
        jbc flag,scan               ;if flag set then not a valid key
        mov newkey,a                ;save key and wait for debounce time
        mov b,#00h
        mov a,#bounce               ;debounce key
        acall softime               ;delay debounce time
        acall keydown               ;see if any key is still down
        jz scan                     ;if not then noise, look again
        acall convert               ;if not noise, then see if a valid key
        jbc flag,scan
        cjne a,newkey,scan          ;if last key = stored key, then valid key
        acall vendit                ;call to vending routine
wait:
        acall keydown               ;wait until keys are all up
        jnz wait                    ;loop until all keys up
        mov b,#00h
        mov a,#next                 ;wait for key up time delay
        acall softime
        acall keydown               ;see if all still up
        jnz wait                    ;if still down then wait
        sjmp scan                   ;get next key
;
;"keydown" gets the contents of the P1 and P3 pins connected to the
;keys and checks that at least one pin is low. No check is made for
;more than one key grounded.
keydown:
        mov r0,p1                   ;use R0 to hold P1 status
        mov a,p3                    ;use A to hold P3 status
        orl a,#0f3h                 ;All bits of A = 1, except 2 and 3
        anl a,r0                    ;if A = FFh then no key is down
        cpl a                       ;if A <> 00h then at least one key down
        ret                         ;if A = 00h then no keys are down
;
;"convert" checks for more than one key down; if more than one key
;is down, then the flag bit is set. If only one key is down, then
;the 1-of-10 pattern is converted to an equivalent number from 00
;to 09 in the A register, and the flag bit is reset. Valid patterns
;are found by a series of CJNE instructions as A is counted up by 1
```

```
;every time a CJNE fails to find a match. Note that a lookup table
;may also be used to relate key patterns to key numbers.
;
convert:
     clr flag                  ;assume that only one key down
     clr a                     ;A = 00h, the first possible key number
     mov r1,p1                 ;use R1 to hold P1 pattern
     mov r3,p3                 ;use R3 to hold P3 pattern
     orl 03h,#0f3h             ;all bits of R3 are 1 except 2 and 3
     cjne r1,#0feh,one         ;if key 0 down, then CJNE will not jump
     sjmp check3               ;check bits 2 and 3 of P3
one:
     inc a                     ;not 0 key, A = 01
     cjne r1,#0fdh,two         ;try bit 1 of P1
     sjmp check3
two:
     inc a                     ;not 1 key, A = 02
     cjne r1,#0fbh,three       ;try bit 2 of P1
     sjmp check3
three:
     inc a                     ;not 2 key, A = 03
     cjne r1,#0f7h,four        ;try bit 3 of P1
     sjmp check3
four:
     inc a                     ;not 3 key, A = 04
     cjne r1,#0efh,five        ;try bit 4 of P1
     sjmp check3
five:
     inc a                     ;not 4 key, A = 05
     cjne r1,#0dfh,six         ;try bit 5 of P1
     sjmp check3
six:
     inc a                     ;not 5 key, A = 06
     cjne r1,#0bfh,seven       ;try bit 6 of P1
     sjmp check3
seven:
     inc a                     ;not 6 key, A = 07
     cjne r1,#07fh,eight       ;try bit 7 of P1
     sjmp check3
eight:
     inc a                     ;not 7 key, A = 08
     cjne r3,#0fbh,nine        ;try bit 2 of P3
     sjmp good
nine:
     inc a                     ;not 8 key, A = 09
     cjne r3,#0f7h,bad         ;try bit 3 of P3
     sjmp good
check3:
```

```
        cjne r3,#0ffh,bad           ;P1 has one low, check P3 for FFh
good:
        ret
bad:
        setb flag                   ;more than one key down, not valid
        ret
;
softime:                            ;time delay routine named "softime"
        push 07h
        push acc
        orl a,b
        cjne a,#00h,ok
        pop acc
        sjmp done
ok:
        pop acc
timer:
        mov r7,#delay
onemil:
        nop
        nop
        nop
        nop
        djnz r7,onemil
        nop
        nop
        dec a
        cjne a,#0ffh,noroll
        dec b
noroll:
        cjne a,#00h,timer
        cjne a,b,timer
done:
        pop 07h
        ret
;
vendit:
        swap a                      ;display valid key on P3 MSN
        orl a,#0fh                  ;leave P3 LSN as 1111
        mov p3,a
        ret                         ;simulate some vending routine
        end
```

◆ COMMENT ◆

- The convert subroutine is looking for a single low bit. The CJNE patterns all have 1 bit low and the rest high.

- Multiple keys are rejected by convert. Held keys are ignored as the program waits for a 100d millisecond all-keys-up period before admitting the next key. The program loops so quickly that it is humanly impossible to hit a key so that it can be missed.
- The main program is predominantly a series of calls to subroutines, which can each be written by different programmers. Agreement on what data is passed to and received from the subroutines is essential for success, as well as a clear understanding of which 8051 registers and memory locations are used.

Interrupt-Driven Programs for Small Keyboards

If the application is so time sensitive that the delays associated with debouncing and awaiting an all-up cannot be tolerated, then some form of interrupt must be used so that the main program can run unhindered.

A compromise may be made by polling the keyboard as the main program loops, but all time delays are done using timers so that the main program does not wait for a software delay. The getkey program can be modified to use a timer to generate the delays associated with the key-down debounce time and the "all-up" delay. The challenge associated with this approach is to have the program remember which delay is being timed out. Remembering which delay is in progress can be handled using a flag bit, or one timer can be used to generate the key-down debounce delay, and another timer to generate the key-up delay. The two-timer approach is examined in the example given in this section.

The important feature of the program is that the main program will check a flag to see whether there is any keyboard activity. If the flag is set, then the program finds the key stored in a RAM location and resets the flag. The getting of the key is transparent to the main program; it is done in the interrupt program.

Inkey

The program *inkey* uses both hardware timers, T0 and T1, to generate the time delays needed to scan and debounce the keyboard.

The subroutines that get a key run in the background by using timer T1 to generate a main program interrupt every 50 milliseconds. At least one 50ms time interval must occur before the T1 interrupt routine will scan the keys. The T1 interrupt routine *scankey* then scans the keys. If a valid key is found down, T0 is started to debounce the key for 20 milliseconds. Timer T1 is stopped, and the T1 interrupt routine returns to the main program. After 20 milliseconds, T0 times out and interrupts the main program.

The T0 interrupt routine *samekey* checks for the same valid key down as was found by scankey. If the same valid key is found, the key's code is stored and a flag, keyflag, is set. T0 is stopped and T1 is restarted for the next 50ms keyboard scan delay.

The main program loop, inkey, checks keyflag and calls routine *key,* which will get each new key and, in turn, call a sample vending machine routine, *vendit,* which shows the key code on port P3.

Register bank 1 is used for all subroutine operations so that the main program registers are not altered. Dedicating a register bank for interrupt subroutines saves stack space normally used to push and pop main program registers.

```
        bncelo equ 0e0h         ;TLO time for a 20d millisecond delay
        bncehi equ 0b1h         ;TH0 time for a 20d millisecond delay
        uplo equ 0b0h           ;TL1 time for a 50d millisecond delay
        uphi equ 3ch            ;TH1 time for a 50d millisecond delay
        newkey equ 7fh          ;RAM variable stores valid new key
        upflag equ 00h          ;signals T1 routine of at least one delay
        keyflag equ 01h         ;if set then a new key
        validflag equ 02h       ;if clear, then valid key
        org 0000h
        mov sp.#30h             ;place stack in GP area
        sjmp over               ;jump over T0 and T1 interrupt addresses
;
sameky:                         ;place T0 interrupt routine here
        org 00bh                ;T0 interrupt
        acall intt0             ;call T0 interrupt routine
        reti
;
scankey:                        ;place T1 interrupt routine here
        org 001bh               ;place T1 routine here
        push acc                ;save A
        push psw                ;save PSW
        mov psw,#08h            ;select register bank 1
        clr tr1                 ;stop T1
        jbc upflag,firstscan    ;upflag set after first 50ms key-up delay
        acall keydown           ;check for keys up
        jnz keysdown
        setb upflag             ;at least one 50ms keys-up delay
keysdown:
        acall startT1           ;restart T1
        pop psw                 ;restore PSW
        pop acc                 ;restore A
        reti                    ;return to main program
firstscan:
        acall keydown           ;look for key(s) down
        jz stillup              ;restart T1 if not, set upflag
        acall convert           ;look for a valid key
        jbc validflag,nextscan  ;if not valid restart T1
        move newkey,a           ;if so then store and start T0
        acall startT0
        pop psw                 ;restore PSW
```

```
        pop acc                 ;restore A
        reti                    ;return with T0 running to debounce
stillup:
        setb upflag             ;no keys down, scan on next T1 interrupt
nextscan:
        acall startT1           ;restart T1 for next scan interval
        pop psw                 ;restore PSW
        pop acc                 ;restore A
        reti
;
intt0:
        push acc                ;save A
        push psw                ;save PSW
        mov psw,#08h            ;select register bank 1
        clr tr0                 ;stop T0
        acall keydown           ;see if key still down
        jz noise                ;if not, then noise, return
        acall convert           ;if not noise, then see if a valid key
        jbc validflag,noise     ;if not a valid key then return
        cjne a,newkey,noise     ;if last key = stored key, then valid key
        setb keyflag            ;signal main program new key present
        clr upflag              ;reset upflag for 50ms keys up delay
noise:
        acall startT1           ;restart T1
        pop psw                 ;restore PSW
        pop acc                 ;restore A
        ret
;
;jump over T0 and T1 interrupt addresses to begin scan processes
over:
        mov tmod,#11h           ;set T1,T0 as 16-bit timers
        mov ie,#8ah             ;enable global, T1, T0 interrupts
        mov p1,#0ffh            ;set P1 and P3 as input ports
        mov p3,#0ffh            ;if no key then P3 MSN = Fh
        acall startT1           ;start T1 key scan timer
        clr upflag              ;start with initial key-up delay
;
;main program loops here
inkey:                          ;simulate main program as it loops
        jbc keyflag,key         ;if flag is set then get key
        sjmp inkey              ;continue looping
key:
        mov a,newkey            ;get new key
        acall vendit            ;call some vending routine
        sjmp inkey              ;get next key
;
;
;store subroutines after this comment
```

```
startT0:
      mov tl0,#bncelo          ;set T0 for a 20ms delay
      mov th0,#bncehi
      setb tr0
      ret
;
startT1:
      mov tl1,#uplo            ;set T1 for a 50ms delay
      mov th1,#uphi
      setb tr1
      ret
;
;"keydown" gets the contents of the P1 and P3 pins connected to the
;keys and checks that at least one pin is low. No check is made for
;more than one key grounded. If A = 00 then keys are up
keydown:
      mov r0,p1               ;use R0 to hold P1 status
      mov a,p3                ;use A to hold P3 status
      orl a,#0f3h             ;All bits of A = 1, except 2 and 3
      anl a,r0                ;if A = FFh then no key is down
      cpl a                   ;if A <> 00h then at least one key down
      ret                     ;if A = 00h then no keys are down
;
;"convert" checks for more than one key down; if more than one key
;is down, then the flag bit is set. If only one key is down, then
;the 1-of-10 pattern is converted to an equivalent number from 00
;to 09 in the A register, and the flag bit is reset. Valid patterns
;are found by a series of CJNE instructions as A is counted up by 1
;every time a CJNE fails to find a match. Note that a lookup table
;may also be used to relate key patterns to key numbers.
;
convert:
      clr validflag           ;assume that only one key down
      clr a                   ;A = 00h, the first possible key number
      mov r1,p1               ;use R1 to hold P1 pattern
      mov r3,p3               ;use R3 to hold P3 pattern. (RBANK 1!)
      orl 0bh,#0f3h           ;all bits of R3 are 1 except 2 and 3
      cjne r1,#0feh,one       ;if key 0 down, then CJNE will not jump
      sjmp check3             ;check bits 2 and 3 of P3
one:
      inc a                   ;not 0 key, A = 01
      cjne r1,#0fdh,two       ;try bit 1 of P1
      sjmp check3
two:
      inc a                   ;not 1 key, A = 02
      cjne r1,#0fbh,three     ;try bit 2 of P1
      sjmp check3
three:
```

```
        inc a                   ;not 2 key, A = 03
        cjne r1,#0f7h,four      ;try bit 3 of P1
        sjmp check3
four:
        inc a                   ;not 3 key, A = 04
        cjne r1,#0efh,five      ;try bit 4 of P1
        sjmp check3
five:
        inc a                   ;not 4 key, A = 05
        cjne r1,#0dfh,six       ;try bit 5 of P1
        sjmp check3
six:
        inc a                   ;not 5 key, A = 06
        cjne r1,#0bfh,seven     ;try bit 6 of P1
        sjmp check3
seven:
        inc a                   ;not 6 key, A = 07
        cjne r1,#07fh,eight     ;try bit 7 of P1
        sjmp check3
eight:
        inc a                   ;not 7 key, A = 08
        cjne r3,#0fbh,nine      ;try bit 2 of P3
        sjmp good
nine:
        inc a                   ;not 8 key, A = 09
        cjne r3,#0f7h,bad       ;try bit 3 of P3
        sjmp good
check3:
        cjne r3,#0ffh,bad       ;P1 has one low, check P3 for FFh
good:
        ret
bad:
        setb validflag          ;more than one key down, not valid
        ret
;
vendit:
        swap a                  ;display valid key on P3 MSN
        orl a,#0fh              ;leave P3 LSN as 1111
        mov p3,a
        ret                     ;simulate some vending routine
end
```

Codekey

The completely interrupt-driven small keyboard example given in this section requires no program action until a key has been pressed. Hardware must be added to attain a completely interrupt-driven event. The circuit of Figure 10.4 is used.

The keyboard is a 2-of-8 type which codes the 16 keys as shown in Table 10.1.

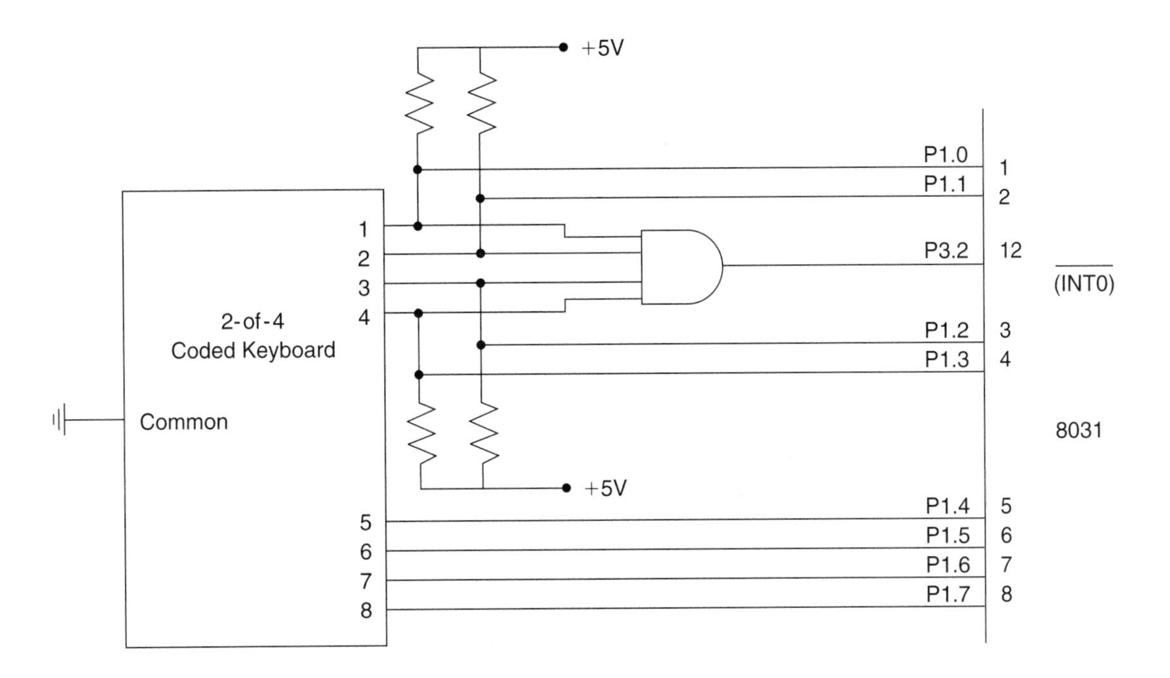

Keyboard Code		
Key	**Pins**	**Low**
0	1	5
1	2	5
2	3	5
3	4	5
4	1	6
5	2	6
6	3	6
7	4	6
8	1	7
9	2	7
A	3	7
B	4	7
C	1	8
D	2	8
E	3	8
F	4	8

FIGURE 10.4 ◆ Keyboard Configuration Used for Codekey Program

TABLE 10.1

Key	Code (hex)	Key	Code (hex)
0	EE	A	BB
1	ED	B	B7
2	EB	C	7E
3	E7	D	7D
4	DE	E	7B
5	DD	F	77
6	DB		
7	D7		
8	BE		
9	BD		

An inspection of the code reveals that each nibble of P1 has only 1 bit that is low for each key and that 2 of the 8 bits are uniquely low for each key. If more than one key is pressed, then 3 or more bits go low, signaling an invalid condition. This popular scheme allows for up to 16 keys to be coded in this manner. Unlike the lead-per-key arrangement, only four of the lines must be active-low ORed to generate an interrupt.

The hardware serves to detect when any number of keys are hit by using an AND gate to detect when any nibble bit goes low. The high-to-low transition then serves to interrupt the microcontroller on port 3.2 ($\overline{\text{INT0}}$). The interrupt program reads the keys connected to port 1 and uses timer T0 to generate the debounce time and T1 for the keys-up delay. The total delay possible at 12 megahertz for the timers is 65.53 milliseconds, which covers the delay times used in the previous examples.

The program *Codekey* is interrupt-driven by a high-to-low transition on $\overline{\text{INT0}}$. Timers T0 and T1 generate the debounce and delay times in an interrupt mode. The $\overline{\text{INT0}}$ interrupt input is disabled until all keys have been seen up for the T1 delay. A lookup table is used to verify that only one key is pressed.

```
;"codekey", an interrupt-driven coded key program. The hex keyboard
;generates the hex code listed below on the P1 port pins. The number
;in parentheses after the hex code is the hex result after the code
;is "zeroed" by subtracting 77h.
;     0 = EE(77), 1 = ED(76), 2 = EB(74), 3 = E7(70)
;     4 = DE(67), 5 = DD(66), 6 = DB(64), 7 = D7(60)
;     8 = BE(47), 9 = BD(46), A = BB(44), B = B7(40)
;     C = 7E(07), D = 7D(06), E = 7B(04), F = 77(00)
;The zeroed codes are used to construct a lookup table that converts
;each code to a hex equivalent.
;Timer T0 is used in an interrupt mode to generate the debounce
```

```
;delay and timer T1 is used in an interrupt mode to generate the
;keys-up delay.
;Port pin 2.3 blinks for each new key pressed. The key code is shown
;on the MSN of P2.
;
    newkey equ 70h          ;store a new key in RAM
    base equ 400h           ;base of lookup table in ROM
    newflg equ 00h          ;if set then a new key has occurred
    org 0000h
codekey:
    sjmp over               ;jump over /INT0 interrupt address
;
;when a key is pressed, the AND gate generates a strobe to /INT0
;which generates an interrupt to this code address
    org 0003h               ;place /INT0 program here
    sjmp keyint             ;jump to /INT0 routine
;
;Timer T0 generates an interrupt to this address after debounce time
    org 000bh               ;place timer T0 program here
    sjmp tim0               ;jump to T0 routine
;
;Timer T1 generates an interrupt to this address after keys-up delay
    org 001bh               ;place timer T1 program here
    sjmp tim1               ;jump to T1 routine
;
keyint:                     ;/INT0 routine
    mov tl0,#0e0h           ;set T0 for a 20ms debounce time delay
    mov th0,#0b1h           ;20 ms = 20,000 usec (4E20h)
    setb tr0                ;start T0
    clr ex0                 ;disable /INT0 during time delay
    reti                    ;return to main program during debounce
;
tim0:                       ;T0 interrupt routine
    push acc                ;save registers used in routine
    push dpl
    push dph
    clr tr0                 ;stop T0
    mov a,p1                ;read key code from keypad
    clr c                   ;zero the Carry flag
    subb a,#77h             ;convert key code to 00 - 77h
    mov dptr,#base          ;point to base of lookup table
    movc a, @a+dptr         ;get key code from offset 00h to 77h
    cjne a,#30h,good        ;all other codes in table are ASCII 0
    sjmp goback
good:
    mov newkey,a            ;store the new key
    setb newflg             ;signal main program new key present
```

```
goback:
        mov tl1,#60h            ;set T1 for a 50,000 usec delay
        mov th1,#3ch            ;before next key will be sensed
        pop dph                 ;restore registers and return
        pop dpl
        pop acc
        setb tr1                ;start T1 for keys-all-up delay
        reti
;
tim1:                           ;T1 routine here
        push acc                ;save A
        clr tr1                 ;stop T1
        mov a,p1                ;see if all keys are now up
        cjne a,#0ffh,wait       ;all keys up = 0FFh
        clr p2.3                ;reset newkey indication on P2.3
        setb ex0                ;enable /INT0
        pop acc                 ;restore A
        reti
wait:                           ;restart T1 and delay until all up
        mov tl1,#60h            ;set T1 for a 50,000 usec delay
        mov th1,#3ch            ;before next key will be sensed
        setb tr1                ;start T1
        pop acc                 ;restore A
        reti
;jump over interrupt routines to this location
over:
        setb p2.3               ;begin with no key
        mov p1,#0ffh            ;set P1 as an input port
        setb it0                ;enable /INT0 as edge triggered
        mov ie,#8bh             ;enable /INT0, T0, T1, global interrupts
        mov tmod,#11h           ;set T0 and T1 as 16-bit counters
;main program loops at simulate label
simulate:
        jbc newflg,key          ;main program checks for a new key
        sjmp simulate           ;loop here as simulated main program
;
key:
        mov a,newkey            ;get key code and display on P2 MSN
        swap a                  ;and show new key flag on P2.3
        orl a,#0fh              ;leave LSN of P2 as Fh
        mov p2,a
        setb p2.3
        clr p2.3                ;blink P2.3
        sjmp simulate
;
;lookup table placed in code memory at this address
        org 400h                ;place lookup table here
```

```
db 0fh                  ;key F
db "000"                ;fill rest of table with 30h (ASCII 0)
org 404h                ;lookup table is scattered from 00 to 77h
db 0eh                  ;key E
db "0"
org 406h
db 0dh                  ;key D
db 0Ch                  ;key C
db"00000000000000000000000000000000000000000000000000000000"
org 440h
db 0bh                  ;key B
db "000"
org 444h
db 0ah                  ;key A
db "0"
org 446h
db 09h                  ;key 9
db 08h                  ;key 8
db "000000000000000000000000"
org 460h
db 07h                  ;key 7
db "000"
org 464h
db 06h                  ;key 6
db "0"
org 466h
db 05h                  ;key 5
db 06h                  ;key 4
db "00000000"
org 470h
db 03h                  ;key 3
db "000"
org 474h
db 02h                  ;key 2
db "0"
org 476h
db 01h                  ;key 1
db 00h                  ;key 0
end
```

◆ COMMENT ◆

Key bounce down is eliminated by the T0 delay, and key bounce up, by the T1 delay. More than two keys down is detected by the self-coding nature of the keyboard. A held key does not interrupt the edge-triggered INT0 input.

Program for a Large Matrix Keyboard

A 64-key keyboard, arranged as an 8-row by 8-column matrix, will be interfaced to the 8051 microcontroller, as shown in Figure 10.5. Port 1 will be used to bring each row low, one row at a time, using an 8-bit latch that is strobed by port 3.2. P1 will then read the 8-bit column pattern by enabling the tri-state buffer from port 3.3. A pressed key will have a unique row-column pattern of one row low, one column low. Multiple key presses are rejected by either an invalid pattern or a failure to match for three complete cycles. Each row is scanned at an interval of 2 milliseconds, or a 16-millisecond cycle for the entire keyboard. A valid key must be seen to be the same key for three cycles (48 milliseconds). There must then be three cycles with no key down before a new key will be accepted. The 2-millisecond delay between scans is generated by timer T0 in an interrupt mode.

Bigkey

The *bigkey* program scans an 8 × 8 keyboard matrix using T0 to generate a periodic 2-ms delay in an interrupt mode. Each row is scanned via an external latch driven by port 1 and strobed by port 3.2. Columns are read via a tri-state buffer under control of port 3.3. Keys found to be valid are passed to the main program by setting the flag newflag and placing the key identifiers in locations newrow and newcol. The main program resets newflag when the new key is fetched. R4 is used as a cycle counter for successful matches and up time cycles. R5 is used to hold the row scan pattern: only 1 bit low.

```
;"bigkey", a large matrix keyboard scanning program.
;T0 is used to generate a 2 ms time delay on an interrupt basis. At
;the end of each delay, the T0 interrupt subroutine scans the
;keyboard and determines if one key is pressed. If the same key is
;seen pressed for 3 scans it is valid, and a new key flag
;"newflag" is set to the main program.
;Bank 0 registers are used for various scan program duties. R5 holds
;the row scan pattern, which brings one row low while all else are
;high. R4 counts up to 3 successful (same key down) succesive scans.
;R7 is used to store A, the column bit pattern read from the keyboard
;matrix via P1. R3 counts A rotates when looking for a single 0 bit
;in a column pattern. R1 converts the 1-bit-low row and column
;patterns to row and column numbers from 0 to 7. P2 displays the row
;and column number of each key pressed.
;
        newrow equ 70h          ;store valid key row number
        newcol equ 71h          ;store valid key column number
        newflg equ 00h          ;main program new key flag
        upflg equ 01h           ;set to signal start of keys-up delay
        org 0000h
bigkey:
```

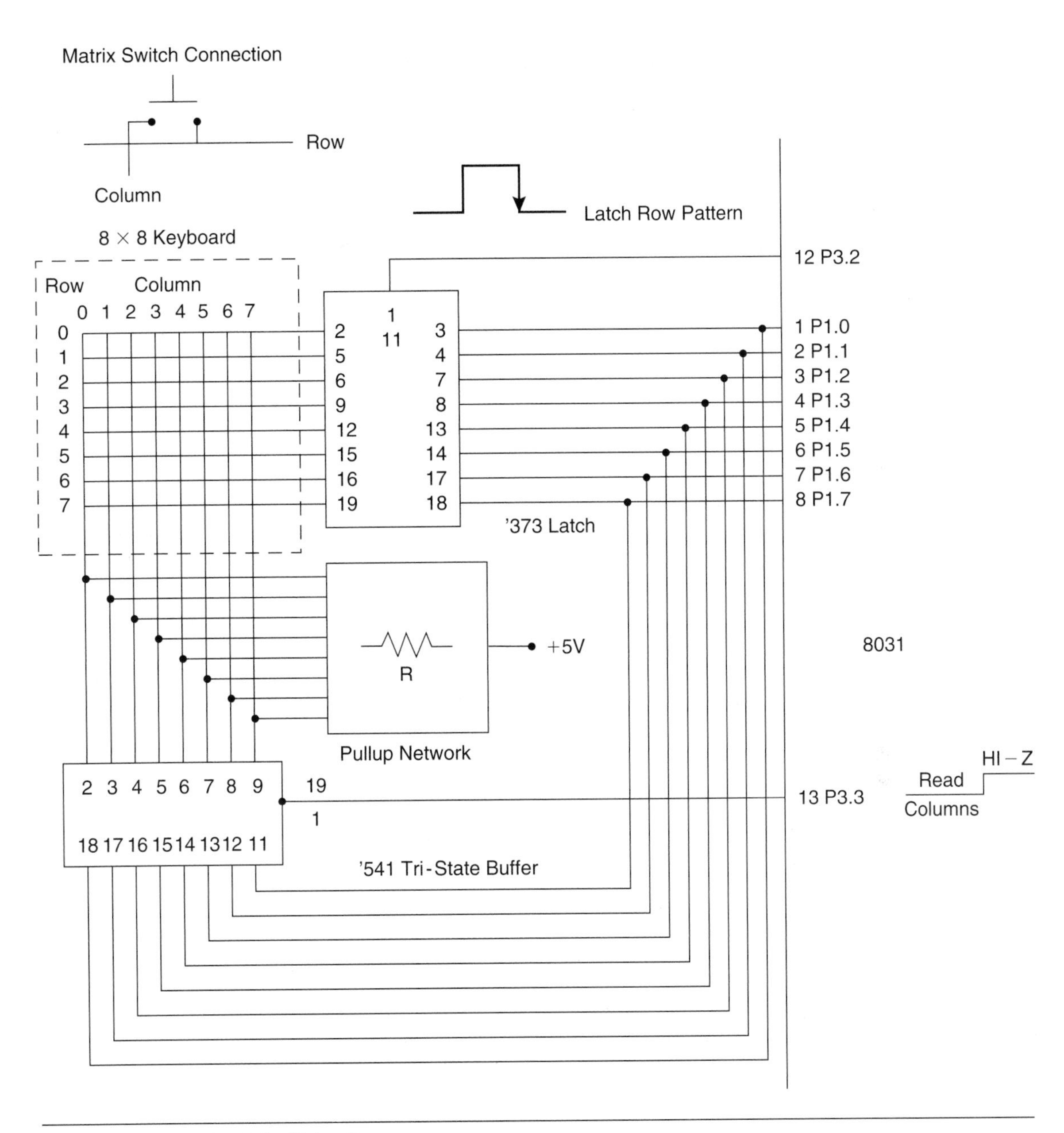

FIGURE 10.5 ◆ Circuit for Bigkey Program

```
       sjmp over                  ;jump over T0 interrupt address
;
;Timer T0 interrupts to this address
;T0 generates a 2ms time delay between each row scan pattern. The
;subroutine then generates a row strobe to the matrix and reads the
;resulting latched column pattern.
       org 000bh                  ;T0 interrupt address
       mov tl0,#30h               ;set for counting 2000 usec
       mov th0,#0f8h
       push acc                   ;save registers
       push psw
       mov p1,r5                  ;latch row scan pattern
       setb p3.2                  ;generate strobe to external latch
       clr p3.2
       mov p1,#0ffh               ;configure P1 as an input port
       clr p3.3                   ;enable external column buffer
       mov a,p1                   ;read columns for this row
       setb p3.3                  ;disable read strobe
       jb upflg,upyet             ;if upflg = 1 then wait for keys all up
       mov r7,a                   ;save A
       setb c                     ;rotate A until first 0 bit found
       mov r3,#08h                ;use R3 as a counter as A is rotated
look:
       rrc a                      ;begin search for first 0 bit (if a key)
       jnc test                   ;if C = 0, then check for A = FFh
       djnz r3,look               ;continue rotates until 0 bit or done
       mov a,r5                   ;no key down now, check if seen last scan
       cjne a,newrow,goback       ;if previous key down then now up
       mov newrow,#00h            ;reset newrow to 00h
       sjmp notdone               ;
test:
       cjne a,#0ffh,notdone       ;if A not FF then more than one key down
       mov a,r5                   ;determine old or new key, check row
       cjne a,newrow,newone       ;if a match, then check column pattern
       mov a,r7                   ;get column pattern
       cjne a,newcol,newone       ;if a match then see how many times seen
       inc r4                     ;3 complete keypad scans = 48 ms
       cjne r4,#03h,goback        ;continue until key seen 3 times in a row
       setb newflg                ;signal main program that new key
       setb upflg                 ;now wait for 3 scans of keys all up
       sjmp notdone
newone:
       mov newcol,a               ;save column pattern of possible new key
       mov newrow,r5              ;save row pattern of possible new key
notdone:
       mov r4,#00h                ;reset R4 scan counter
       sjmp goback
```

```
upyet:
      cjne a,#0ffh,notdone          ;look for keys up for 3 scans
      inc r4                        ;count keys up for 3 scans in R4
      cjne r4,#03h,goback
      clr upflg                     ;when up for 3 scans enable next key
goback:
      mov a,r5                      ;rotate row low bit to next position
      rl a
      mov r5,a
      pop psw                       ;restore registers used
      pop acc
      reti
;the T0 interrupt program ends here. The main program initializes
;the scan and then waits for newflg to be set. The row and column
;patterns then are coded into a key number from 01 to 64d. P2 is
;used to display the row and column number for each key pressed.
;
over:
      mov r5,#0feh                  ;initialize row 0 low
      mov tmod,#01h                 ;set T0 as a 16-bit counter
      mov tl0,#30h                  ;set for counting 2000 usec
      mov th0,#0f8h
      mov ie,#82h                   ;enable global and T0 interrupts
      setb tr0                      ;start T0
      mov r4,#00h                   ;clear R4 to 00 successive scans
      setb upflg                    ;wait for keys all up
      clr newflag                   ;no new key at start
      clr p3.2                      ;start latch strobe low
      setb p3.3                     ;start buffer strobe high
;the main program loops at the main label until the key flag is set
main:
      jbc newflg,getkey             ;loop until key pressed
      sjmp main
;getkey converts 1-of-8 low row and column patterns to row
;and column numbers between 0 and 7.
getkey:
      mov a,newrow                  ;determine row number from 00 to 07
      clr c                         ;count R1 up until row bit found
      mov r1,#00h
rownum:
      rrc a                         ;rotate right carry until 0 found
      jnc showrow
      inc r1                        ;count until low bit found
      sjmp rownum
showrow:
      mov a,r1                      ;display row on MSN of P2
      swap a
```

```
        anl a,#0f0h
        mov p2,a
        mov a,newcol              ;determine column number
        clr c
        mov r1,#00h
colnum:
        rrc a                     ;rotate right carry until 0 found
        jnc showcol
        inc r1                    ;count until low bit found
        sjmp colnum
showcol:
        mov a,r1                  ;display col on LSN of P2
        anl a,#0fh
        orl p2,a
        clr p3.0                  ;blink P3.0 to indicate a new key
        setb p3.0
        sjmp main                 ;loop back to main program
        end
```

◆ COMMENT ◆

- Once begun by the main program, T0 continues to time out and generate the row scan pattern in the interrupt program. To the main program, the keys appear in some unknown way; the interrupt program is said to run in the *background.*
- There is considerable adjustment (tweak) in this program to accommodate keys with various bounce characteristics. The debounce time can be altered in a gross sense by changing the number of cycles (R4) for acceptance and in a fine way by changing the basic row scan time (T0).
- This same program can be used to monitor any multipoint array of binary data points. The array can be expanded easily to a 16 × 16 matrix by adding one more latch and tri-state buffer and using two more port 3 pins to generate the latch and enable strobes.
- Note that only A can compare against memory contents in a CJNE instruction.

10.2 Displays

If keyboards are the predominant means of interface to human input, then visible displays are the universal means of human output. Displays may be grouped into three broad categories:

1. Single light(s)
2. Single character(s)
3. Intelligent alphanumeric

Single light displays include incandescent and, more likely, LED indicators that are treated as single binary points to be switched off or on by the program. *Single character* displays include numeric and alphanumeric arrays. These may be as simple as a seven-segment numeric display up to intelligent dot matrix displays that accept an 8-bit ASCII character and convert the ASCII code to the corresponding alphanumeric pattern. *Intelligent alphanumeric* displays are equipped with a built-in microcontroller that has been optimized for the application. Inexpensive displays are represented by multicharacter LCD windows, which are becoming increasingly popular in handheld wands, factory-floor terminals, and automotive dashboards. The high-cost end is represented by LCD ASCII terminals of the type commonly used to interface to a multi-user computer.

The individual light and intelligent single-character displays are easy to use. A port presents a bit or a character and then strobes the device. The intelligent ASCII terminals are normally serial devices, which are the subject of Chapter 11.

The two examples in this section—seven-segment and intelligent LCD displays—require programs of some length.

Seven-Segment Numeric Display

Seven-segment displays commonly contain LED segments arranged as an 8, with one common lead (anode or cathode) and seven individual leads for each segment. Figure 10.6 shows the pattern and an equivalent circuit representation of our example, a common cathode display. If more than one display is to be used, then they can be time multiplexed; the human eye cannot detect the blinking if each display is relit every 10 milliseconds or so. The 10 milliseconds is divided by the number of displays used to find the interval between updating each display.

The example examined here uses four seven-segment displays; the segment information is output on port 1 and the cathode selection is done on ports 3.2 to 3.5, as shown in Figure 10.7. A segment will be lit *only if* the segment line is brought high *and* the common cathode is brought low.

Transistors must be used to handle the currents required by the LEDs, typically 10 milliamperes for each segment and 70 milliamperes for each cathode. These are average current values; the peak currents will be four times as high for the 2.5 milliseconds each display is illuminated.

The program is interrupt driven by T0 in a manner similar to that used in the program bigkey. The interrupt program goes to one of four 2-byte character locations and finds the cathode segment pattern to be latched to port 1 and the anode pattern to be latched to port 3. The main program uses a lookup table to convert from a hex number to the segment pattern for that number. In this way, the interrupt program automatically displays whatever number the main program has placed in the character locations. The main program loads the character locations and is not concerned with how they are displayed.

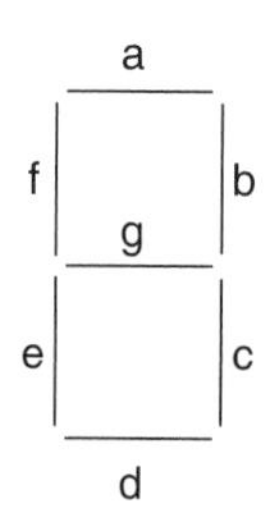

Segment Pattern

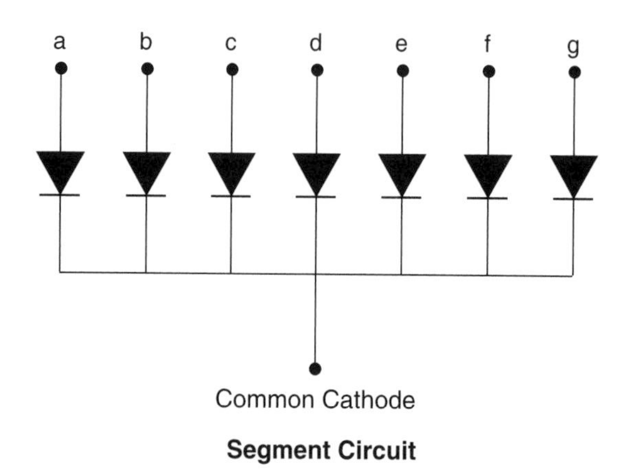

Segment Circuit

FIGURE 10.6 ◆ Seven-Segment LED Display and Circuit

Svnseg

The program *svnseg* displays characters found in locations ch1 to ch4 on four common-cathode seven-segment displays. Port 1 holds the segment pattern from the low byte of chx; port 3 holds the cathode pattern from the high byte of chx. T0 generates a 2.5 milliseconds delay interval between characters in an interrupt mode. The main program uses a lookup table to convert from hex to a corresponding pattern. R0 of bank 1 is dedicated as a pointer to the displayed character.

```
ch1 equ 50h         ;assign RAM character locations
ch2 equ 52h         ;2 bytes per character
ch3 equ 54h
ch4 equ 56h
org 0000h           ;jump over T0 interrupt location
```

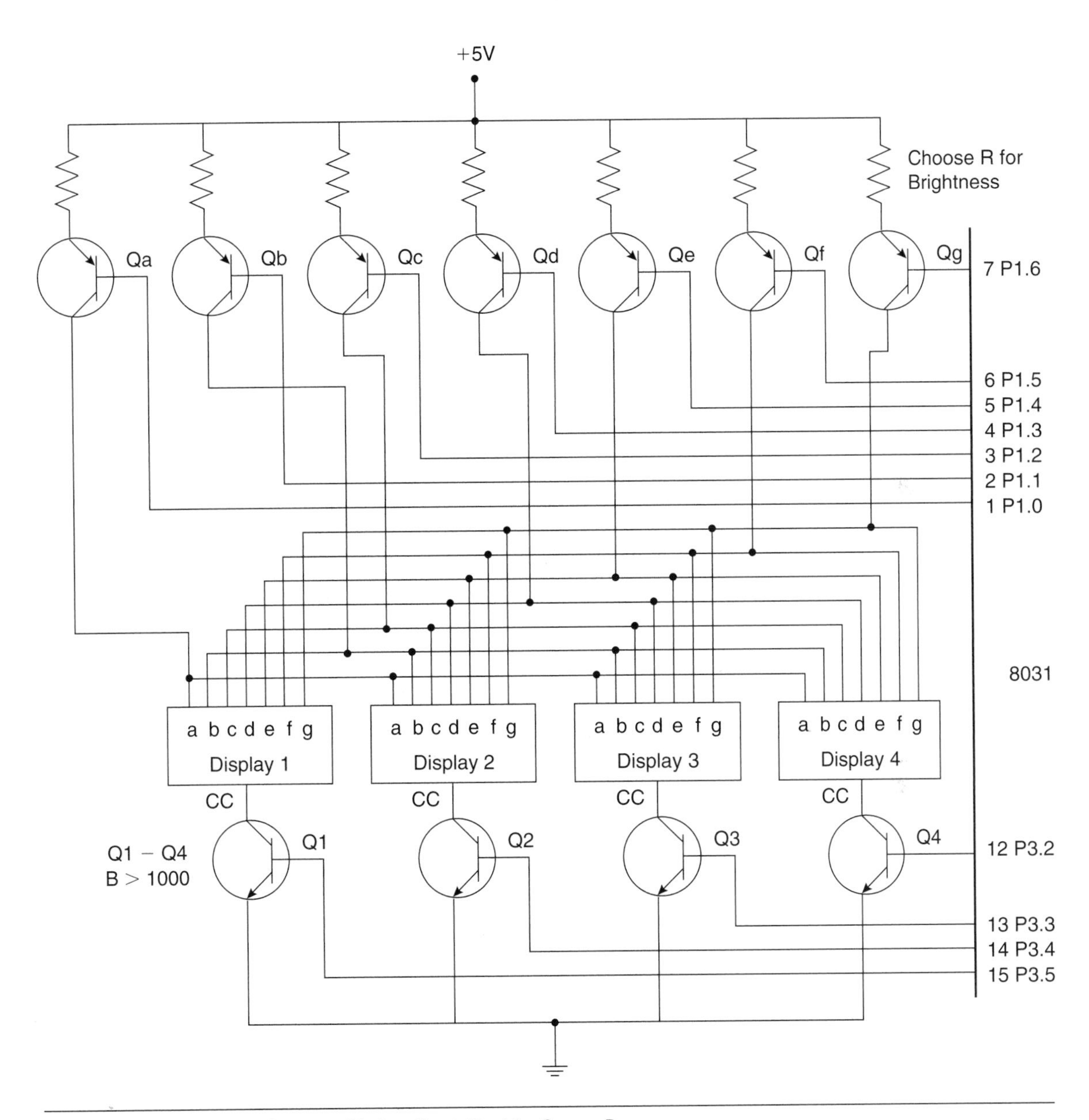

FIGURE 10.7 ◆ Seven-Segment Display Circuit Used for Svnseg Program

```
svnseg:     mov sp,#0fh          ;get the stack above bank one
            sjmp over
;
;begin the interrupt-driven program at the T0 interrupt location
;
            org 000bh
            mov tl0,#0fbh        ;reload T0 for next interrupt
            mov th0,#0f2h
            setb psw.3           ;select bank 1
            mov p1,@r0           ;place segment pattern on port 1
            inc r0               ;point to accompanying cathode pattern
            mov p3,@r0           ;place cathode patten on port 3
            inc r0               ;check for fourth character
            cjne r0,#58h,nxt
            mov r0,#ch1          ;if ch4 just displayed go to ch1
nxt:        clr psw.3            ;return to register bank 0
            reti                 ;return to main program
;the main program loads sample characters and starts the T0
;interrupt.
            mov a,#00h           ;use an example sequence of 0, 1, 2, 3
            acall convert        ;convert to segment pattern and store
            mov ch1,a
            mov a,#01h
            acall convert
            mov ch2,a
            mov a,#02h
            acall convert
            mov ch3,a
            mov a,#03h
            acall convert
            mov ch4,a            ;last segment pattern stored
            setb psw.3           ;select register bank 1
            mov r0,#ch1          ;set R0 to point to ch1 RAM location
            inc r0               ;now load anode pattern for ch1
            mov @r0,#20h         ;set anode for character 1 only high
            inc r0               ;point to next character and continue
            inc r0               ;load ch2 pattern
            mov @r0,#10h
            inc r0
            inc r0               ;load ch3 pattern
            mov @r0,#08h
            inc r0
            inc r0               ;load ch4 pattern
            mov @r0,#04h
            mov r0,#ch1          ;point to RAM address for ch1
            mov tl0,#0fbh        ;load T0 for first interrupt
            mov th0,#0f2h
            mov tmod,#01h        ;set T0 to mode 1
```

```
          mov ie,#82h        ;enable T0 interrupt
          setb tcon.4        ;start timer
          clr psw.3          ;return to register bank 0
here:     sjmp here          ;loop and simulate rest of program
;
;convert uses the PC to point to the base of the 16-byte table
;
convert:  inc a              ;compensate for RET byte
          mov a,@pc+a        ;get byte
          ret                ;return with segment pattern in A
          db c0h             ;0
          db f9h             ;1
          db a4h             ;2
          db b0h             ;3
          db 99h             ;4
          db 92h             ;5
          db 82h             ;6
          db f8h             ;7
          db f0h             ;8
          db 98h             ;9
          db 88h             ;A
          db 83h             ;b
          db c6h             ;C
          db b1h             ;d
          db 86h             ;E
          db 8eh             ;F
          end
```

◆ COMMENT ◆

- Using bank 1 as a dedicated bank for the interrupt routine cuts down on the need for pushes and pops. Bank 1 may be selected quickly, giving access to the eight registers while saving the bank 0 registers. Note that the stack, at reset, points to R0 of bank 1, so that it must be relocated.
- The intensity of the display may also be varied by blanking the displays completely for some interval using the program.

Intelligent LCD Display

In this section, we examine an intelligent LCD display of two lines, 20 characters per line, that is interfaced to the 8051. The protocol (handshaking) for the display is shown in Figure 10.8, and the interface to the 8051 in Figure 10.9.

The display contains two internal byte-wide registers, one for commands (RS = 0) and the second for characters to be displayed (RS = 1). It also contains a user-programmed RAM area (the character RAM) that can be programmed to generate any desired character that can be formed using a dot matrix. To distin-

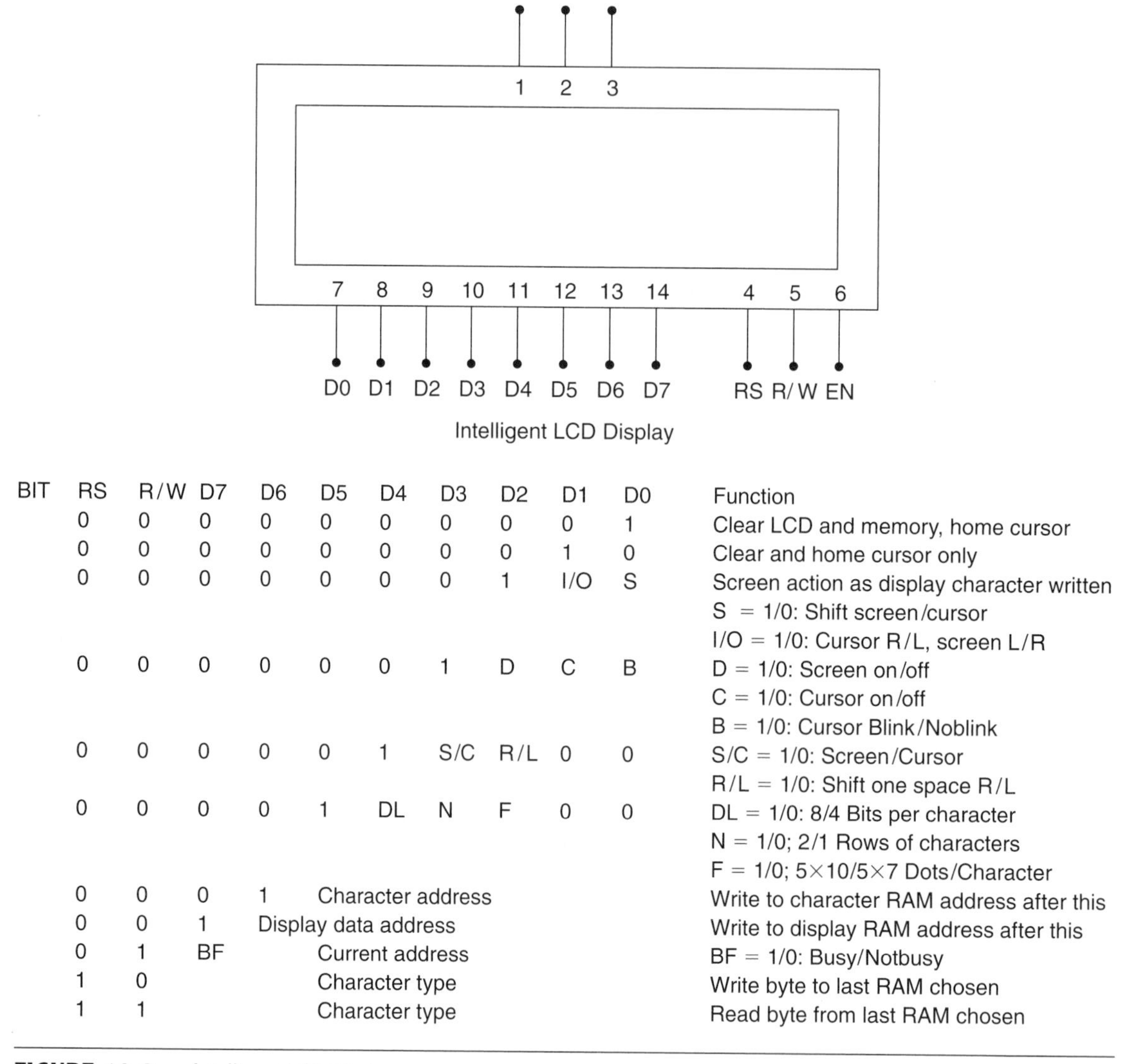

BIT	RS	R/W	D7	D6	D5	D4	D3	D2	D1	D0	Function
	0	0	0	0	0	0	0	0	0	1	Clear LCD and memory, home cursor
	0	0	0	0	0	0	0	0	1	0	Clear and home cursor only
	0	0	0	0	0	0	0	1	I/O	S	Screen action as display character written
											S = 1/0: Shift screen/cursor
											I/O = 1/0: Cursor R/L, screen L/R
	0	0	0	0	0	0	1	D	C	B	D = 1/0: Screen on/off
											C = 1/0: Cursor on/off
											B = 1/0: Cursor Blink/Noblink
	0	0	0	0	0	1	S/C	R/L	0	0	S/C = 1/0: Screen/Cursor
											R/L = 1/0: Shift one space R/L
	0	0	0	0	1	DL	N	F	0	0	DL = 1/0: 8/4 Bits per character
											N = 1/0; 2/1 Rows of characters
											F = 1/0; 5×10/5×7 Dots/Character
	0	0	0	1	Character address						Write to character RAM address after this
	0	0	1	Display data address							Write to display RAM address after this
	0	1	BF		Current address						BF = 1/0: Busy/Notbusy
	1	0			Character type						Write byte to last RAM chosen
	1	1			Character type						Read byte from last RAM chosen

FIGURE 10.8 ◆ Intelligent LCD Display

guish between these two data areas, the hex command byte 80 will be used to signify that the display RAM address 00h is chosen.

Port 1 is used to furnish the command or data byte, and ports 3.2 to 3.4 furnish register select and read/write levels.

The display takes varying amounts of time to accomplish the functions listed in Figure 10.8. LCD bit 7 is monitored for a logic high (busy) to ensure the

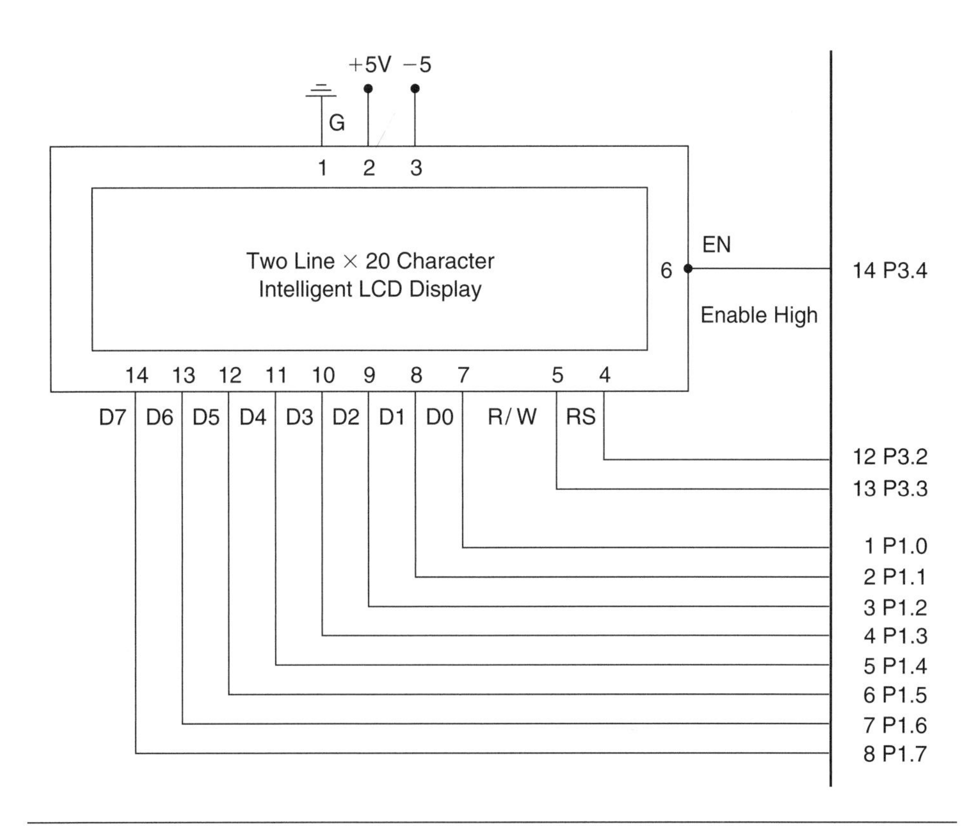

FIGURE 10.9 ◆ Intelligent LCD Circuit for Lcdisp Program

display is not overwritten. A slightly more complicated LCD display (4 lines × 40 characters) is currently being used in medical diagnostic systems to run a very similar program.

Lcdisp

The program *lcdisp* sends the message "Hello There" to an intelligent LCD display shown in Figure 10.8. Port 1 supplies the data byte. Port 3.2 selects the command (0) or data (1) registers. Port 3.3 enables a read (0) or write (1) level, and port 3.4 generates an active high-enable strobe.

```
;"lcdisp", a program that initializes an intelligent LCD display and
;displays a message on the display. Display line 1 begins at address
;80h, display line 2 begins at address C0h.
;
     org 0000h
lcdisp:
```

```
        mov a,#3ch          ;8 bits/char, 2 rows, 5×10 dots/char
        acall command       ;strobe command to display
        mov a,#0eh          ;screen and cursor on, no blink
        acall command       ;strobe command to display
        mov a,#06h          ;shift cursor right
        acall command       ;strobe command to display
        mov a,#01h          ;clear memory and home cursor
        acall command       ;strobe command to display
        mov a,#87h          ;move cursor to space 7, line 1
        acall command
        mov a,#'H'          ;say "Hello There" on 2 lines
        acall display
        mov a,#'e'
        acall display
        mov a,#'l'
        acall display
        acall display
        mov a,#'o'
        acall display
        mov a,#0c7h         ;switch to address 7 of line 2
        acall command
        mov a,#'T'
        acall display
        mov a,#'h'
        acall display
        mov a,#'e'
        acall display
        mov a,#'r'
        acall display
        mov a,#'e'
        acall display
here:
        sjmp here           ;loop here after displaying message
;
;LCD strobe subroutines
command:
        acall ready         ;write when display not busy
        mov p1,a            ;command character in P1
        clr p3.2            ;command register chosen
        clr p3.3            ;write enabled
        setb p3.4           ;strobe character to display
        clr p3.4
        ret
display:
        acall ready         ;write when display ready
        mov p1,a            ;data character in P1
        setb p3.2           ;data register chosen
```

```
        clr p3.3                ;write enabled
        setb p3.4               ;strobe character to display
        clr p3.4
        ret
ready:
        clr p3.4                ;display disabled
        mov p1,#0ffh            ;configure P1 for input
        clr p3.2                ;select command register
        setb p3.3               ;read enabled
wait:
        clr p3.4                ;strobe display
        setb p3.4               ;read busy status of display
        jb p1.7,wait            ;wait for not busy (BF = 0)
        clr p3.4                ;end display strobe
        ret
        end
```

Lcdisp2

The program lcdisp is very cumbersome when many messages must be displayed. A better way to display messages is to make up message strings that have embedded in them control characters. The display program can recognize the control characters, command the display, and then display the message.

A flexible LCD display program, named *lcdisp2,* is shown next. A message string is made up of displayable ASCII characters and nondisplayable ASCII control characters. The control characters, chosen at random, are FCh to denote that LCD programming codes follow, FDh to indicate that data (displayable) characters follow, and FEh to show end of message.

```
;"lcdisp2", a more flexible program to display messages.
;
;equates for control characters
        com equ 0fch            ;command(s) follow this header
        dat equ 0fdh            ;data follows this header
        eot equ 0feh            ;end-of-message character
        org 0000h
lcdisp2:
        mov dptr,#msg1          ;initialize LCD and display message
        acall message           ;
;loop here at end of message display
here:
        sjmp here
;
;subroutine is passed address of message to be displayed in DPTR
message:
        acall ready             ;wait until display not busy
        clr a                   ;use the DPTR only
```

```
        movc, a,@a+dptr         ;get character
        inc dptr                ;point to next character
        cjne a,#eot,comd        ;see if end-of-message string or command
        ret                     ;return if end of message
comd:
        cjne a,#com,datta       ;if command header then set RS = 0
        clr p3.2                ;set RS = 0 until data control character
        sjmp message            ;go until done
datta:
        cjne a,#dat,sendit      ;if data header then set RS = 1
        setb p3.2               ;if data then RS = 1
        sjmp message            ;go until done
sendit:                         ;write character to display
        mov p1,a                ;character in P1
        clr p3.3                ;write enabled
        setb p3.4               ;strobe character to display
        clr p3.4
        sjmp message            ;go until done
;
;check busy flag from LCD display
ready:
        mov r0,p3               ;save state of P3
        clr p3.4                ;display disabled
        mov p1,#0ffh            ;configure P1 for input
        clr p3.2                ;select command register
        setb p3.3               ;read enabled
wait:
        clr p3.4                ;strobe display
        setb p3.4               ;read busy status of display
        jb p1.7,wait            ;wait for not busy (BF = 0)
        clr p3.4                ;end display strobe
        mov p3,r0               ;restore state of P3
        ret
;
;many messages may be stored, each given a different label name
msgl:
        db com,3ch,06h,0eh,01h,87h,dat,"Hello",com,0c7h,dat,"Y'all",eot
end
```

10.3 Pulse Measurement

Sensors used for industrial and commercial control applications frequently produce pulses that contain information about the quantity sensed. Varying the sensor output frequency, using a constant duty cycle but variable frequency pulses to indicate changes in the measured variable, is most common. Varying the duration of the pulse width, resulting in constant frequency but variable duty cycle, is also used. In this section, we examine programs that deal with both techniques.

Measuring Frequency

Timers T0 and T1 can be used to measure external frequencies by configuring one timer as a counter and using the second timer to generate a timing interval over which the first can count. The frequency of the counted pulse train is then:

Unknown frequency = Counter/timer

For example, if the counter counts 200 pulses over an interval of .1 second generated by the timer, the frequency is:

UF = 200/.1 = 2000 Hz

Certain fundamental limitations govern the range of frequencies that can be measured. An input pulse must make a 1-to-0 transition lasting two machine cycles, or f/24, to be counted. This restriction on pulse deviation yields a maximum frequency of 500 kilohertz using our 12 megahertz crystal (assuming a square wave input).

The lowest frequency that can be counted is limited by the duration of the time interval generated, which can be exceedingly long using all the RAM to count timer rollovers (65.54 milliseconds × 2^32768). There is no practical limitation on the lowest frequency that can be counted.

Happily, most frequency-variable sensors generate signals that fall inside of 0 to 500 kilohertz. Usually the signals have a range of 1,000 to 10,000 hertz.

A timing interval of 1 second generates a frequency count accurate to the nearest 1 hertz; an interval of .1 second yields a count accurate to the nearest 10 hertz.

Freq

A program that measures an unknown frequency on pin 3.4 (T0), named *freq,* is shown next. The unknown frequency is counted in T0 configured as a 16-bit counter. T0 is reset and begins counting the unknown frequency. A delay of .1 second is done, and T0 stopped. The count in T0 is the unknown frequency divided by 10.

T1 is used, in an interrupt mode, as a timer with an exact delay of .0001 seconds. Bank 1 registers R0 and R1 count 1,000 T1 interrupts to yield a delay of .1 second. T0 and T1 are reset and enabled to count and time at T = 0 in the program. After the .1 second delay is up, T0 is stopped, and the contents of T0 displayed on P1 (LSB) and P2 (MSB).

```
;Freq, a program to measure an unknown frequency on the T0 pin.
;Timer T1, in the 8-bit auto-reload mode, generates an exact time
;delay of .000100 seconds between interrupts from a 12MHz crystal.
;Registers R0 and R1 of register bank 1 count 1,000d of the T1 timer
;interrupts to yield an exact time delay of .1 seconds. Timer T0,
;operating as a 16-bit counter, counts the pulses from external pin
;T0. The count in T0 is then the external frequency divided by 10.
;T0 may count from 0000(0)h or 0 Hz up to C350(0)h or 50000(0) Hz.
;Note that frequencies above 500,000 Hz, which exceed the f/24
;requirements of the 8051 counting circuits, are not measured
;accurately. Note also that approximately 12 cycles or 12 microseconds
;for this example are required to stop T0. This stopping time could
;lead to a false count in T0 of from 1 count at 83,333 Hz to 6
;counts at an input frequency of 500,000 Hz.
;
        frqflg equ 0fh              ;use a bit flag to signal main program
freq:
        org 0000h
        mov sp,#30h                 ;set stack above registers/bit area
        sjmp over                   ;jump over T1 interrupt address
;
;T1 will overflow and the T1 Interrupt flag will cause the program
;to vector to this address. R0 and R1 of register bank 1 are used to
;count from 0000d to 1,000d. R0 is the least significant byte
;counter, R1 the most significant byte counter. When the count
;reaches 1,000d, or 03E8H, then "frqflg", the main program flag, is
;set, and the timers stopped. 12 cycles (12 usec for this example)
;are required to stop T0, or from 1 additional count at 83,333Hz
;to 6 false counts at 500,000 Hz.
;                                                                       usec
        org 001bh                   ;T1 overflow flag interrupt to here   3
        setb psw.3                  ;switch to register bank 1            1
        inc r0                      ;count R0 up until overflow at 00h    1
        cjne r0,#00h,checktime      ;check to see if time is up           2
        inc r1                      ;or increment R1 when R0 rolls over
checktime:
        cjne r1,#03h,goback         ;check R1 for terminal count          2
        cjne r0,#0e8h,goback        ;check R0 for terminal count          2
        clr tr0                     ;stop T0                              1
                                    ;
        clr tr1                     ;stop T1. Time before T0 stopped =   12
        setb frqflg                 ;signal main program that T0 = frequency
goback:
        clr psw.3                   ;return to bank 0 registers
        reti                        ;return to main program
;
;The main program sets up T0 to count external pulses, and T1 to
```

```
;count 100 microseconds in the autoreload mode 2. The main program
;flag, frqflg, is then monitored by the main program until it is
;set. When frqflg is set, the main program displays the unknown
;frequency (divided by 10) on P1 (LSB) and P2 (MSB).
;
over:
        mov tcon,#00h           ;All timers stopped – flags reset
        setb psw.3              ;select register bank 1 and reset R0, R1
        mov r0,#00h
        mov r1,#00h
        clr psw.3               ;return to bank 0
        mov tmod,#25h           ;T1 a mode 2 timer; T0 a mode 1 counter
        mov tl1,#9ch            ;start TL1 at 9CH
        mov th1,#9ch            ;TH1 = 156d, overflows in 100 clocks
        mov tl0,#00h            ;zero T0
        mov th0,#00h
        clr frqflg              ;reset the frequency measured flag
        mov tcon,#50h           ;start timer T1 and counter T0
        mov ie,#88h             ;enable global and T1 overflow interrupts
simulate:
        jbc frqflg,getfrq       ;have main program test "frqflg"
        sjmp simulate           ;loop here until frequency is measured
getfrq:
        mov p1,tl0              ;show frequency count (in hex) on P1 & P2
        mov p2,th0
        sjmp over               ;make next measurement
        end
```

Pulse Width Measurement

Theoretically, if the input pulse is known to be a perfect square wave, the pulse frequency can be measured by finding the time the wave is high (Th). The frequency is then

$$UF = \frac{1}{\text{TH} \times 2}$$

If Th is 200 microseconds, for example, then UF is 2500 hertz. The accuracy of the measurement will fall as the input wave departs from a 50% duty cycle.

Timer X may be configured so that the internal clock is counted only when the corresponding $\overline{\text{INTX}}$ pin is high by setting the GATE X bit in TMOD. The accuracy of the measurement is within approximately one timer clock period, or 1 microsecond for a 12 megahertz crystal. This accuracy can only be attained if the measurement is started when the input wave is low and stopped when the input next goes low. Pulse widths greater than the capacity of the counter, which is

65.54 milliseconds for a 12 megahertz crystal, can be measured by counting the overflows of the Timer flag and adding the final contents in the counter.

The width of an unknown pulse is measured by enabling a timer when the pulse generates an interrupt on one of the $\overline{\text{INTX}}$ pins. The interrupt is programmed to occur on a high-to-low edge on the $\overline{\text{INTX}}$ pin. The counter begins counting when the unknown pulse goes high, enabling the counter to count when $\overline{\text{INTX}}$ goes high. The next pulse edge stops the counter. The counter will contain the width of the pulse to the nearest microsecond for a 12 MHz crystal.

Width

A program, named *width* measures the width of a pulse fed to pin 3.3 ($\overline{\text{INT1}}$). Timer T1 is enabled to count on the first pulse edge, counts when the pulse is high, and stops on the second edge. Ports P1 (LSB) and P2 (MSB) show, in hex, the width of the pulse in microseconds.

```
;"Width", measures the width of a pulse fed to pin INT1.
;T1 is enabled to count the 1 MHz clock when enabled to do so by a
;high-to-low transition interrupt on INT1. T1 is gated by the level
;on  INT1 so there is a delay between enabling T1 (edge of INT1) and
;then counting of the width of the positive portion of the pulse on
;INT1. This routine will measure pulse widths to the nearest
;microsecond, which is the uncertainty as to when the level of INT1
;will go high relative to the clock pulse edges that feed counter T1.
;
      wflg equ 00h             ;bit flag used to notify main program
      org 0000h
      mov sp,#30h              ;set stack above bit addressable area
      sjmp over                ;jump over INT1 interrupt area
;
;place /INT1 interrupt routine here
      org 13h                  ;INT1 interrupt at this program location
      jb int1,noise            ;if not low then noise – leave
      jbc tr1,stop             ;if T1 is running then stop it
      setb tr1                 ;if T1 is not running then start it
noise:
      reti                     ;return with T1 enabled to time pulse
stop:
      setb wflg                ;set flag to indicate measurement done
      clr ex1                  ;disable INT1 until next measurement
      reti                     ;return with T1 stopped and wflg set
;
;the main program is placed here. The program monitors wflg and
;displays the high pulse width on P1 (LSBY) and P2 (MSBY).
;The width is accurate to the nearest microsecond.
over:
      mov tmod,#90h            ;T1 to count when INT1 pin high
      setb it1                 ;INT1 interrupt on negative edge
```

```
        mov tl1,#00h          ;reset T1 to 00h
        mov th1,#00h
        mov ie,#84h           ;enable global and INT1 interrupts
simulate:
        jbc wflg,getwidth     ;test flag until measurement made
        sjmp simulate         ;loop until done
getwidth:
        mov p1,tl1            ;display LS Byte on P1
        mov p2,th1            ;display MS Byte on P2
        sjmp over             ;get next width measurement
        end
```

◆ COMMENT ◆

If there is a considerable amount of electrical noise present on the $\overline{INTX}$ pin, an average value of the pulse width could be found by measuring the widths of a number of consecutive pulses. A counter could be incremented at the end of each cycle and the sum of the widths divided by the counter contents. The noise should average to zero.

10.4 D/A and A/D Conversions

Conversion between the analog and digital worlds requires the use of integrated circuits that have been designed to interface with computers. Highly intelligent converters are commercially available that all have the following essential characteristics:

- Parallel data bus: tri-state, 8-bit
- Control bus: enable (chip select), read/write, ready/busy

The choice the designer must make is whether to use the converter as a RAM memory location connected to the memory busses or as an I/O device connected to the ports. Once that choice is made, the set of instructions available to the programmer becomes limited. The memory location assignment is the most restrictive, having only MOVX available. The design could use the additional 32K RAM address space with the addition of circuitry for A15. By enabling the RAM when A15 is low, and the converter when A15 is high, the designer could use the upper 32K RAM address space for the converter, as was done to expand port capacity by memory mapping in Chapter 9. All of the examples examined here are connected to the ports.

D/A Conversions

A generic R-2R type D/A converter, based on several commercial models, is connected to ports 1 and 3 as shown in Figure 10.10. Port 1 furnishes the digi-

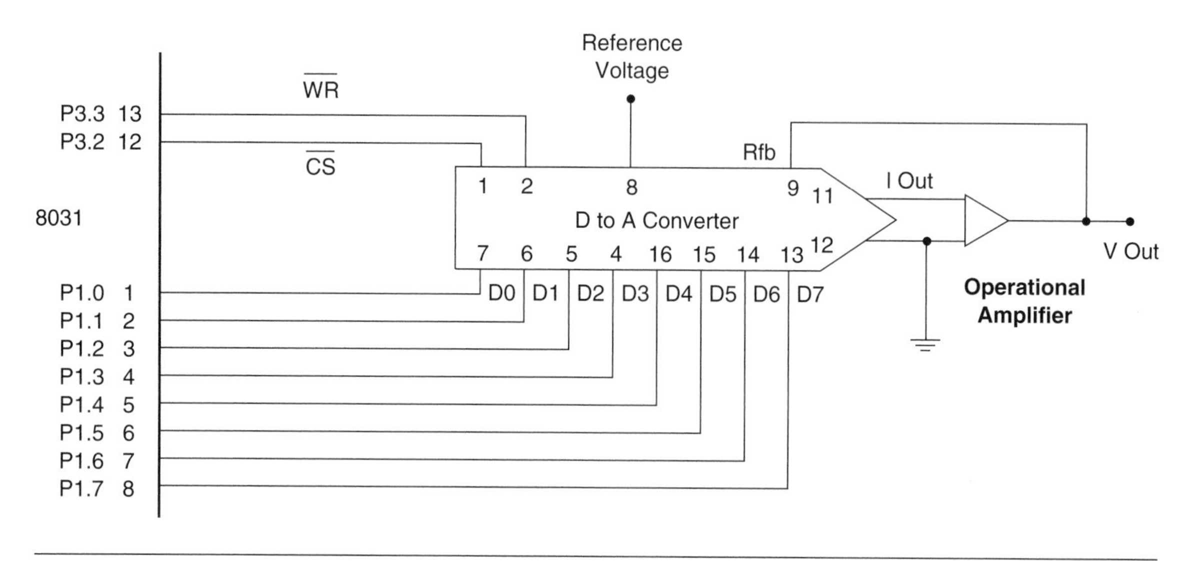

FIGURE 10.10 ◆ D/A Converter Circuit for Davcon Program

tal byte to be converted to an analog voltage; port 3 controls the conversion process. The converter has these features:

- Vout = $-$Vref $\times$ (byte in/100H), Vref = $\pm$10 V
- Conversion time: 5 μs
- Control sequence: $\overline{\text{CS}}$ then $\overline{\text{WR}}$

For this example, a 1,000 hertz sine wave that will be generated can have a programmable frequency. Vref is chosen to be -10 volts, and the wave will swing from $+9.96$ volts to 0 volt around a midpoint of 4.48 volts. The program uses a lookup table to generate the amplitude of each point of the sine wave; the time interval at which the converter is fed bytes from the table determines the wave frequency. Note that a 16MHz crystal is used.

The conversion time limits the highest frequency that can be generated using S sample point. In this example, the shortest period that can be used is

$$\text{Tmin} = \text{S} \times 5\ \mu\text{s} = 5\text{S}\ \mu\text{s}, \qquad \text{fmax} = \frac{200{,}000}{\text{S}}$$

The design tension is high frequency versus high resolution. For a 1,000-hertz wave, S could be 200d samples. In reality, we cannot use this many samples; the program cannot fetch the data, latch it to port 1, and strobe port 3.3 in 5 microseconds. An inspection of the program will show that the time needed for a single wave point is 6 microseconds, and setting up for the next wave takes another 2.25 microseconds. S becomes 166d samples using the 6-microsecond interval, and the addition of 2.25 microseconds at the end of every wave yields a true frequency of 1001.75 hertz.

Davcon

The D/A converter program *davcon* generates a 1,000-hertz sine wave using an 8-bit converter. 166d samples are stored in a lookup table and fed to the converter at a rate of one sample every 6 microseconds. The lookup table is pointed to in external ROM by the DPTR, and R1 is used to count the samples. Numbers in parentheses indicate the number of cycles.

```
            org 0000h
davcon:     clr p3.2                ;enable chip select to converter
            mov dptr,#table         ;get base address to DPTR
repeat:     mov r1,#0a6h            ;initialize R1 to 166d (1)
next:       mov a,r1                ;offset into table (1)
            movc a,@a+dptr          ;get sample (2)
            mov p1,a                ;sample to port 1 (1)
            clr p3.3                ;write strobe low (1)
            setb p3.3               ;write strobe high (1)
            djnz r1,next            ;loop for 166D samples (2)
            sjmp repeat             ;reload R1 and generate next wave (2)
;
;the lookup table begins here; a cosine wave is chosen to make the
;table readable; the first 83 samples cover the wave from maximum to
;1 more than 0; the next 83 cover the wave from 0 to maximum. 83
;samples per half-cycle means a sample every 2.17 degrees
;
table:      db 00h                  ;no entry at A = 00h
            db ffh                  ;FFhcos 0 = FFh. s1
            db feh                  ;7Fh + 7Fhcos 2.17 = FEh. s2
            db feh                  ;7Fh + 7Fhcos 4.34 = FEh. s3
            db fdh                  ;sample 4
            db fdh                  ;sample 5
            ;and so on until we near 90 degrees:
            db 81h                  ;7Fh + 7Fhcos 88.9 = 81h. s42
            db 7ch                  ;7Fh + 7Fhcos 91.1 = 7Ch. s43
            ;near 180 degrees we have:
            db 01h                  ;7Fh + 7Fh cos 173.5 = 01h. s81
            db 00h                  ;7Fh + 7Fh cos 175.7 = 00h. s82
            db 00h                  ;7Fh + 7Fh cos 177.8 = 00h. s83
            db 00h                  ;7Fh + 7Fh cos 180 = 00h. s84.
            db 00h                  ;7Fh + 7Fh cos 182.2 = 00h. s85
            db 00h                  ;7Fh + 7Fh cos 184.33 = 00h. s86
            db 01h                  ;7Fh + 7Fh cos 186.5 = 01h. s87
;finally, close to 360 degrees the table contains:
            db fbh                  ;s 161
            db fch                  ;s 162
```

```
db fdh          ;s 163
db fdh          ;s 164
db feh          ;s 165
db feh          ;s 166
end
```

◆ COMMENT ◆

The program retrieves the data from the highest to the lowest address.

A/D Conversion

The easiest A/D converters to use are the *flash* types, which make conversions based on an array of internal comparators. The conversion is very fast, typically in less than 1 microsecond. Thus, the converter can be told to start, and the digital equivalent of the input analog value will be read one or two instructions later. Modern successive approximation register (SAR) converters do not lag far behind, however, with conversion times in the 2 – 4 microsecond range for 8 bits.

At this writing, flash converters are more expensive (by a factor of two) than the traditional SAR types, but this cost differential should disappear (within time). Typical features of an 8-bit flash converter are:

- Data: Vin = Vref(−), data = 00h; Vin = Vref(+), data = FFh
- Conversion time: 1 μs
- Control sequence: $\overline{CS}$ then $\overline{WR}$ then $\overline{RD}$

An example circuit, using a generic flash converter, is shown in Figure 10.11. Port 1 is used to read the byte value of the input analog voltage, and port 3 controls the conversion. A conversion is started by pulsing the write line low, and the data is read by bringing the read line low.

Our example involves the digitizing of an input waveform every 100d microseconds until 1,000d samples have been stored in external RAM.

Adconv

The program *adconv* will digitize an input voltage by sampling the input every 100 μs and storing the digitized values in external RAM locations 4000h to 43E7h (1000d samples). Numbers in parentheses are cycles. The actual delay between samples is 99.75 microseconds.

```
         begin equ 4000h      ;start storage at 4000h
         delay equ 74h        ;delay in DJNZ loop for 87 usec
         end1 equ 43h         ;high byte of ending address
         end2 equ 0e8h        ;low byte of ending address
         org 0000h
adconv:  mov dptr,#begin      ;point to starting address in RAM
```

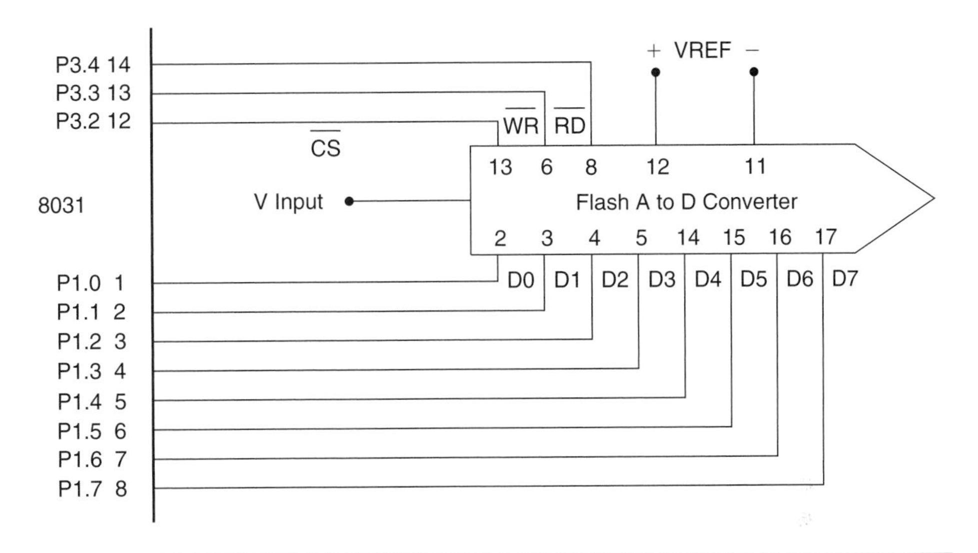

FIGURE 10.11 ◆ A/D Converter Circuit for Adconv Program

```
         clr p3.2             ;generate CS to ADC
next:    clr p3.3             ;generate WR pulse (1)
         setb p3.3            ;(1)
         clr p3.4             ;generate RD pulse (1)
         mov a,p1             ;get data (1)
         setb p3.4            ;end of RD pulse (1)
         movx @dptr,a         ;store in external RAM (2)
         inc dptr             ;point to next and see if done (2)
         mov a,dph            ;(1)
         cjne a,#end1,wait    ;(2)
         mov a,dpl            ;(1)
         cjne a,#end2,wait    ;(2)
         sjmp done            ;finished if both tests pass
wait:    mov r1,#delay        ;delay for 87d µs
here:    djnz r1,here         ;(2) × .75 µs × 116d = 87 µs
         sjmp next            ;(2) 17d cycles (12.75 µs)
done:    sjmp done            ;simulate rest of program
         end
```

◆ **COMMENT** ◆

Using this program, we could fill up the RAM in 3.2 seconds, which illustrates the volumes of data that can be gathered quickly by such a circuit. Realistic applications would feature some data reduction at the microcontroller before the reduced (massaged) data were relayed to a host computer.

10.5 Multiple Interrupts

The 8051 is equipped with two external interrupt input pins: $\overline{\text{INT0}}$ and $\overline{\text{INT1}}$ (P3.2 and P3.3). These are sufficient for small systems, but the need may arise for more than two interrupt points. There are many schemes available to multiply the number of interrupt points; they all depend on the following strategies:

- Connect the interrupt sources to a common line
- Identify the interrupting source using software

Because the external interrupts are active-low, the connections from the interrupt source to the $\overline{\text{INTX}}$ pin must use open-collector or tri-state devices.

An example of increasing the $\overline{\text{INT0}}$ from one to eight points is shown in Figure 10.12. Each source goes to active-low when an interrupt is desired. A corresponding pin on port 1 receives the identity of the interrupter. Once the interrupt program has handled the interrupt situation, the interrupter must receive

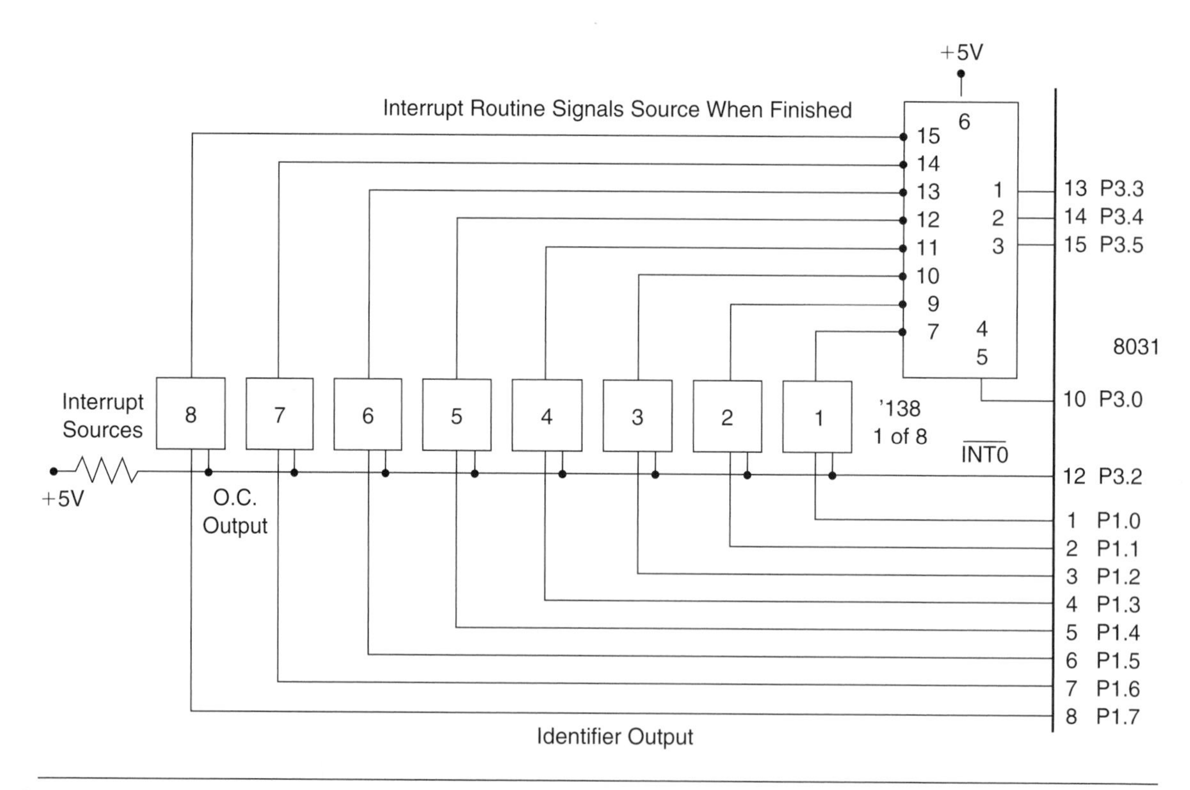

FIGURE 10.12 ◆ Multiple-Source Interrupt Circuit Used in Lopri and Hipri Programs

an acknowledgment so that the interrupt line for that source can be brought back to a high state. Port 3 pins 3.3, 3.4, and 3.5 supply, via a 3-to-8 decoder, the acknowledgment feedback signal to the proper interrupt source. The decoder is enabled by port pin 3.0.

Multiple and simultaneous interrupts can be handled by the program in as complex a manner as is desired. If there is no particular urgency attached to any of the interrupts then they can be handled as the port 1 pins are scanned sequentially for a low.

A simple priority system can be established whereby the most important interrupt sources are examined in the priority order, and the associated interrupt program is run until finished. An elaborate priority system involves ordering the priority of each source. The elaborate system acknowledges an interrupt immediately, thus resetting that source's interrupt line, and begins executing the particular interrupt program for that source. A new interrupt from a higher priority source forces the current interrupt program to be suspended and the new interrupter to be serviced.

To acknowledge the current interrupt in anticipation of another, it is necessary to also re-arm the $\overline{\text{INTX}}$ interrupt by issuing a "dummy" RETI instruction. The mechanism for accomplishing this task is illustrated in the program named *hipri*. First, a low priority scheme is considered.

Lopri

The program *lopri* scans port P1 for the source of an interrupt that has triggered $\overline{\text{INT0}}$. The pins are scanned for a low and the scan resumed after any interrupt is found and serviced. The interrupt source is acknowledged prior to a RETI instruction. R5 of bank 1 is used to store the next pin to be scanned, and R6 is used to scan the pins for a low. A jump table is used to select the interrupt routine that matches the particular interrupt. Each interrupt routine supplies the 3-to-8 decoder a unique acknowledge pattern before a RETI.

```
            ack equ 70h          ;each interrupt routine loads its
                                 ;unique acknowledge byte in ack
            org 0000h
lopri:      sjmp over            ;jump over the INT0 interrupt address
;
;The INT0 interrupt will vector the program here
;
            org 0003h            ;INT0 vector address
            mov ack,#0ffh        ;place enable pattern in ack
            push acc             ;save A
            push dpl             ;save DPTR
            push dph
            setb psw.3           ;select register bank 1
            mov a,r5             ;get pattern in R5 to A
```

```
            orl a,p1            ;OR the single 0 in A with P1
            mov r6,#08h         ;rotate A through C eight times
which:      rrc a               ;find the 0 starting at P1.0
            jnc loww            ;keep rotating until low found
            djnz r6,which       ;if not found then it was not this pin
            sjmp goback         ;return with no action taken
loww:       mov a,r6            ;convert from 1-of-8 low to number
            subb a,#01h         ;A was 8 to 1, now 07 to 00
            rl a                ;A is now 0Eh to 00 (2 bytes/sjmp)
            mov dptr,#jmptbl    ;DPTR points to the base of jump table
            jmp @a+dptr         ;jump to the matching interrupt
                                ;routine
goback:     mov a,r5            ;rotate r5 to the next pin position
            rl a
            mov r5,a
            clr psw.3           ;select register bank 0
            pop dph             ;restore register used in subroutine
            pop dpl
            pop acc
            mov p3,ack          ;each routine loads proper P3 pattern
            nop                 ;give the interrupt circuit a few
            nop                 ;microseconds to respond and remove
            nop                 ;the low level before returning
            mov p3,#0ffh        ;enable the next interrupt to occur
            reti
;
;the main program starts here followed by the interrupt routine jump
;table (simulated)
;
over:       mov sp,#0fh         ;move stack above register bank 1
            mov p3,#0ffh        ;set port 3 to disable 3/8 all high
            setb psw.3          ;select bank 1 and set R5 to 1 low
            mov r5,#0feh        ;port pin 1.0 selected
            clr psw.3           ;return to bank 0o
            mov ie,#81h         ;enable INT0 interrupt
            mov tcon,#00h       ;enable level trigger for INT0
simulate:   sjmp simulate       ;simulate main program
jmptbl:     sjmp goback         ;simulate interrupt programs; pin 7
            sjmp goback         ;6
            sjmp goback         ;5
            sjmp goback         ;4
            sjmp goback         ;3
            sjmp goback         ;2
            sjmp goback         ;1
            sjmp goback         ;0
            end
```

◆ COMMENT ◆

- The instruction JMP @A+DPTR has been used to select one of a number of jump addresses, depending on the number found in A. The simulated subroutines could be an SJMP to the actual interrupt handling subroutine. Because each SJMP takes 2 bytes to execute, A has to be doubled to point to every other byte in the jump table. When this action is not convenient, A can use a lookup table to get a new A, which then accesses a jump address.
- R5 has 1 bit low, and that bit acts as a mask when ORed with P1 to find the low bit in P1. When the low pin does not match the R5 pattern, the RETI will immediately cause $\overline{\text{INT0}}$ to interrupt again, and R5 will be set to the next pin position. The worst-case response time, if eight pins must be searched before the low pin is found, will be in the order of 600 microseconds.
- If $\overline{\text{INT0}}$ is triggered by noise, the routine returns after the first fruitless search with no action taken and re-arms the interrupt structure.
- The external interrupt flags are cleared when the program vectors to the interrupt address *only* when the external interrupt is *edge* triggered. *Level*-triggered interrupts must have the low level removed *before* the RETI, or an immediate interrupt is regenerated. Each interrupt routine loads the internal RAM location ack with the proper bit pattern to the decoder to enable and decode the proper line to reset the interrupting source.

Hipri

Suppose that we wish to have a priority system by which the priority of each input pin is assigned at a different level—that is, there are eight priority levels, and each higher level can interrupt one at a lower level. Theoretically, this leads to at least nine return addresses being pushed on the stack (plus any other registers saved), so the stack should be expected to grow more than 18d bytes; it is set above the addressable bits at location 2Fh.

In order to enable the interrupt structure in anticipation of a higher level interrupt, it is necessary to issue a RETI instruction without actually leaving the interrupt routine that currently has the highest priority. One way to accomplish this task is to push on the stack the address of the current interrupt routine to be done. Then, use a RETI that will return to the address on the stack, the desired current interrupt subroutine, and also re-arm the interrupt structure should another interrupt occur. The addresses of each subroutine can be known before assembly by originating each at a known address, or the program can find each address in a lookup table and push it on the stack, as illustrated in the example program.

For this example, the priority of each interrupt source is equivalent to the port 1 pin to which its identity line is connected. P1.0 has the highest priority, and P1.7, the lowest. A lookup table is used to find the address of the subroutine to be pushed on the stack.

External interrupt $\overline{\text{INT0}}$ is connected to the common interrupt line from all sources. It is enabled edge triggered whenever an interrupt routine is running so that any higher priority interrupt will be immediately acknowledged. If a lower priority interrupt occurs, it will interrupt the program in progress long enough to determine the priority. The interrupted subroutine will resume, and the lower level interrupt source priority will be saved until the subroutine in progress is finished. All interrupting sources maintain their identity lines low until they are acknowledged. The common interrupt line is reset immediately to enable any other source to interrupt the 8051.

If a higher level source interrupts a lower priority interrupt, then the high priority routine will interrupt the lower priority routine. The priority of the lower level interrupt will be saved.

The program *hipri* assigns eight levels of priority to the interrupt sources connected to port 1. A lookup table is used to find the address of the interrupt handling subroutine that is pushed on the stack. A RETI instruction is then used to "return" to the desired subroutine and re-arm the interrupt hardware on the 8051.

```
                org 0000h
hipri:          ljmp over               ;jump over the INT0 routine
;the INT0 interrupt will vector here to find the identity and
;priority of the interrupt source
;
                org 0003h               ;INT0 interrupt vectors here
int:            push dph                ;save registers used
                push dpl
                push acc
                setb psw.3              ;use register bank 1
                clr p3.0                ;reset common INT line by strobing
                setb p3.0               ;pin 3.0
                mov dptr,#base          ;get base address of address table
                mov a,R5                ;get priority of current interrupt
                orl a,P1                ;determine if new interrupt is
                                        ;higher
                cjne a,#0ffh,higher     ;A will be FFh if new < old
                pop acc                 ;not higher priority; return to
                                        ;current
                pop dpl
                pop dph
                reti
higher:         push 0dh                ;higher priority; save old (R5)
                jnb acc.0,first         ;find higher priority interrupt
                jnb acc.1,second
```

```
          jnb acc.2,third
          jnb acc.3,fourth
          jnb acc.4,fifth
          jnb acc.5,sixth
          jnb acc.6,seventh
          jnb acc.7,eighth
          sjmp goback          ;noise; return with no new interrupt
first:    mov r5,#0ffh         ;highest priority
          mov a,#00h           ;load A with offset into lookup
                               ;table
          sjmp pushadd         ;"pushadd" will push the address
second:   mov r5,#0feh         ;may only be interrupted by P1.0
          mov a,#02h           ;load A with offset for next program
          sjmp pushadd
third:    mov r5,#0fch         ;interrupt by 0 – 1
          mov a,#04h
          sjmp pushadd
fourth:   mov r5,#0f8h         ;interrupt by 0 – 2
          mov a,#06h
          sjmp pushadd
fifth:    mov r5,#0f0h         ;interrupt by 0 – 3
          mov a,#08h
          sjmp pushadd
sixth:    mov r5,#0e0h         ;interrupt by 0 – 4
          mov a,#0ah
          sjmp pushadd
seventh:  mov r5,#0c0h         ;interrupt by 0 – 5
          mov a,#0ch
          sjmp pushadd
eighth:   mov r5,#80h          ;interrupt by 0 – 6
          mov a,#0eh
pushadd:  mov r6,a             ;save A for second byte fetch
          inc a                ;point to the low byte of the
                               ;address
          movc a,@a+dptr       ;get first program address low byte
          push acc             ;push the low byte
          mov a,r6             ;get A back
          movc a,@a+dptr       ;get the high byte of the address
          push acc             ;push the high byte
          reti                 ;execute subroutine; enable
                               ;interrupt
goback:   pop 0dh              ;restore old priority mask
          mov a,p1             ;look at P1 for more interrupts
          cjne a,#0ffh,old     ;see if any are waiting, or in
                               ;progress
          pop acc              ;if none waiting then return to main
          pop dpl              ;program
```

```
                pop dph
                clr psw.3               ;return to register bank 0
                reti                    ;return to main program
old:            orl a,r5                ;A = FFh if next interrupt
                                        ;waiting was
                cjne a,#0ffh,next       ;itself interrupted
                pop acc                 ;get old interrupt values
                pop dpl
                pop dph
                reti                    ;return to old interrupt in progress
next:           pop acc                 ;the waiting interrupt is a new one
                pop dpl                 ;that has never begun to execute
                pop dph                 ;jump to "int" as if an INT0 has
                ljmp int                ;occurred
;
;the lookup table that contains the addresses of the eight interrupt
;programs is assembled here; the assembler knows all the actual
;numbers at assembly time
;
base:           dw prog1                ;"progx" is the actual interrupt
                                        ;routine
                dw prog2
                dw prog3
                dw prog4
                dw prog5
                dw prog6
                dw prog7
                dw prog8
prog1:          nop                     ;simulate interrupt program.
                ljmp goback             ;be sure to acknowledge before ljmp
prog2:          nop                     ;after subroutine has finished
                ljmp goback
prog3:          nop
                ljmp goback
prog4:          nop
                ljmp goback
prog5:          nop
                ljmp goback
prog6:          nop
                ljmp goback
prog7:          nop
                ljmp goback
prog8:          nop
                ljmp goback
;
;the main program starts here; progx could have been assembled
;after the main program if desired
;
```

```
over:     mov sp,#2fh      ;set stack above addressable bits
          setb tcon.0      ;enable INT0 edge-triggered
          setb psw.3       ;choose register bank one
          mov r5,#00h      ;set for interrupt at all levels
          mov ie,#81h      ;enable INT0
here:     clr psw.3        ;return to bank 0
          sjmp here        ;simulate main program
          end
```

◆ COMMENT ◆

- The dw assembler directive will store the *high* byte of the 2-byte word at the *lower* address in memory. For the RETI in pushadd to work properly, the *low* address byte must be placed on the stack first.
- If interrupt A has just gone low, and interrupt B, which is of a higher priority, occurs after the system has vectored to the $\overline{\text{INT0}}$ address, interrupt B will be accessed if the B line goes low before the polling software starts (JNB ACC.x). If the polling has caused A to be chosen, then B will be recognized after the RETI in pushadd causes the A address to be POPed from the stack. One instruction of A will be executed, then the IE0 flag in TCON will cause an interrupt.
- The 8051 interrupt system will generate an interrupt unless any of the following conditions is true:
  - Another routine of equal or greater priority is running.
  - The current instruction is not finished.
  - The instruction is a RETI or any IE/IP access.
- The edge-triggered interrupt sets the IE0 flag, and the interrupt that generated the edge serviced after any of the listed conditions are cleared.

Hardware Circuits for Multiple Interrupts

Solutions to the expanded interrupt problem proposed to this point have emphasized using a minimal amount of external circuitry to handle multiple, overlapping interrupts. A hardware strategy, which can be expanded to cover up to 256 interrupt sources, is shown in Figure 10.13. This circuit is a version of the *daisy chain* approach, which has long been popular.

The overall philosophy of the design is as follows:

1. The most important interrupt source is physically connected first in the chain, with those of lesser importance next in line. Lower priority interrupt sources are "behind" (connected further from $\overline{\text{INT0}}$) those of a higher priority.
2. Each interrupting source can disable all signals from sources that are wired behind it. All sources that lose the INACTOUT signal (a low level) from the

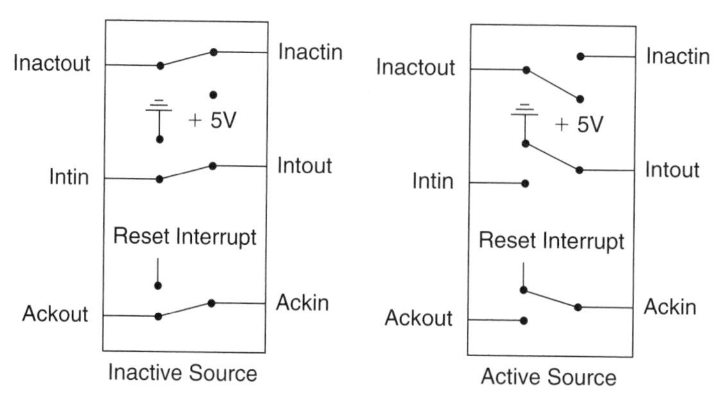

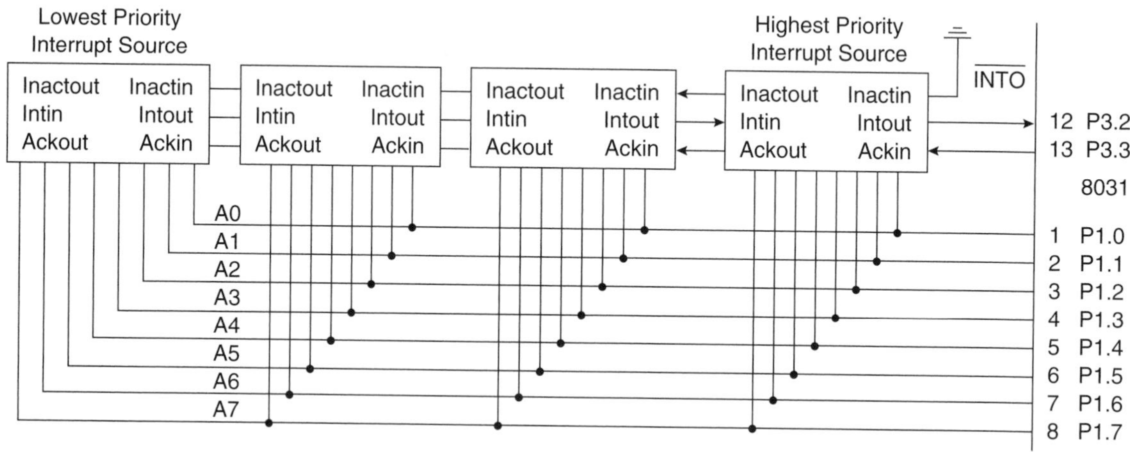

FIGURE 10.13 ◆ Daisy Chain Circuit Used for Hardint

source(s) ahead of it will place their source address buffer in a tri-state mode until INACTOUT is restored.

3. A requesting source pulls its $\overline{\text{INTOUT}}$ line low and places its 8-bit identifier on the tri-state bus connected to port 1. The interrupt routine at the $\overline{\text{INT0}}$ vector location reads P1 and, using a lookup table, finds the address of the subroutine that handles that interrupt. The address is placed on the stack and a RETI executed to go to that routine and re-arm the interrupt structure.
4. The interrupt subroutine generates an ACKIN signal (a low-level pulse) to the source from the 8051 at the end of the subroutine; the source then removes $\overline{\text{INTOUT}}$ and the 8-bit source address. When an interrupt is acknowledged, the interrupting source must bring the $\overline{\text{INTOUT}}$ line high for

at least one machine cycle so that the 8051 interrupt structure can recognize the next high-to-low transition on $\overline{INT0}$.

The software is very simple for this scheme. Any interrupt received is always of higher priority than the one now running, and the source address on port 1 enables rapid access to the interrupt subroutine.

Accomplishing this interrupt sequence requires that the source circuitry be complex or that the source contain some intelligence such as might be provided by a microcontroller. The additional source hardware will entail considerable relative expense for each source. As the number of interrupt sources increases, system costs rise rapidly. At some point the designer should consider another microcontroller that has extensive interrupt capability.

Hardint

The program *hardint* is used with daisy-chained interrupt sources to service 16 interrupt sources. An interrupt is falling-edge triggered on $\overline{INT0}$ and the interrupt address read on P1. A lookup table then finds the address of the interrupt routine that is pushed on the stack and the RETI "returns" to the interrupt subroutine. The interrupt subroutine issues an acknowledgment on port 3.3, which resets the interrupting source.

```
          org 0000h
hardint:  ljmp over
;
;the interrupt program located at the INT0 vector address will read
;the source address on port 1, and push that address for a RETI to
;the interrupt subroutine for that address
;
          org 0003h
          setb psw.3          ;choose register bank 1
          push acc            ;save registers used
          push dpl
          push dph
          mov a,p1            ;read port 1 for source address
          cjne a,#10h,less    ;valid addresses are 00h to 0Fh
less:     jnc goback          ;invalid, return
          rl a                ;valid address, adjust A for addresses
          mov r6,a            ;save A for low-byte fetch
          mov dptr,#base      ;point to program address lookup table
          inc a               ;point to low byte
          movc a,@a+dptr      ;get low byte
          push acc
          mov a,r6            ;get high-byte offset
          movc a,@a+dptr
          push acc
          reti                ;execute subroutine; enable interrupt
```

```
goback:   pop dph         ;return to program in progress
          pop dpl
          pop acc
          clr psw.3       ;back to register bank 0
          reti
base:     dw prog0        ;make lookup table for subroutines
          dw prog1
          dw prog2
          dw prog3
          dw prog4
          dw prog5
          dw prog6
          dw prog7
          dw prog8
          dw prog9
          dw proga
          dw progb
          dw progc
          dw progd
          dw proge
          dw progf
prog0:    nop             ;simulate interrupt subroutine
          ljmp goback
prog1:    nop             ;simulate interrupt subroutine
          ljmp goback
prog2:    nop             ;simulate interrupt subroutine
          ljmp goback
prog3:    nop             ;simulate interrupt subroutine
          ljmp goback
prog4:    nop             ;simulate interrupt subroutine
          ljmp goback
prog5:    nop             ;simulate interrupt subroutine
          ljmp goback
prog6:    nop             ;simulate interrupt subroutine
          ljmp goback
prog7:    nop             ;simulate interrupt subroutine
          ljmp goback
prog8:    nop             ;simulate interrupt subroutine
          ljmp goback
prog9:    nop             ;simulate interrupt subroutine
          ljmp goback
proga:    nop             ;simulate interrupt subroutine
          ljmp goback
progb:    nop             ;simulate interrupt subroutine
          ljmp goback
progc:    nop             ;simulate interrupt subroutine
          ljmp goback
```

```
progd:      nop                 ;simulate interrupt subroutine
            ljmp goback
proge:      nop                 ;simulate interrupt subroutine
            ljmp goback
progf:      nop                 ;simulate interrupt subroutine
            ljmp goback
;
;place the main routine here
;
over:       mov sp,#2fh         ;set stack above addressable bits
            setb tcon.0         ;set INT0 for edge triggered
            mov ie,#81h         ;enable INT0
here:       sjmp here           ;simulate main program
            end
```

◆ COMMENT ◆

- If the lookup table goes beyond 128 addresses, or 256 bytes, then DPH is incremented by 1 to point to a second complete table.
- Each interrupt subroutine must contain an acknowledge byte that is placed on P3 to reset each source.
- Note the use of CJNE and the Carry flag to determine relative sizes of 2 bytes at label less.

10.6 Putting It All Together

All of the examples presented to this point have used the free ports (P1 and parts of P3) that the "cheap" design affords. It is clear that to do a real-world design requires the use of additional port chips to enable several functions to be interfaced to the 8051 at one time. Such a design is illustrated in this section, using an 8255 programmable port chip memory-mapped at external RAM location 8000h to 8003h. A review of memory mapping found in Chapter 9 shows that the required address decoding can be done using an inverter to enable external RAM whenever A15 is low, and the 8255 whenever A15 is high. Actually, *any* address that begins with A15 high can address the 8255; 8000h seems convenient.

Ant

The example program uses the intelligent LCD display, a coded 16-key keypad, and is capable of serial data communications. This type of design is suitable for many applications where a small, inexpensive, alphanumeric terminal (dubbed the *ANT*) is needed for the factory floor or the student lab.

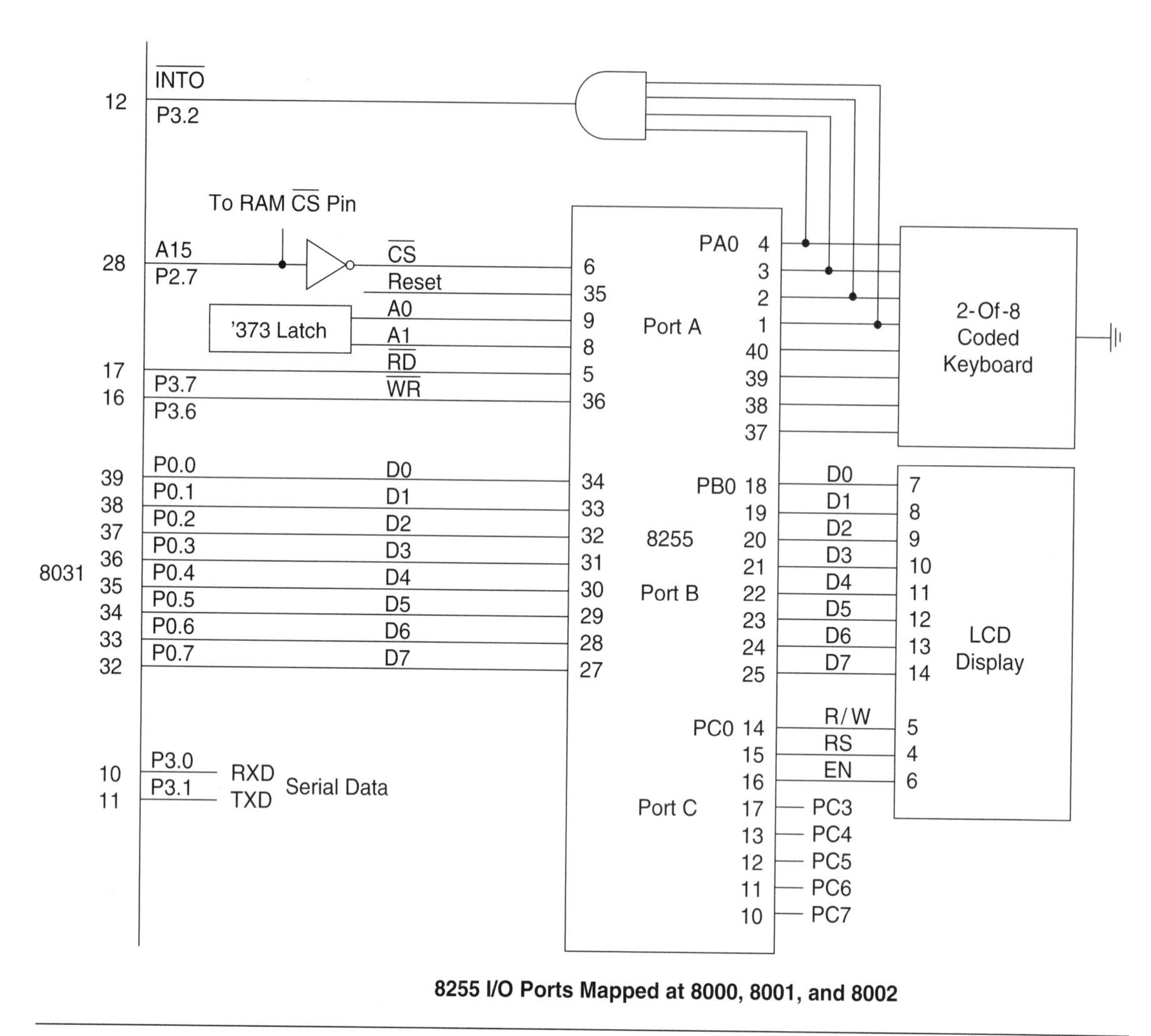

FIGURE 10.14 ◆ A Multi-Tasking Circuit Using Memory-Mapped I/O

The design is shown in Figure 10.14. Port A of the 8255 is connected to the keypad; port B supplies data bytes to the LCD and the lower half of port C controls the display. The program is interrupt-driven by the keypad and the serial port. $\overline{\text{INT0}}$ is used to detect a keypress via the AND gate array and the serial interrupt is internal to the 8051. The serial port has the highest priority. This type of program is often called *multi-tasking* because the routines are called by the interrupt structure, and the computer appears to be doing many things simultaneously.

A keypad program developed in this chapter combined with a serial communication program from Chapter 11 completes the design.

The program *ant* controls the actions of an 8051 configured as a terminal with a LCD display and hexadecimal keypad. The serial port is enabled and has the highest priority of any function. The coded keyboard is a 2-of-8 type that can use a lookup table to detect valid key presses. A shift key capability is possible because unique patterns are possible if one key is held down while another is pressed.

```
         con equ 8003h       ;address of 8255 mode control register
         prta equ 8000h      ;address of 8255 port A
         prtb equ 8001h      ;address of 8255 port B
         prtc equ 8002h      ;address of 8255 port C
         conant equ 98h      ;A = input, B and lower C = output
         bf equ 05h          ;C pattern to read LCD busy flag
         wrd equ 06h         ;C pattern to write data to LCD
         wrc equ 04h         ;C pattern to write control to LCD
         setlcd equ 3fh      ;initialize LCD to 2 lines, 5 × 10 dots
         curs equ 06h        ;LCD cursor blinks, moves left
         lcdon equ 0eh       ;LCD on
         klr equ 01h         ;clear LCD and home cursor
         org 0000h
ant:     ljmp over           ;jump over the interrupt locations
;
;when a key is pressed, or a serial data character is sent or
;received, the program vectors to the interrupt address locations;
;dummy routines will be written here; refer to the key routines in
;this chapter and the serial data routines in Chapter 11 for examples
;of these programs
;
         org 0003h           ;origin the keypad program here
         sjmp keypad         ;jump to keypad handling program
         org 0023h           ;origin serial interrupt program here
         sjmp serial         ;jump to serial program
keypad:  push dph            ;dummy keypad program, get the key
         push dpl
         mov dptr,#prta      ;read the key value
         movx a,@dptr        ;insert a key-handling routine next
         pop dpl
         pop dph
         reti
serial:  nop                 ;dummy serial program
         reti
;
```

```
;the main program begins here; All the interrupts are initialized,
;the main program sends a "hello" to the display and waits for an
;interrupt
;
over:      setb tcon.0          ;set INT0 for edge-triggered operation
           setb ip.4            ;set serial interrupt high priority
           acall serset         ;call the serial port setup routine
           mov dptr,#con        ;initialize 8255 mode to basic I/O
           mov a,#conant        ;set A = input, B and C = output
           movx @dptr,a         ;initialize 8255 mode register
           mov a,#setlcd        ;initialize the LCD and say "hello"
           lcall lcdcon
           mov a,#curs
           lcall lcdcon
           mov a,#lcdon
           lcall lcdcon
           mov a,#klr
           lcall lcdcon         ;LCD is now initialzed and blank
           mov dptr,#msg        ;use DPTR to point to "hello"
           lcall lcddta         ;send the message to the LCD
           mov ie,#91h          ;enable serial and INT0 interrupts
here:      sjmp here            ;simulate the rest of the program
serset:    ret                  ;dummy serial setup routine
;
;The subroutine lcddta sends data characters to the LCD until the
;character ~ is found; the beginning of the message is passed to the
;subroutine in the DPTR by the calling program
;
lcddta:    mov a, #00h          ;get character of message
           movc a, @a+dptr
           cjne a,#7eh,modd     ;stop when ~ (7Eh) is found
           ret                  ;message sent
modd:      acall datta          ;send data character
           inc dptr
           sjmp lcddta          ;loop until done
;
;the subroutine that sends control characters passed in A to the LCD
;display, via the 8255
;
lcdcon:    push dph             ;save registers used
           push dpl
           mov dptr,#prtb       ;get control data in A to port B
           movx @dptr,a
           mov dptr,#prtc       ;point to port C for LCD control
           mov a,#wrc           ;strobe character to LCD using port C
           movx @dptr,a
```

```
            mov a,#00h          ;end strobe
            movx @dptr,a
            acall dun           ;wait for LCD to finish
            pop dpl             ;restore registers
            pop dph
            ret
;
;the subroutine "datta" sends data characters passed in A to the
;LCD screen for display
;
datta:      push dph            ;save registers used
            push dpl
            mov dptr,#prtb      ;get character data in A to port B
            movx @dptr,a
            mov dptr,#prtc
            mov a,#wrd          ;strobe character to LCD using port C
            movx @dptr,a
            mov a,#00h          ;end strobe
            movx @dptr,a
            acall dun           ;wait for LCD to finish
            pop dpl             ;restore registers
            pop dph
            ret
;
;"dun" reads the busy flag on the LCD and returns if the flag is low
;
dun:        mov dptr,#con       ;configure port B as an input
            mov a,#9ah
            movx @dptr,a
            mov dptr,#prtc      ;set port C for a read command
            mov a,#bf
            movx @dptr,a        ;send command to read flag
            mov dptr,#prtb      ;read port B
            movx a,@dptr        ;the busy flag is bit 7
            jnb acc.7,go        ;done when BF = 0
            mov dptr,#prtc      ;if still busy then read again
            mov a,#00h
            movx @dptr,a
            sjmp dun
go:         mov a,#00h          ;finished, remove strobe
            mov dptr,#prtc
            movx @dptr,a
            mov dptr,#con       ;reset port B as an output
            mov a,#98h
            movx @dptr,a
            ret
```

```
;the message "hello" is assembled here; a great number of
;messages, each with a unique label, can be sent in this way
;
msg:          db "hello~"
end
```

◆ COMMENT ◆

- The LCD example shows the extensive use of the DPTR and MOVX commands when dealing with a memory-mapped external port.
- Forgetting to terminate every message with a ~ results in a very confused LCD because the remainder of ROM is written to the LCD.
- There will be no interference between any of these programs if the serial interrupts always have priority. Serial data is received as it occurs, and the keypad program and any messages to the LCD are suspended for the few microseconds it takes to read the serial port. The suspended programs can resume until the next serial character, which is normally an interval of 1 or more milliseconds.

10.7 Summary

Hardware designs and programs have been illustrated to solve several common application problems that are especially suitable for solution using a microcontroller. These hardware circuits are:

- Keyboards: lead-per-key; X-Y matrix; coded
- Displays: 7-segment LED; intelligent LCD
- Pulse measurement: frequency; pulse width
- Data converters: R/2R digital to analog; flash analog to digital
- Interrupts: multi-source; daisy chain
- Expanded 8051 system: memory-mapped I/O

The programs in this chapter interface the 8051 to these circuits. New programming concepts introduced are:

- Interrupt handling
- Register bank switching in *svnseg*
- Jump tables in *lopri*
- Stack RETI in *hipri*
- Using CJNE for relative size in *hardint*
- Multi-tasking in *ant*

These programs can be used as the kernels for more comprehensive applications.

10.8 Questions and Problems

1. List the most likely effects if a keyboard program does not accomplish the following:

   a. Debounce keys when pressed down

   b. Check for a valid key code

   c. Wait for all keys up before ending keyboard routine

   d. Debounce keys when released

2. A keyboard has two keys: run and stop. Write a program that is interrupt-driven by these two keys using $\overline{\text{INT0}}$ for the run key and $\overline{\text{INT1}}$ for the stop key. If run is selected, set pin P3.0 high; if stop is selected, set the pin low. Bounce time is 10 milliseconds for the keys.
3. Determine why it is important to employ some kind of debounce subroutine in a keyboard program, particularly for interrupt-driven programs, even if keys with absolutely no bounce are used.
4. The lookup table used in the program *codekey* is very inefficient, using 120 bytes to form a table for the valid keys. Write a subroutine using a series of CJNE instructions that will obtain the same result.
5. Write a lookup table subroutine for the program *bigkey* that will convert the row and column bytes for each key to a single byte number.
6. Expand the lookup table convert in the program *svnseg* to include these characters: G, H, I, J, L, O, P, S, T, and U.
7. Write a program that will display the following message on the intelligent display:

   "Hello!

   Please Enter Command."

   Center each line of the display.

8. Write a subroutine that is passed the starting address of an ASCII string in ROM and then displays the string on the intelligent display. The string length is fixed.
9. Repeat Problem 8 for a string of any length.

10. Write a program for the LCD display that will display the contents of register R1 as follows:

    R1 = XX

    XX is the R1 contents in hex. Center the display. (Hint: Remember the contents are in hex, and the display speaks ASCII.)
11. Write a program using timer 0 that will delay exactly .100000 milliseconds ± 1 microsecond from the time the timer starts until it is stopped. (The crystal frequency is 12 megahertz.)
12. Make a table that shows the accuracy of pulse width measurements as a function of multiples of count periods. The table should be arranged as follows:

Pulse Width ($\times 1\ \mu s$)	Accuracy (%)
2	
3	
4	
5	
6	
7	
8	
9	
10	
20	
50	
100	

13. Write a program that can use the stack to "return" to any of 256 subroutines pointed to by the number 00 to FFh in A.
14. Compose a 40-value lookup table that will generate a sawtooth wave using a D/A converter.
15. Repeat Problem 14 without using a lookup table of any kind.
16. Repeat Problem 14 for a rectified sine wave.
17. Outline a method of measuring the frequency of a sine wave using a flash A/D converter. Estimate the highest frequency that can be measured to an accuracy of 1%.
18. Assume that the highest frequency that may be accurately measured by an 8051 system is 45,000 Hz. Determine the number of cycles of a 45-kHz wave that must be measured in order to guarantee a measurement accuracy of 1%.

19. Write a program that performs all of the functions of an intelligent daisy chain interrupt source controller.

20. Write a lookup table program for the *ant* program that will allow the F key of a 2-of-8 coded keypad to be used as a shift key. A shift key makes possible 31 valid key combinations. The key codes are

Key	Output Pin								
1	1	2	3	4	5	6	7	8	9
0	x				x				x
1	x					x			x
2	x						x		x
3	x							x	x
4		x			x				x
5		x				x			x
6		x					x		x
7		x						x	x
8			x		x				x
9			x			x			x
A			x				x		x
B			x					x	x
C				x	x				x
D				x		x			x
E				x			x		x
F				x				x	x

An x means a connection is made; pin 9 is the common pin for all codes.

11

Serial Data Communication

Chapter Outline

CHAPTER OBJECTIVES

On successful completion of this chapter you will be able to:

- Discuss several typical communication configurations.
- Use 8051 communication modes 0, 1, 2, and 3.
- Write mode 0, 1, 2, and 3 communication programs.

11.0 Introduction

Chapter 3 contains an extensive review of serial data communication concepts and the hardware and software that are built into the 8051 for enabling serial data transfers. Chapter 9 contains some brief programming examples of how this capability may be used. Serial data transmission has become so important to the overall computing strategy of industrial and commercial applications that a separate chapter on this crucial subject is appropriate.

One hallmark of contemporary computer systems is interconnectivity: the joining of computers via data networks that link the computers to each other and to shared resources, such as disk drives, printers, and other I/O devices. The beginning of the "computer age" saw isolated CPUs connected to their peripherals using manufacturer-specific data transmission configurations. One of the peripherals, however, was the teletype that had been borrowed from the telephone industry for use as a human interface to the computer, using the built-in keyboard and printer.

The teletype was designed to communicate using standard voice-grade telephone lines via a *modem* (modulator demodulator) that converts digital signals to analog frequencies and analog frequencies to digital signals. The data, by the very nature of telephone voice transmission, is sent and received serially. Various computer manufacturers adapted their equipment to fit the teletype, and, perhaps, the first "standard" interface in the industry was born.

This standard was enhanced in the early 1960s with the establishment of an electrical/mechanical specification for serial data transmission that was assigned the number RS 232 by the Electronics Industry Association. A standard data code was also defined for all the characters in the alphabet, decimal numbers, punctuation marks, and control characters. Based on earlier telephonic codes, the standard became known as the American Standard Code for Information Interchange (ASCII).

The establishment of RS 232 and ASCII coincided with the development of multi-user computer organizations wherein a number of users were linked to a host mainframe via serial data links. By now, the CRT terminal had replaced the slower teletype, but the RS 232 serial plug remained, and serial data was encoded in ASCII. Peripheral devices, such as printers, adopted the same standards in order to access the growing market for serial devices.

Serial data transmission using ASCII became so universal that specialized integrated circuits, Universal Asynchronous Receiver Transmitters (UARTs), were developed to perform the tasks of converting an 8-bit parallel data byte to a 10-bit serial stream and converting 10-bit serial data to an 8-bit parallel byte. When the second-generation 8051 microcontroller was designed, the UART became part of the circuit.

Chapter 9 introduces the basic programming concepts concerning transmitting and receiving data using the serial port of the 8051. In this chapter, we study the serial data modes available to the programmer and develop programs that use these modes. The four modes are as follows:

- Mode 0: Shift register mode
- Mode 1: Standard UART mode
- Mode 2: Multiprocessor fixed mode
- Mode 3: Multiprocessor variable mode

In this chapter, we also identify multiprocessor configurations that are appropriate for each mode and write sample programs to enable data communication between 8051 microcontrollers.

11.1 Network Configurations

The first problem faced by the network system designer is how to physically hook the computers together. The two possible basic configurations are the star and the loop, which are shown in Figure 11.1.

The star features one line from a central computer to each remote computer, or from *host* to *node*. This configuration is often used in time-sharing applications when a central mainframe computer is connected to remote terminals or personal computers using a dedicated line for each node. Each node sees only the data on its line; all communication is private from host to node.

The loop uses one communication line to connect all of the computers together. There may be a single host that controls all actions on the loop, or any computer may be enabled to be the host at any given time. The loop configuration is often used in data-gathering applications where the host periodically interrogates each node to collect the latest information about the monitored process. All nodes see all data; the communication is public between host and nodes.

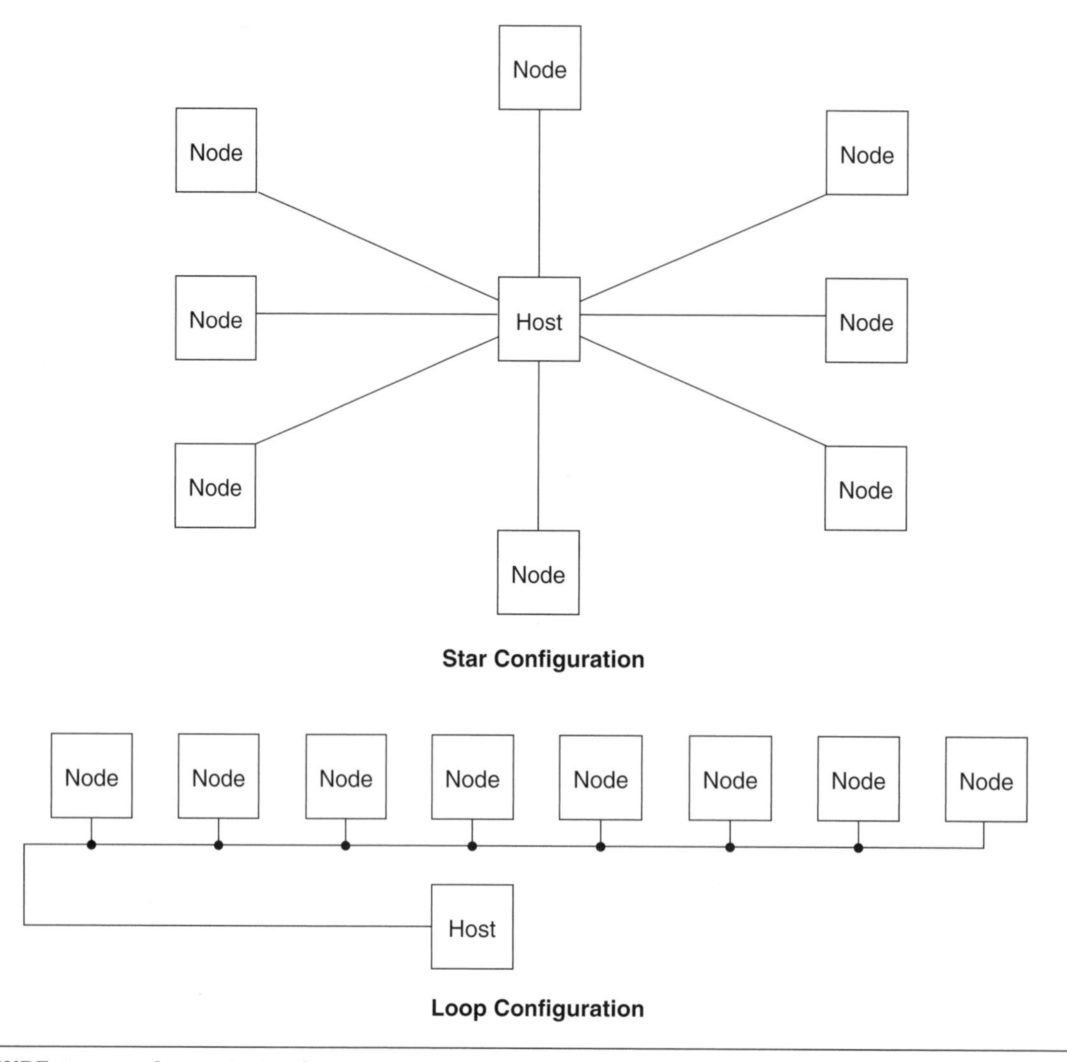

FIGURE 11.1 ◆ Communication Configurations

Choosing the configuration to use depends on many external factors that are often beyond the control of the system designer. Some general guidelines for selection are shown in Table 11.1.

The star is a good choice when the number of nodes is small, or the physical distance from host to node is short. But, as the number of nodes grows, the cost and physical space represented by the cables from host to nodes begins to represent the major cost item in the system budget. The loop configuration becomes attractive as cost constraints begin to outweigh other considerations.

TABLE 11.1

Objective	Network	Comments
Reliability	Star	Single node loss per line loss
Fault isolation	Star	Fault traceable to node and line
Speed	Star	Each node has complete line use
Cost	Loop	Single line for all nodes

Microcontrollers are usually applied in industrial systems in large numbers distributed over long distances. Loop networks are advantageous in these situations, often with a host controlling data transmission on the loop. Host software is used to expedite fault isolation and, thus, to improve system reliability. High speed data transmission schemes can be employed to enhance system response time where necessary.

The old racing adage "Speed costs money: How fast do you want to go?" should be kept in mind when designing a loop system. Successors to RS 232, most notably RS 485, have given the system designer 100 kilobaud rates over 4000-foot distances using inexpensive twisted-pair transmission lines. Faster data rates are possible at shorter distances, or more expensive transmission lines, such as coaxial cable or fiber-optic bundles, can be employed. Remember that wiring costs are often the major constraint in the design of large distributed systems.

Many hybrid network arrangements have evolved from the star and the loop. Figure 11.2 shows two of the more popular types that contain features found in both basic configurations.

11.2 8051 Data Communication Modes Example Programs

The 8051 has one serial port—port pins 3.0 (RXD) and 3.1 (TXD)—that receives and transmits data. All data is transmitted or received in two registers with one name: SBUF. Writing to SBUF results in data transmission; reading SBUF accesses received data. Transmission and reception can take place simultaneously, and the receiver can be in the process of receiving a byte while a previous byte is still in SBUF. The first byte must be read before the reception is complete, or the second byte will be lost.

Physically the data is a series of voltage levels that are sampled, in the center of the bit period, at a frequency that is determined by the serial data mode and the program that controls that mode. All devices that wish to communicate must use the same voltage levels, mode, character code, and sampling frequency (baud rate). The wires that connect the ports must also have the same polarity so that the idle state, logic high, is seen by all ports.

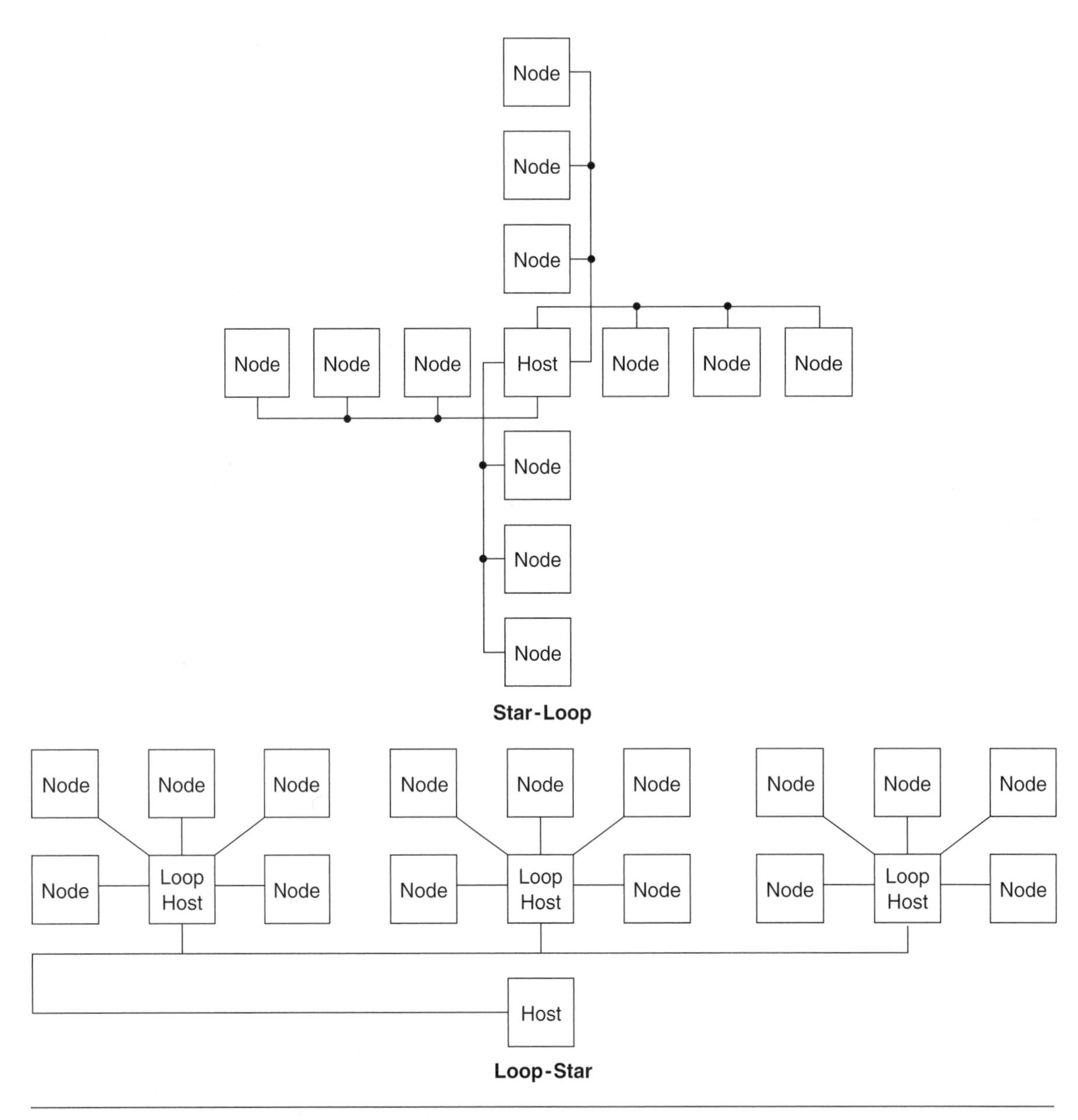

FIGURE 11.2 ◆ Hybrid Communication Configurations

The installation and checkout of a large distributed system are subject to violations of all of the "same" constraints listed previously. Careful planning is essential if cost and time overruns are to be avoided.

The four communication modes possible with the 8051 present the system designer and programmer with opportunities to construct very sophisticated data communication networks.

Mode 0: Shift Register Mode

Mode 0 is not suitable for the interchange of data between 8051 microcontrollers. Mode 0 uses SBUF as an 8-bit shift register that transmits and receives data on port pin 3.0, while using pin 3.1 to output the shift clock. The data and the shift clock are synchronized using the six internal machine states, and even for microcontrollers using the same crystal frequency, they can be slightly out of phase because of differences in reset and start-up times.

Figure 11.3 shows the timing for the transmission and reception of a data character. Remember that the shift clock is generated internally and is always *from* the 8051 *to* the external shift register. The clock runs at the machine cycle frequency of f/12. Note that transmission is enabled any time SBUF is the destination of a write operation, regardless of the state of the Transmitter Empty Interrupt flag, SCON bit 1 (TI).

Data is transmitted, LSB first, when the program writes to SBUF. Data is shifted right during S6P2 (state 6, phase 2 of a machine cycle), or 24/f seconds after the rising edge of the shift clock at S6P1. Data is stable from just after S6P2

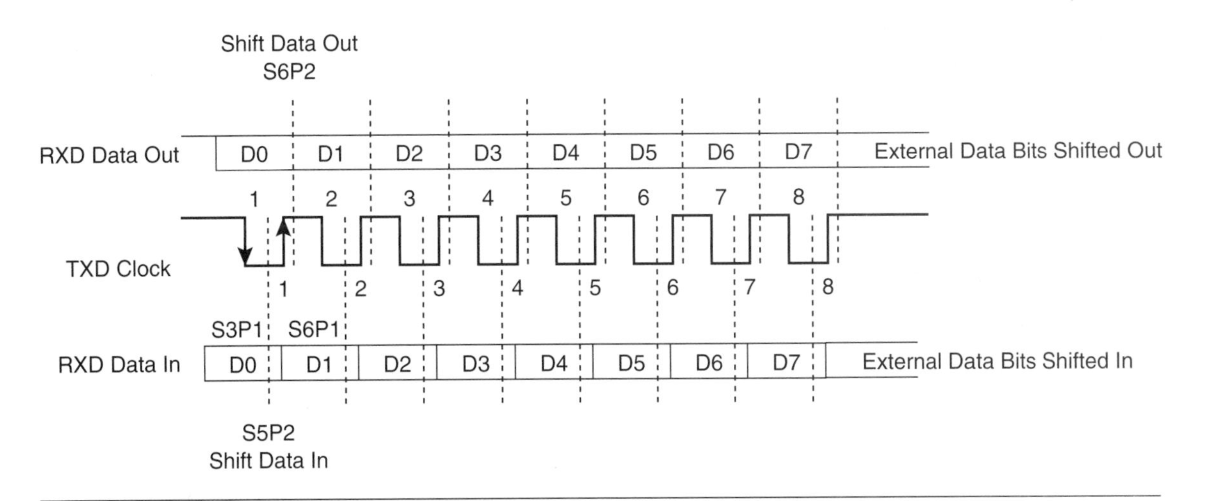

FIGURE 11.3 ◆ Mode 0 Timing

for 1 cycle. Good design practice dictates that the data be shifted into the external shift register during the high-to-low transition of the shift clock, at S3P1, to avoid problems with clock skew.

The receiver is enabled when SCON bit 5 (REN) is enabled by software and SCON bit 0 (RI) is set to 0. At the end of reception RI will set, inhibiting any form of character reception until reset by the software. The condition of RI cleared to 0 is unique for mode 0; all other modes are enabled to receive when REN is set without regard as to the state of RI. The reason is clear: Mode 0 is the only mode that controls when reception can take place. Enabling reception also enables the clock pulses that shift the received data into the receiver.

Reception begins, LSB first, with the data that is present during S5P2, or 24/f seconds before the rising edge of the shift clock at S6P1. The incoming data is shifted to the right. Incoming data should be stable during the low state of the shift clock, and good design practice indicates that the data be shifted from the external shift register during the low-to-high transition of the shift clock, at S6P1, so that the data is stable up to 1 clock period before it is sampled.

A serial data transmission interrupt is generated at the end of the transmission or reception of bit 8 if enabled by the ES interrupt bit EI.4 of the Enable Interrupt register. Software must reset the interrupting bit RI or TI. Because the same physical pin is used for transmission and reception, simultaneous interrupts are not possible.

Mode 0 is well suited for rapid data collection and control of multi-point systems that use a simple two-wire system for data interchange. Multiple external shift registers can expand the external points to an almost infinite number, limited only by the response time desired for the application. For instance, at f = 16 megahertz, each point of a 10,000 point system could be monitored every 60 milliseconds. Common industrial systems do not require rates this high, and a reasonable rate of one point per second would leave adequate time for processing by the program.

Modezero

A small system that features 16 points of monitored data and 16 points of control is shown in Figure 11.4. Data from the process is converted from parallel to serial in the '166 type registers. Data to the process is converted from serial to parallel in the type '164 registers and latched into the '373 latches.

It is important that the data be "frozen" before the shifting begins. The bits shifted in could be changed before reaching the microcontroller, or a control bit might be changed, momentarily, as it shifts through the output shift registers. Port pin 3.2 is used to disable the input registers from the process when high and to enable loading input values when low.

To read the inputs, P3.2 is brought high and the receiver is enabled (twice) to generate 16 input shift clocks. The high level on P3.2 prevents the shift clocks from reaching the output registers. At the end of the read, P3.2 is brought low to enable loading input values into the input registers. No clock pulses are generated, so the output control registers do not change state.

Control bits to the output registers are transmitted when P3.2 is low and SBUF has 2 data bytes written in succession. The 2 bytes generate 16 clock pulses that fill the output registers with the SBUF data. Port pin 3.3 is used to latch the newly shifted control data to the process by strobing the output data latches. A program that monitors and controls the points follows.

The program *modezero* monitors 16 bits and controls the state of 16 bits. The system can be expanded indefinitely by expanding the shift register configurations shown in Figure 11.4. In this example program, whatever data is

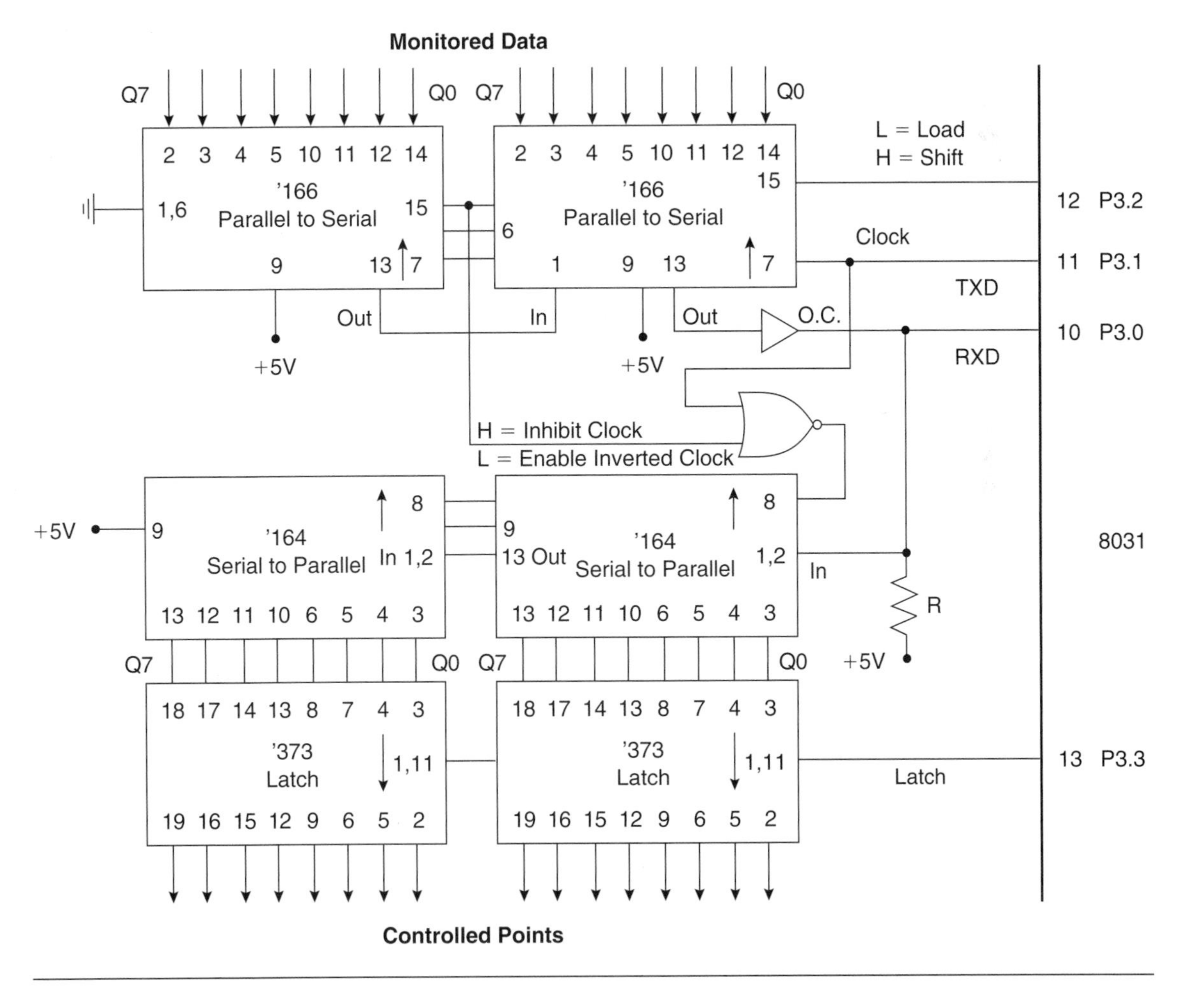

FIGURE 11.4 ◆ Shift Register Circuit Used with Modezero Program

read on the monitored points is written to the control points. The direction of data flow to/from the 8051 is controlled by P3.2, (high = in). P3.3 latches new data to the process.

```
                org 0000h
                mon1 equ 70h        ;store first 8 monitored points
                mon2 equ 71h        ;store second 8 monitored points
modezero:       clr p3.2            ;load data from process to input
                                    ;registers
                setb p3.2           ;enable data shift in
                acall monit         ;get first byte
                mov mon1,a          ;store first byte
                acall monit         ;get second byte
                mov mon2,a          ;store second byte
                clr p3.2            ;enable data to be shifted out
                acall conit         ;start sending data, second byte first
                mov a,mon1          ;get first byte
                acall conit         ;send first byte
                setb p3.3           ;latch data to output latches
                clr p3.3            ;end latch strobe
                sjmp modezero       ;loop for any new input
;
;the routine that reads the monitored points follows
;
monit:          mov scon,#10h       ;set mode 0 and enable reception
                                    ;reset RI
here:           jnb ri,here         ;wait for end of reception
                mov scon,#00h       ;clear receive enable and interrupt bit
                mov a,sbuf          ;read byte received
                ret                 ;return to calling program
;
;the routine that sends the control data follows
;
conit:          mov scon,#00h       ;set mode 0 and clear all interrupt bits
                mov sbuf,a          ;start transmission
wait:           jnb ti,wait         ;wait until transmission complete
                ret                 ;return to calling program
                end
```

◆ COMMENT ◆

- Note that in both the transmit and receive cases the interrupt bit must go high before the subroutine can be ended.
- The data transmission and reception time is so short that interrupt-driven schemes are not efficient.

Mode 1: Standard 8-Bit UART Mode

In Chapter 9, several simple communication programs are studied that use the serial port configured as mode 1, the standard UART mode normally used to communicate in 8-bit ASCII code. Only 7 bits are needed to encode the entire set of ASCII characters. The eighth bit can be used for even or odd parity or ignored completely. Asynchronous data transmission requires a start and stop bit to enable the receiving circuitry to detect the start and finish of a complete character. A total of 10 bits is needed to transmit the 7-bit ASCII character, as shown in Figure 11.5.

Transmission begins whenever data is written to SBUF. It is the responsibility of the programmer to ensure that any previous character has been transmitted by inspecting the TI bit in SCON for a set condition. Data transmission begins with a high-to-low start bit transition on TXD that signals receiving circuitry that a new character is about to arrive. The 8-bit character follows, LSB first and MSB parity bit last, and then the stop bit, which is high for 1 bit period. If another character follows immediately, a new start bit is signaled by a high-to-low transition; otherwise, the line remains high. The width of each transmitted data bit is controlled by the baud rate clock used. The receiver must use the same baud rate as the transmitter, or it reads the data at the wrong time in the character stream.

Reception begins if the REN bit is set in SCON and a high-to-low transition is sensed on RXD. Data bits are sampled at the baud rate in the center of the bit duration period. The received character is loaded into SBUF and the stop bit into SCON bit 2 (RB8) *if* the RI bit in SCON is cleared, indicating that the program has read the previous character, *and* either SM2 in SCON is cleared or SM2 is set and the received stop bit is high, which is the normal state for stop bits.

If these conditions are met, then SBUF is loaded with the received character, and RI is set. If the conditions are not met, the character is ignored, RI is not set, and the receive circuitry awaits the next start bit.

The restriction that a new character is not accepted unless RI is cleared seems reasonable. Data is lost if either the previous byte is overwritten or the new byte discarded, which is the action taken by the 8051. The restriction on

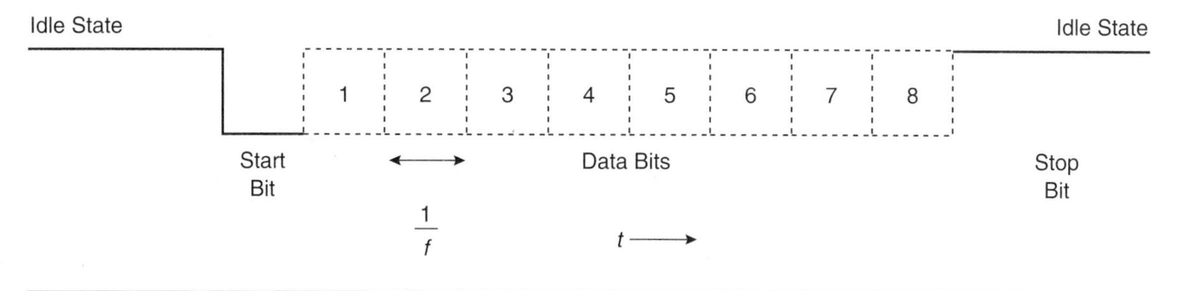

FIGURE 11.5 ◆ Asynchronous 8-Bit Character Used in Mode 1

SM2 and the stop bit are not as obvious. Normally, SM2 will be set to 0, and the character will be accepted no matter what the state of the stop bit. Software can check RB8 to ensure that the stop bit is correct before accepting the character if that is thought to be important.

Possible reasons for setting SM2 to force reception only when the stop bit is a 1 could be useful if the transmitter has the ability to change the stop bit from the normal high state. If the transmitter has this capability, then the stop bit can serve as an address bit in a multiprocessor environment where many loop microcontrollers are all receiving the same transmission. Only the microcontroller that has SM2 cleared can receive characters ending in either of the stop bit states. If all the microcontrollers but one have SM2 set, then all data transmissions ending in a low stop bit interrupt the unit with SM2 = 0; the rest ignore the data. Transmissions ending in a high stop bit can interrupt all microcontrollers.

Transmitters with the capability to alter the stop bit state are not standard. The 8051 communication modes 2 and 3 use the SM2 bit for multiprocessing. Mode 1 is not needed for this use.

In summary, mode 1 should be used with SM2 cleared, as a standard 8-bit UART, with software checks for proper stop bit magnitude if needed. As discussed in Chapter 9, the baud rate for modes 1 and 3 are determined by the overflow rate of timer 1, which is usually configured as an auto-reload timer. PCON bit 7 (SMOD) can double the baud rate when set.

Modeone

Mode 1 is most likely to be used in a dedicated system where the 8051 serial port is connected to a single similar port. A program that transmits and receives large blocks of data on an interrupt-driven basis is developed to investigate some problems common to data interchange programs.

To the main program, interrupt-driven communication routines are transparent: Data appears in RAM as it is received and disappears from RAM as it is transmitted. In both cases, the link between the main program and the interrupt-driven communication subroutines are areas of RAM called *buffers.* These buffers serve to store messages that are to be sent and messages that are received.

A prototype communication program, named *modeone,* is shown next. Modeone relies on interrupt-driven reception and transmission of data stored in RAM buffers. The main program reads received data from one buffer, and stores data in another buffer to be transmitted. Interrupt-driven receive and transmit subroutines do the actual work of receiving data and storing it in a receive buffer, or getting data from a buffer and transmitting it.

The receive and transmit buffers are in external RAM and are addressed using pointing registers R0 and R1 in different register banks. Each buffer can hold up to 128 bytes of data. Larger buffers, up to 256 bytes each, may be constructed using pointing registers, or up to 64K using DPTR.

The receive buffer, *inbuf,* is pointed to by registers R0 and R1 of register bank 1. Inbuf occupies RAM address 00h to 7Fh. As characters are received they are stored in inbuf, and R1 incremented to point to the next space in the buffer. If inbuf reaches 126 (7Eh) stored bytes, P3.4 is set to signal the other transmitting device to stop further transmission until inbuf can be emptied. If inbuf reaches 128 (80h), further reception is disabled until room is available in inbuf. The main program gets characters from inbuf using R0 to point to the next character to read. When R0 and R1 are equal, then inbuf is empty, and the pointers are reset to 00h.

The transmit buffer, *outbuf,* is pointed to by registers R0 and R1 in register bank 2. Outbuf occupies RAM addresses 80h to FFh. R0 points to the next character to be transmitted from outbuf, and R1 points to the next character to be stored in outbuf by the main program. When R0 and R1 are equal then outbuf is empty, and both registers are reset to 80h. If R1 reaches 256 (00h), the main program stops storing characters until room is available in outbuf.

```
;Modeone, receives and transmits data. A 12MHz crystal is used to
;generate a nominal baud rate of 2400, or 2403.8 actual.
;
     baudnum equ 0f3h      ;number loaded into TH1 for 2403.8 baud
     org 0000h
     sjmp over             ;jump over serial interrupt at 0023h
;
;place serial data interrupt subroutine here
     org 0023h
     push psw              ;save machine status
     push acc              ;save A register used in subroutines
     push b                ;save B register used in subroutines
     jbc ri,rcve           ;received data has priority
     jbc ti,xmit           ;then send data when time permits
     sjmp go               ;fail safe instruction
rcve:
     mov psw,#08           ;choose register bank 1 for inbuf pointers
     mov a,sbuf            ;get received character
     movx @r1,a            ;store character at top of input buffer stack
     inc r1                ;increment to point to next location
     cjne r1,#7eh,filled   ;check to see if input buffer almost full
     setb p3.4             ;signal sending device that almost full
filled:
     cjne r1,#80h,tran     ;check to see if input buffer filled
     clr ren               ;disable further data reception
tran:
     jbc ti,xmit           ;check if transmission waiting
     sjmp go               ;else return to main program
xmit:
     mov psw,#10h          ;choose register bank 2 for output buffer
     mov a,r0              ;check R0 and R1 for equality
     mov b,r1              ;if A = B then reset output buffer pointers
     cjne a,b,mor
```

```
      mov r0,#80h            ;pointers are equal, buffer empty, reset
      mov r1,#80h            ;both buffers to bottom of output buffer
      sjmp go                ;return to main routine
mor:
      movx a,@r0             ;get next character for transmission
      mov sbuf,a             ;transmit
      inc r0                 ;point to next character for transmission
go:
      pop b
      pop acc                ;return main program to original state
      pop psw
      reti                   ;return to main program
;
;jump to main program at this address
over:
      mov sp,#30h            ;set stack above bit addressable area
      mov p2,#00h            ;use bank 1 of the external SRAM
      anl pcon,#7fh          ;set SMOD bit to 0 for Baud 3 32 rate
      anl tmod,#0fh          ;alter timer T1 configuration only
      orl tmod,#20h          ;set timer T1 as an 8-bit autoload
      mov th1,#baudnum       ;TH1 set for divide clock by 13d
      setb tr1               ;T1 running at 1E6/13 5 76923 Hz
      mov scon,#50h          ;UART mode 1; Baud 5 76923/32 5 2403.8
      mov psw,#08h           ;set R0 and R1 of bank 1 to point to inbuf
      mov r0,#00h            ;inbuf extends from 00h to 7Fh
      mov r1,#00h
      mov psw,#10h           ;select register bank 2 to point to outbuf
      mov r0,#80h
      mov r1,#80h            ;outbuf extends from 80h to FFh
      mov psw,#00h           ;use register bank 0 for main program
      clr p3.4               ;signal talker that input buffer not full
      orl ie,#90h            ;enable global and serial interrupts
;
;main program loops here
loop:
      mov psw,#08h           ;use bank 1 for inbuf pointers
      mov a,r0               ;inbuf is empty when R0 = R1
      mov b,r1               ;R0 is current place, R1 is top of inbuf
      cjne a,b,rdit          ;if current <> top then read character
      clr ea                 ;disable interrupts as pointers reset
      mov r0,#00h            ;buffer empty, reset to start
      mov r1,#00h
      clr p3.4               ;signal talker that buffer empty
      setb ren               ;enable data reception, if disabled
      setb ea                ;enable interrupts
      sjmp send              ;send characters
rdit:
```

```
      movx a,@r0            ;read character from inbuf
      inc r0                ;point to next character in inbuf
send:
      mov psw,#10h          ;use bank 2 registers to point to outbuf
      cjne r1,#81h,fullup   ;If buffer has 1 character then set TI
      setb ti               ;initiate first character transmission
fullup:
      cjne r1,#00h,sd       ;R1 points to top of outbuf, rolls over
      sjmp loop             ;if R1 = 00 then outbuf is full
sd:
      mov a,r1              ;send the characters in R1 as a test
      movx @r1,a            ;store character in outbuf at R1 address
      inc r1                ;increment R1 to point to next outbuf slot
      sjmp loop             ;stay in main program loop
      end
```

◆ COMMENT ◆

- Note that the program has to initiate the first interrupt for the first character that is stored in a previously empty outbuf. If the first interrupt action were not done, transmission would never take place, as the TI bit would remain a 0. The 0 state of the TI bit is ambiguous: It can mean that the transmitter is busy sending a byte or that no activity is taking place at all. The 1 state of TI is specific: A byte has been transmitted, and SBUF can receive the next byte.
- The example program fills outbuf quickly, until R1 rolls over to 00h. Outbuf is emptied until R1 rolls over also, and outbuf is re-initialized to 80h. Received data is always read before inbuf can fill up, as there is very little for the program to do. Adding a time delay in the program ensures that inbuf grows beyond 1 byte.
- The data source should cease sending data to the 8051 until port 3.4 goes low. In this example, "full" is arbitrarily set at 1 byte below the maximum capacity of inbuf. The actual number for a full condition should be set at maximum capacity less the response time of the source expressed in characters.
- No feedback from the source to the 8051 has been provided for halting transmission of data from the 8051. Feedback can be accomplished by using one of the $\overline{INT}$ lines as an input from the source to signal a full condition.

Modes 2 and 3: Multiprocessor

Modes 2 and 3 are identical except for the baud rate. Mode 2 uses a baud rate of f/32 if SMOD (PCON.7) is cleared or f/64 if SMOD is set. For our 12 megahertz example, this results in baud rates of 375,000 and 187,500 bits per second, respectively. Pulse rates of these frequencies require care in the selection and installation of the transmission lines used to carry the data.

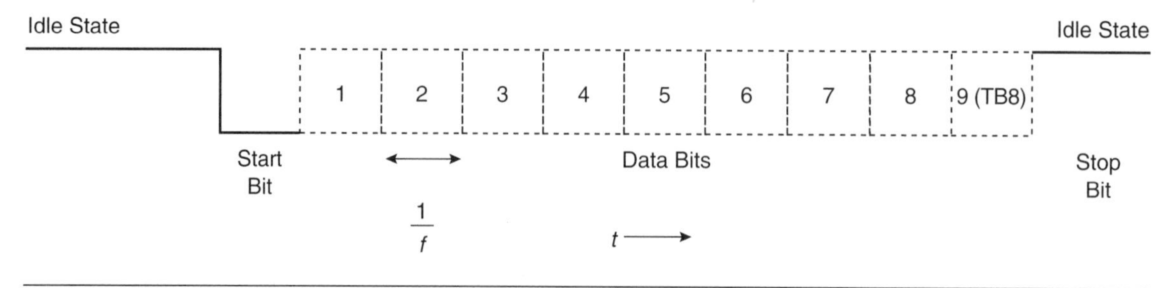

FIGURE 11.6 ◆ Asynchronous 9-Bit Character Used in Modes 2 and 3

Baud rates for mode 3 are programmable using the overflows of timer 1 exactly as for data mode 1. Baud rates as high as 83333 bits per second are possible using a 16 megahertz crystal. These rates are compatible with RS 485 twisted-pair transmission lines.

Data transmission using modes 2 and 3 features 11 bits per character, as shown in Figure 11.6. A character begins with a start bit, which is a high-to-low transition that lasts 1 bit period, followed by 8 data bits, LSB first. The tenth bit of this character is a programmable bit that is followed by a stop bit. The stop bit remains in a high state for a minimum of 1 bit period.

Inspection of Figures 11.5 and 11.6 reveals that the only difference between mode 1 and mode 2 and 3 data transmission is the addition of the programmable tenth bit in modes 2 and 3.

When the 8051 transmits a character in modes 2 and 3, the 8 data bits are whatever value is loaded in SBUF. The tenth bit is the value of bit SCON.3, named TB8. This bit can be cleared or set by the program. Interrupt bit TI (SCON.1) is set after a character has been transmitted and must be reset by program action.

Characters received using modes 2 and 3 have the 8 data bits placed in SBUF and the tenth bit is in SCON.2, called RB8, if certain conditions are true. Two conditions apply to receive a character. First, interrupt bit RI (SCON.0) must be cleared before the last bit of the character is received, and second, bit SM2 (SCON.5) must be a 0 or the tenth bit must be a 1. If these conditions are met, then the 8 data bits are loaded in SBUF, the tenth bit is placed in RB8, and the receive interrupt bit RI is set. If these conditions are not met, the character is ignored, and the receiving circuitry awaits the next start bit.

The significant condition is the second. If RI is set, then the software has not read the previous data (or forgot to reset RI), and it would serve no purpose to overwrite the data. Clearing SM2 to 0 allows the reception of multiprocessor characters transmitted in modes 2 and 3. Setting SM2 to 1 prevents the reception of those characters that have bit 10 equal to 0. Put another way, if bit 10 is a 1, then reception always takes place; SM2 is ignored. *If* bit 10 is a 0 then *only* those receivers with SM2 set to 0 are interrupted.

Modes 2 and 3 have been included in the 8051 specifically to enhance the use of multiple 8051s that are connected to a common loop in a multiprocessor configuration. The term *multiprocessing* implies many processors acting in some unified manner and connected so that data can be interchanged between them. When the processors are connected in a loop configuration, then there is generally a controlling or *talker* processor that directs the activities of the remainder of the loop units, or *listeners.*

One particular characteristic of a talker-listener loop is the frequent transmission of data between the talker and individual listeners. All data broadcast by the talker is received by all the listeners, although often the data is intended only for one or a few listeners. At times, data is broadcast that is meant to be used by all the listeners.

There are many ways to handle the addressing problem. Systems that use standard UART technology, such as mode 1, can assign unique addresses to all the listeners. Each message from the talker can begin with the address of the particular listener for which it is intended. When a message is sent, all the listeners process the message and react only if the address that begins the message matches their assigned addresses. If messages are sent frequently, the listeners will waste a lot of processing time rejecting those messages not addressed to them.

Modes 2 and 3 reduce listener processing time by enabling character reception based on the state of SM2 in a listener and the state of bit 10 in the transmitted character. A single strategy is used to enable a few listeners to receive data while the majority ignore the transmissions.

All listeners initially have SM2 set to 0, the normal reset state, and receive all multiprocessor messages. Each listener has a unique address. The talker addresses each of the listeners that are not of interest and commands them to set SM2 to 1, leaving the listeners to which communication is desired with SM2 cleared to 0. All characters from the talker to the unique listeners are then sent with bit 10 set to 0. The listeners with SM2 cleared receive the data; those with SM2 set ignore the data because of the condition of bit 10. Communication with all listeners is done by setting bit 10 to 1, which enables reception of characters with no regard as to the state of SM2.

A variation of this strategy is to have all listeners set SM2 to 1 on power-up. All address messages have a 1 in bit position 10, so all listeners receive and process any address message to see whether action is required. Listeners chosen are commanded in the address message to set SM2 to 0, and data communication proceeds with bit 10 cleared to 0.

The multiprocessing strategy works best when there is extensive data interchange between the talker and each individual listener. Frequent changes of listeners with little data flow results in heavy address usage and subsequent interruption of all listeners to process the address messages.

Modethree

A multiprocessor configuration that demonstrates the use of mode 3 is shown in Figure 11.7. An RS 485 twisted-pair transmission line is used to form a loop that has 15d 8051 microprocessors connected to the lines so that all data on the loop is common to all serial ports. The 8051 has been programmed to be the talker, and the rest are listeners.

The purpose of the loop is to collect 10 data bytes from each listener, in sequential order. All listeners initialize SM2 to 1 after power-up, and the talker configures all address messages using a 1 in bit 10. Addressed listeners transmit 10 data characters to the talker with bit 10 set to 0. The talker has SM2 set to 0 so that all communications from listeners are acknowledged. Data characters from a listener to the talker are ignored by the remaining listeners. At the end of the 10 data bytes, the addressed listener resets SM2 to 1. The data rate is set by timer 1 in the auto-reload mode to be 83,333 baud. That portion of the talker and listener program that has to do with setting up the multiprocessor environment will be programmed. Note f = 16MHz.

The messages that are sent from the talker to the listeners are called *canned* because the contents of each is known when the program is written; the messages can be placed in ROM for later use. The subroutine *sendit* in the talker program can send canned messages of arbitrary length, as long as each message ends in the character $.

Message contents from the listeners to the talker are not known when the program is written. A version of sendit, *sndat,* can still be used if the message

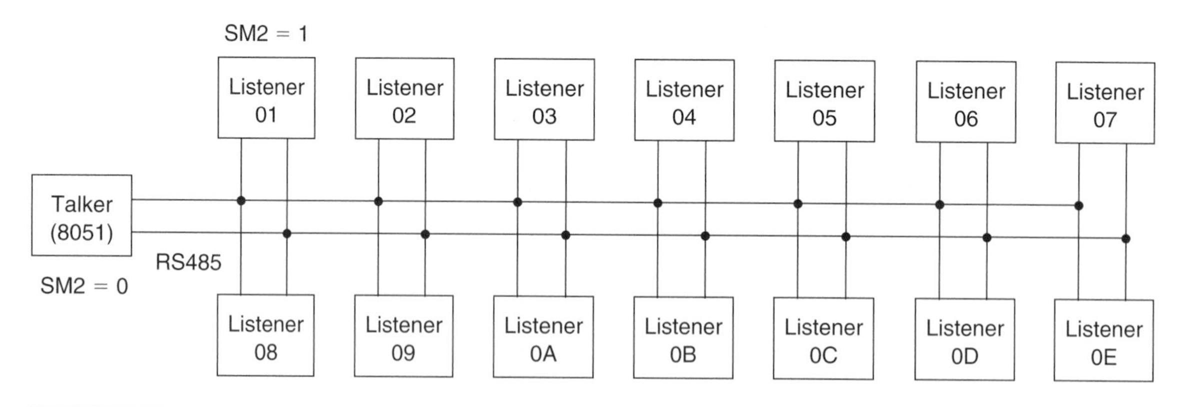

FIGURE 11.7 ◆ Communication Loop Used for Modethree Program

is constructed in the same manner as the canned messages in the ROM of the talker program.

The program *modethree* sends a canned address message to each of Fh listeners on a party-line loop using serial data mode 3. All canned messages are transmitted with bit 10 set to 1; all received data from the addressed listener has bit 10 set to 0. SM2 is set in all listeners and reset in the talker.

```
            org 0000h
modethree:  mov scon,#0dah      ;set mode 3, REN, TB8 and TI,
                                ;clear SM2
            mov th1,#0ffh       ;set TH1 for 83,333d overflow rate
            orl pcon,#80h       ;set SMOD
            mov tmod,#20h       ;set timer 1 to auto-reload mode
            mov tcon,#40h       ;start T1 to generate baud clock
            mov dptr,#add1      ;send first listener message
talker:     acall sendit
            acall getit         ;"getit" is a data reception
                                ;routine
            mov dptr,#add2      ;send second listener message
            acall sendit
            acall getit         ;continue until all data is gathered
            ; .....
            sjmp talker
getit:      ret                 ;dummy routine for this example
;
;the subroutine "sendit" will transmit characters starting at the
;address passed in DPTR until a $ character is found
;
sendit:     clr a               ;zero offset for MOVC
here:       jnb scon.1,here     ;wait for transmitter not busy
            movc a,@a+dptr      ;get character of message
            mov sbuf,a          ;send character
            cjne a,#'$',out     ;if a $ then return to calling
                                ;program
            ret
out:        inc dptr            ;point to next character
            sjmp sendit         ;continue until done
;
;the canned address messages are assembled in ROM next
;
add1:       db "01$"            ;address message for listener 1
add2:       db "02$"            ;address message for listener 2
; ............                  ;continue for all listeners
add15:      db "0f$"
            end
```

```
;the program "listener" recognizes its address and responds with
;10 data characters; the data message is built in RAM, and ends
;with a $ character; for this example, the data is gotten by reading
;port 1 ten times and storing the data; this is the program for
;listener 01
;
listener:   org 0000h
            mov scon,#0f2h     ;set mode3, SM2, REN, TI; clear TB8,
                               ;RI, RB8
            mov th1,#0ffh      ;set TH1 for 83,333d overflow rate
            orl pcon,#80h      ;set SMOD to double baud rate
            mov tmod,#20h      ;set timer 1 to auto-reload mode
            mov tcon,#40h      ;start T1 to generate baud clock
who:        jnb ri,who         ;look for the first address
                               ;character
            clr ri             ;first character, clear receive flag
            mov a,sbuf         ;get character
            cjne a,#'0',no     ;compare against expected address
nxt:        jnb ri,nxt         ;first character correct, get second
            clr ri             ;second character, clear receive
                               ;flag
            mov a,sbuf         ;check next character
            cjne a,#'1',no
ok:         jnb ri,ok          ;wait for $ and then send data
            clr ri
            mov a,sbuf
            cjne a,#'$',no     ;if not $ then reset
            sjmp sendata
no:         jnb ri,no          ;wait for $ and then loop
            clr ri
            mov a,sbuf         ;get character
            cjne a,#'$',no     ;loop until $ found
            sjmp who           ;loop until proper address sent
sendata:    mov r0,#50h        ;build the message in RAM starting
                               ;at 50h
            mov r1,#0ah        ;set R1 to count data bytes from
                               ;port 1
indat:      mov @r0,p1         ;get data from port 1 to RAM
            inc r0             ;point to next RAM location
            djnz r1,indat      ;continue until 10d bytes are stored
            mov @r0,#'$'       ;finish data string with a
                               ;$ character
            mov r0,#50h        ;reset R0 to point to start of
                               ;message
sndat:      jnb ti,sndat       ;wait for transmitter empty
            mov a,@r0          ;get character from message
            mov sbuf,a         ;load SBUF for transmission
```

```
clr ti                ;clear ti
inc r0                ;point to next character
cjne a,#'$',sndat     ;look for $ then stop transmission
sjmp who              ;loop for next cycle
end
```

◆ COMMENT ◆

- The inclusion of the $ character in each message is useful both as a check for the end of a message and to reset a listener that somehow misses one of the three characters expected in an address. If a listener misses a character, as a result of noise for example, it will get to the no label within one or two characters. The next $ will reset the listener program back to the who label.
- Programs that interchange data must be written to eliminate any chance of a receiving unit getting caught in a trap waiting for a predetermined number of characters. Common schemes that accomplish this goal use special *end-of-message* characters, as in the case of modethree, or set timers to interrupt the receiving program if the data is not received within a certain period of time.
- Much more elaborate protocols than those used here in this example would be used by the listeners when sending data to the talker. There is always the possibility that errors will occur as a result of noise or interference from the improper operation of another listener. The talker may store these errors. Error-checking bytes may be added to the data stream so that the talker can verify that the string of characters is error-free.

11.3 Summary

Four serial data communication modes for the 8051 are covered in this chapter:

- Mode 0: High-speed, 8-bit shift register; one baud rate of f/12
- Mode 1: Standard 8-bit UART; variable baud rate using timer 1 overflows
- Mode 2: Multiprocessor 9-bit UART; two baud rates of f/32 and f/64
- Mode 3: Multiprocessor 9-bit UART; variable baud rate using timer 1 overflows

Programs in this chapter use these modes and feature several standard communication techniques:

- High-speed shift register data gathering
- Interrupt-driven transmit and receive buffers
- Sending preprogrammed, or canned, messages

11.4 Questions and Problems

1. Explain why mode 0 is not suitable for 8051 communications.
2. How much clock skew, in terms of clock period, can transmitted data using mode 0 have before data is shifted in error?
3. Repeat Problem 2 for data reception.
4. Assume you are determined to use mode 0 as a communication mode from one 8051 to another. Outline a system of hardware and software that would allow this. (Hint: A "buffer" is needed.)
5. Sketch the mode 1 no-parity ASCII serial characters U, 0, and w.
6. Many communication terminals can determine the baud rate of standard (mode 1) characters by making measurements on the first few *fill* characters received. Outline a program strategy that would set the 8051 baud rate automatically based on the first character received.
7. Character transmission can be done by using a time delay greater than the character time before moving a new byte to SBUF. Explain why character reception must use an Interrupt flag if all characters are to be received.
8. ASCII characters can have even (number of 1s), odd, or no parity using bit 7 as a parity bit. Write a program that checks the incoming data for odd parity and sets a flag if the parity is incorrect.
9. Write a program that converts odd parity bytes to even parity bytes (bit 7 is the parity bit).
10. An overrun is said to occur in data reception whenever a new byte of data is received before the previously received byte has been read. Discuss two methods by which overruns might be detected by the 8051 program.
11. List two reasons why stop bits are used in asynchronous communications.
12. A framing error is said to have occurred if the stop bit is not a logic high. What mode(s) can detect a framing error?
13. Why is it necessary for the main program (see *modeone*) to set the TI bit to begin the transmission of a string of characters using interrupt-driven routines? Name another way for the main program to initiate transmission.
14. Determine if an 8051 in mode 1 can communicate with an 8051 in mode 3.
15. Modify the modeone program to use 4K buffers.

A

8051 Operational Code Mnemonics

A.0 Introduction

Appendix A lists two arrangements of mnemonics for the 8051: by function, and alphabetically. The mnemonic definitions here differ from those of the original manufacturer (Intel Corporation) in the names used for addresses or data; for example, add is used to represent an address in internal RAM, whereas Intel uses the name direct. The author believes that the names used here are clearer than those used by Intel. This appendix also includes an alphabetical listing of the mnemonics using Intel names. There is **no** difference between the mnemonics when real numbers replace the names. For example: MOV add,#n and MOV direct,#data become MOV 10h,#40h when the number 10h replaces the internal RAM address (add/direct) and 40h replaces the number (#n/# data).

A.1 Mnemonics, Arranged by Function

Arithmetic

Mnemonic	Description	Bytes	Cycles	Flags
ADD A,Rr	A+Rr → A	1	1	C OV AC
ADD A,add	A+(add) → A	2	1	C OV AC
ADD A,@Rp	A+(Rp) → A	1	1	C OV AC
ADD A,#n	A+n → A	2	1	C OV AC
ADDC A,Rr	A+Rr+C → A	1	1	C OV AC
ADDC A,add	A+(add)+C → A	2	1	C OV AC
ADDC A,@Rp	A+(Rp)+C → A	1	1	C OV AC
ADDC A,#n	A+n+C → A	2	1	C OV AC
DA A	Abin → Adec	1	1	C
DEC A	A−1 → A	1	1	
DEC Rr	Rr−1 → Rr	1	1	
DEC add	(add)−1 → (add)	2	1	
DEC @Rp	(Rp)−1 → (Rp)	1	1	
DIV AB	A/B → AB	1	4	0 OV
INC A	A+1 → A	1	1	
INC Rr	Rr+1 → Rr	1	1	
INC add	(add)+1 → (add)	2	1	
INC @Rp	(Rp)+1 → (Rp)	1	1	
INC DPTR	DPTR+1 → DPTR	1	2	
MUL AB	A×B → AB	1	4	0 OV
SUBB A,Rr	A−Rr−C → A	1	1	C OV AC
SUBB A,add	A−(add)−C → A	2	1	C OV AC
SUBB A,@Rp	A−(Rp)−C → A	1	1	C OV AC
SUBB A,#n	A−n−C → A	2	1	C OV AC

Logic

Mnemonic	Description	Bytes	Cycles	Flags
ANL A,Rr	A AND Rr → A	1	1	
ANL A,add	A AND (add) → A	2	1	
ANL A,@Rp	A AND (Rp) → A	1	1	
ANL A,#n	A AND n → A	2	1	
ANL add,A	(add) AND A → (add)	2	1	
ANL add,#n	(add) AND n → (add)	3	2	
CLR A	00 → A	1	1	
CPL A	A → A	1	1	
ORL A,Rr	A OR Rr → A	1	1	
ORL A,add	A OR (add) → A	2	1	
ORL A,@Rp	A OR (Rp) → A	1	1	
ORL A,#n	A OR n → A	2	1	
ORL add,A	(add) OR A → (add)	2	1	
ORL add,#n	(add) OR n → (add)	3	2	
XRL A,Rr	A XOR Rr → A	1	1	
XRL A,add	A XOR (add) → A	2	1	
XRL A,@Rp	A XOR (Rp) → A	1	1	
XRL A,#n	A XOR n → A	2	1	
XRL add,A	(add) XOR A → (add)	2	1	
XRL add,#n	(add) XOR n → (add)	3	2	
NOP	PC+1 → PC	1	1	
RL A	A0←A7←A6..←A1←A0	1	1	
RLC A	C←A7←A6..←A0←C	1	1	C
RR A	A0→A7→A6..→A1→A0	1	1	
RRC A	C→A7→A6..→A0→C	1	1	C
SWAP A	Alsn ↔ Amsn	1	1	

Data Moves

Mnemonic	Description	Bytes	Cycles	Flags
MOV A,Rr	Rr → A	1	1	
MOV A,add	(add) → A	2	1	
MOV A,@Rp	(Rp) → A	1	1	
MOV A,#n	n → A	2	1	
MOV Rr,A	A → Rr	1	1	
MOV Rr,add	(add) → Rr	2	2	
MOV Rr,#n	n → Rr	2	1	
MOV add,A	A → (add)	2	1	
MOV add,Rr	Rr → (add)	2	2	
MOV add1,add2	(add2) → (add1)	3	2	
MOV add,@Rp	(Rp) → (add)	2	2	
MOV add,#n	n → (add)	3	2	

Mnemonic	Description	Bytes	Cycles	Flags
MOV @Rp,A	A → (Rp)	1	1	
MOV @Rp,add	(add) → (Rp)	2	2	
MOV @Rp,#n	n → (Rp)	2	1	
MOV DPTR,#nn	nn → DPTR	3	2	
MOVC A,@A+DPTR	(A+DPTR) → A	1	2	
MOVC A,@A+PC	(A+PC) → A	1	2	
MOVX A,@DPTR	(DPTR)^ → A	1	2	
MOVX A,@Rp	(Rp)^ → A	1	2	
MOVX @Rp,A	A → (Rp)^	1	2	
MOVX @DPTR,A	A → (DPTR)^	1	2	
POP add	(SP) → (add)	2	2	
PUSH add	(add) → (SP)	2	2	
XCH A,Rr	A ↔ Rr	1	1	
XCH A,add	A ↔ (add)	2	1	
XCH A,@Rp	A ↔ (Rp)	1	1	
XCHD A,@Rp	Alsn ↔ (Rp)lsn	1	1	

Calls and Jumps

Mnemonic	Description	Bytes	Cycles	Flags
ACALL sadd	PC+2 → (SP); sadd → PC	2	2	
CJNE A,add,radd	[A<>(add)]: PC+3+radd → PC	3	2	C
CJNE A,#n,radd	[A<>n]: PC+3+radd → PC	3	2	C
CJNE Rr,#n,radd	[Rr<>n]: PC+3+radd → PC	3	2	C
CJNE @Rp,#n,radd	[(Rp)<>n]: PC+3+radd → PC	3	2	C
DJNZ Rr,radd	[Rr−1<>00]: PC+2+radd → PC	2	2	
DJNZ add,radd	[(add)−1<>00]: PC+3+radd → PC	3	2	
LCALL ladd	PC+3 → (SP); ladd → PC	3	2	
AJMP sadd	sadd → PC	2	2	
LJMP ladd	ladd → PC	3	2	
SJMP radd	PC+2+radd → PC	2	2	
JMP @A+DPTR	DPTR+A → PC	1	2	
JC radd	[C=1]: PC+2+radd → PC	2	2	
JNC radd	[C=0]: PC+2+radd → PC	2	2	
JB b,radd	[b=1]: PC+3+radd → PC	3	2	
JNB b,radd	[b=0]: PC+3+radd → PC	3	2	
JBC b,radd	[b=1]: PC+3+radd → PC; 0 → b	3	2	
JZ radd	[A=00]: PC+2+radd → PC	2	2	
JNZ radd	[A>00]: PC+2+radd → PC	2	2	
RET	(SP) → PC	1	2	
RETI	(SP) → PC; EI	1	2	

Boolean

Mnemonic	Description	Bytes	Cycles	Flags
ANL C,b	C AND b → C	2	2	C
ANL C,$\overline{b}$	C AND $\overline{b}$ → C	2	2	C
CLR C	0 → C	1	1	C=0
CLR b	0 → b	2	1	
CPL C	$\overline{C}$ → C	1	1	C
CPL b	$\overline{b}$ → b	2	1	
ORL C,b	C OR b → C	2	2	C
ORL C,$\overline{b}$	C OR $\overline{b}$ → C	2	2	C
MOV C,b	b → C	2	1	C
MOV b,C	C → b	2	2	
SETB C	1 → C	1	1	C=1
SETB b	1 → b	2	1	

A.2 Mnemonics, Arranged Alphabetically

Mnemonic	Description	Bytes	Cycles	Flags
ACALL sadd	PC+2 → (SP); sadd → PC	2	2	
ADD A,add	A+(add) → A	2	1	C OV AC
ADD A,@Rp	A+(Rp) → A	1	1	C OV AC
ADD A,#n	A+n → A	2	1	C OV AC
ADD A,Rr	A+Rr → A	1	1	C OV AC
ADDC A,add	A+(add)+C → A	2	1	C OV AC
ADDC A,@Rp	A+(Rp)+C → A	1	1	C OV AC
ADDC A,#n	A+n+C → A	2	1	C OV AC
ADDC A,Rr	A+Rr+C → A	1	1	C OV AC
AJMP sadd	sadd → PC	2	2	
ANL A,add	A AND (add) → A	2	1	
ANL A,@Rp	A AND (Rp) → A	1	1	
ANL A,#n	A AND n → A	2	1	
ANL A,Rr	A AND Rr → A	1	1	
ANL add,A	(add) AND A → (add)	2	1	
ANL add,#n	(add) AND n → (add)	3	2	
ANL C,b	C AND b → C	2	2	C
ANL C,$\overline{b}$	C AND $\overline{b}$ → C	2	2	C
CJNE A,add,radd	[A<>(add)]: PC+3+radd → PC	3	2	C
CJNE A,#n,radd	[A<>n]: PC+3+radd → PC	3	2	C
CJNE @Rp,#n,radd	[(Rp)<>n]: PC+3+radd → PC	3	2	C
CJNE Rr,#n,radd	[Rr<>n]: PC+3+radd → PC	3	2	C
CLR A	0 → A	1	1	
CLR b	0 → b	2	1	
CLR C	0 → C	1	1	C=0
CPL A	$\overline{A}$ → A	1	1	

Mnemonic	Description	Bytes	Cycles	Flags
CPL b	$\overline{b}$ → b	2	1	
CPL C	$\overline{C}$ → C	1	1	C
DA A	Abin → Adec	1	1	C
DEC A	A−1 → A	1	1	
DEC add	(add)−1 → (add)	2	1	
DEC @Rp	(Rp)−1 → (Rp)	1	1	
DEC Rr	Rr−1 → Rr	1	1	
DIV AB	A/B → AB	1	4	0 OV
DJNZ add,radd	[(add)−1<>00]: PC+3+radd → PC	3	2	
DJNZ Rr,radd	[Rr−1<>00]: PC+2+radd → PC	2	2	
INC A	A+1 → A	1	1	
INC add	(add)+1 → (add)	2	1	
INC DPTR	DPTR+1 → DPTR	1	2	
INC @Rp	(Rp)+1 → (Rp)	1	1	
INC Rr	Rr+1 → Rr	1	1	
JB b,radd	[b=1]: PC+3+radd → PC	3	2	
JBC b,radd	[b=1]: PC+3+radd → PC; 0 → b	3	2	
JC radd	[C=1]: PC+2+radd → PC	2	2	
JMP @A+DPTR	DPTR+A → PC	1	2	
JNB b,radd	[b=0]: PC+3+radd → PC	3	2	
JNC radd	[C=0]: PC+2+radd → PC	2	2	
JNZ radd	[A>00]: PC+2+radd → PC	2	2	
JZ radd	[A=00]: PC+2+radd → PC	2	2	
LCALL ladd	PC+3 → (SP); ladd → PC	3	2	
LJMP ladd	ladd → PC	3	2	
MOV A,add	(add) → A	2	1	
MOV A,@Rp	(Rp) → A	1	1	
MOV A,#n	n → A	2	1	
MOV A,Rr	Rr → A	1	1	
MOV add,A	A → (add)	2	1	
MOV add1,add2	(add2) → (add1)	3	2	
MOV add,@Rp	(Rp) → (add)	2	2	
MOV add,#n	n → (add)	3	2	
MOV add,Rr	Rr → (add)	2	2	
MOV b,C	C → b	2	2	
MOV C,b	b → C	2	1	C
MOV @Rp,A	A → (Rp)	1	1	
MOV @Rp,add	(add) → (Rp)	2	2	
MOV @Rp,#n	n → (Rp)	2	1	
MOV DPTR,#nn	nn → DPTR	3	2	
MOV Rr,A	A → Rr	1	1	
MOV Rr,add	(add) → Rr	2	2	
MOV Rr,#n	n → Rr	2	1	
MOVC A,@A+DPTR	(A+DPTR) → A	1	2	
MOVC A,@A+PC	(A+PC) → A	1	2	
MOVX A,@DPTR	(DPTR)^ → A	1	2	

Mnemonic	Description	Bytes	Cycles	Flags
MOVX A,@Rp	(Rp)^ → A	1	2	
MOVX @DPTR,A	A → (DPTR)^	1	2	
MOVX @Rp,A	A → (Rp)^	1	2	
MUL AB	A×B → AB	1	4	0 OV
NOP	PC+1 → PC	1	1	
ORL A,add	A OR (add) → A	2	1	
ORL A,@Rp	A OR (Rp) → A	1	1	
ORL A,#n	A OR n → A	2	1	
ORL A,Rr	A OR Rr → A	1	1	
ORL add,A	(add) OR A → (add)	2	1	
ORL add,#n	(add) OR n → (add)	3	2	
ORL C,b	C OR b → C	2	2	C
ORL C,$\overline{b}$	C OR $\overline{b}$ → C	2	2	C
POP add	(SP) → (add)	2	2	
PUSH add	(add) → (SP)	2	2	
RET	(SP) → PC	1	2	
RETI	(SP) → PC; EI	1	2	
RL A	A0←A7←A6..←A1←A0	1	1	
RLC A	C←A7←A6..←A0←C	1	1	C
RR A	A0→A7→A6..→A1→A0	1	1	
RRC A	C→A7→A6..→A0→C	1	1	C
SETB b	1 → b	2	1	
SETB C	1 → C	1	1	C=1
SJMP radd	PC+2+radd → PC	2	2	
SUBB A,add	A−(add)−C → A	2	1	C OV AC
SUBB A,@Rp	A−(Rp)−C → A	1	1	C OV AC
SUBB A,#n	A−n−C → A	2	1	C OV AC
SUBB A,Rr	A−Rr−C → A	1	1	C OV AC
SWAP A	Alsn ↔ Amsn	1	1	
XCH A,add	A ↔ (add)	2	1	
XCH A,@Rp	A ↔ (Rp)	1	1	
XCH A,Rr	A ↔ Rr	1	1	
XCHD A,@Rp	Alsn ↔ (Rp)lsn	1	1	
XRL A,add	A XOR (add) → A	2	1	
XRL A,@Rp	A XOR (Rp) → A	1	1	
XRL A,#n	A XOR n → A	2	1	
XRL A,Rr	A XOR Rr → A	1	1	
XRL add,A	(add) XOR A → (add)	2	1	
XRL add,#n	(add) XOR n → (add)	3	2	

Mnemonic Acronyms

add	Address of the internal RAM from 00h to FFh.
ladd	Long address of 16 bits from 0000h to FFFFh.
radd	Relative address, a signed number from −128d to +127d.
sadd	Short address of 11 bits; complete address = PC11–PC15 and sadd.

b	Addressable bit in internal RAM or a SFR.
C	The Carry flag.
lsn	Least significant nibble.
msn	Most significant nibble.
n	Any immediate 8-bit number from 00h to FFh.
Rr	Any of the eight registers, R0 to R7, in the selected bank.
Rp	Either of the pointing registers R0 or R1 in the selected bank.
[ ]:	IF the condition inside the brackets is *true,* THEN the action listed will occur; ELSE go to the next instruction.
^	External memory location.
( )	Contents of the location inside the parentheses.

Note that flags affected by each instruction are shown where appropriate; any operations that affect the PSW address may also affect the flags.

A.3 Intel Corporation Mnemonics, Arranged Alphabetically

ACALL addr11	PC+2 → (SP); addr11 → PC	2	2	
ADD A,direct	A+(direct) → A	2	1	C OV AC
ADD A,@Ri	A+(Ri) → A	1	1	C OV AC
ADD A,#data	A+#data → A	2	1	C OV AC
ADD A,Rn	A+Rn → A	1	1	C OV AC
ADDC A,direct	A+(direct)+C → A	2	1	C OV AC
ADDC A,@Ri	A+(Ri)+C → A	1	1	C OV AC
ADDC A,#data	A+#data+C → A	2	1	C OV AC
ADDC A,Rn	A+Rn+C → A	1	1	C OV AC
AJMP addr11	addr11 → PC	2	2	
ANL A,direct	A AND (direct) → A	2	1	
ANL A,@Ri	A AND (Ri) → A	1	1	
ANL A,#data	A AND #data → A	2	1	
ANL A,Rn	A AND Rn → A	1	1	
ANL direct,A	(direct) AND A → (direct)	2	1	
ANL direct,#data	(direct) AND #data → (direct)	3	2	
ANL C,bit	C AND bit → C	2	2	C
ANL C,$\overline{bit}$	C AND $\overline{bit}$ → C	2	2	C
CJNE A,direct,rel	[A<>(direct)]: PC+3+rel → PC	3	2	C
CJNE A,#data,rel	[A<>n]: PC+3+rel → PC	3	2	C
CJNE @Ri,#data,rel	[(Ri)<>n]: PC+3+rel → PC	3	2	C
CJNE Rn,#data,rel	[Rn<>n]: PC+3+rel → PC	3	2	C
CLR A	0 → A	1	1	
CLR bit	0 → bit	2	1	
CLR C	0 → C	1	1	C=0
CPL A	$\overline{A}$ → A	1	1	
CPL bit	$\overline{bit}$ → bit	2	1	
CPL C	$\overline{C}$ → C	1	1	C
DA A	Abin → Adec	1	1	C

DEC A	A−1 → A	1	1	
DEC direct	(direct)−1 → (direct)	2	1	
DEC @Ri	(Ri)−1 → (Ri)	1	1	
DEC Rn	Rn−1 → Rn	1	1	
DIV AB	A/B → AB	1	4	0 OV
DJNZ direct,rel	[(direct)−1<>00]: PC+3+rel → PC	3	2	
DJNZ Rn,rel	[Rn−1<>00]: PC+2+rel → PC	2	2	
INC A	A+1 → A	1	1	
INC direct	(direct)+1 → (direct)	2	1	
INC DPTR	DPTR+1 → DPTR	1	2	
INC @Ri	(Ri)+1 → (Ri)	1	1	
INC Rn	Rn+1 → Rn	1	1	
JB bit,rel	[b=1]: PC+3+rel → PC	3	2	
JBC bit,rel	[b=1]: PC+3+rel → PC; 0 → bit	3	2	
JC rel	[C=1]: PC+2+rel → PC	2	2	
JMP @A+DPTR	DPTR+A → PC	1	2	
JNB bit,rel	[b=0]: PC+3+rel → PC	3	2	
JNC rel	[C=0]: PC+2+rel → PC	2	2	
JNZ rel	[A>00]: PC+2+rel → PC	2	2	
JZ rel	[A=00]: PC+2+rel → PC	2	2	
LCALL addr16	PC+3 → (SP); addr16 → PC	3	2	
LJMP addr16	addr16 → PC	3	2	
MOV A,direct	(direct) → A	2	1	
MOV A,@Ri	(Ri) → A	1	1	
MOV A,#data	#data → A	2	1	
MOV A,Rn	Rn → A	1	1	
MOV direct,A	A → (direct)	2	1	
MOV direct,direct	(direct) → (direct)	3	2	
MOV direct,@Ri	(Ri) → (direct)	2	2	
MOV direct,#data	#data → (direct)	3	2	
MOV direct,Rn	Rn → (direct)	2	2	
MOV bit,C	C → bit	2	2	
MOV C,bit	bit → C	2	1	C
MOV @Ri,A	A → (Ri)	1	1	
MOV @Ri,direct	(direct) → (Ri)	2	2	
MOV @Ri,#data	#data → (Ri)	2	1	
MOV DPTR,#data16	#data16 → DPTR	3	2	
MOV Rn,A	A → Rn	1	1	
MOV Rn,direct	(direct) → Rn	2	2	
MOV Rn,#data	#data → Rn	2	1	
MOVC A,@A+DPTR	(A+DPTR) → A	1	2	
MOVC A,@A+PC	(A+PC) → A	1	2	
MOVX A,@DPTR	(DPTR)^ → A	1	2	
MOVX A,@Ri	(Ri)^ → A	1	2	
MOVX @DPTR,A	A → (DPTR)^	1	2	
MOVX @Ri,A	A → (Ri)^	1	2	
MUL AB	A×B → AB	1	4	0 OV

NOP	PC+1 → PC	1	1	
ORL A,direct	A OR (direct) → A	2	1	
ORL A,@Ri	A OR (Ri) → A	1	1	
ORL A,#data	A OR #data → A	2	1	
ORL A,Rn	A OR Rn → A	1	1	
ORL direct,A	(direct) OR A → (direct)	2	1	
ORL direct,#data	(direct) OR #data → (direct)	3	2	
ORL C,bit	C OR bit → C	2	2	C
ORL C,$\overline{\text{bit}}$	C OR $\overline{\text{bit}}$ → C	2	2	C
POP direct	(SP) → (direct)	2	2	
PUSH direct	(direct) → (SP)	2	2	
RET	(SP) → PC	1	2	
RETI	(SP) → PC; EI	1	2	
RL A	A0←A7←A6..←A1←A0	1	1	
RLC A	C←A7←A6..←A0←C	1	1	C
RR A	A0→A7→A6..→A1→A0	1	1	
RRC A	C→A7→A6..→A0→C	1	1	C
SETB bit	1 → bit	2	1	
SETB C	1 → C	1	1	C=1
SJMP rel	PC+2+rel → PC	2	2	
SUBB A,direct	A−(direct)−C → A	2	1	C OV AC
SUBB A,@Ri	A−(Ri)−C → A	1	1	C OV AC
SUBB A,#data	A−#data−C → A	2	1	C OV AC
SUBB A,Rn	A−Rn−C → A	1	1	C OV AC
SWAP A	Alsn ↔ Amsn	1	1	
XCH A,direct	A ↔ (direct)	2	1	
XCH A,@Ri	A ↔ (Ri)	1	1	
XCH A,Rn	A ↔ Rn	1	1	
XCHD A,@Ri	Alsn ↔ (Ri)lsn	1	1	
XRL A,direct	A XOR (direct) → A	2	1	
XRL A,@Ri	A XOR (Ri) → A	1	1	
XRL A,#data	A XOR #data → A	2	1	
XRL A,Rn	A XOR Rn → A	1	1	
XRL direct,A	(direct) XOR A → (direct)	2	1	
XRL direct,#data	(direct) XOR #data → (direct)	3	2	

Acronyms

addr11	Page address of 11 bits, which is in the same 2K page as the address of the following instruction.
addr16	Address for any location in the 64K memory space.
bit	The address of a bit in the internal RAM bit address area or a bit in an SFR.
C	The Carry flag.
#data	An 8-bit binary number from 00 to FFh.
#data16	A 16-bit binary number from 0000 to FFFFh.
direct	An internal RAM address or an SFR byte address.
lsn	Least significant nibble.

msn	Most significant nibble.
rel	Number that is added to the address of the next instruction to form an address +127d or −128d from the address of the next instruction.
Rn	Any of registers R0 to R7 of the current register bank.
@Ri	Indirect address using the contents of R0 or R1.
[ ]:	IF the condition inside the brackets is *true*, THEN the action listed will occur; ELSE go to the next instruction.
^	EXTERNAL memory location.
( )	Contents of the location inside the parentheses.

Note that flags affected by each instruction are shown where appropriate; any operations that affect the PSW address may also affect the flags.

A.4 8051 Instruction Hexadecimal Codes

Each 8051 program instruction is converted into a binary machine code when the instruction is assembled. In the bad old days, program assembly was done manually by looking up and constructing each code by hand. Manual assembly is tedious, error-prone, and slow. Today we have assembler programs such as the one supplied with this book, A51, that assemble a program text file into a machine code binary file in seconds.

There may be those rare instances, however, when the programmer may wish to check the contents of an assembler or EPROM file that contains the hexadecimal operational codes (opcodes) of a program. The table that follows lists each hex opcode, in numerical order, and the corresponding instruction for that opcode. Note that the instruction mnemonics ACALL and AJMP have eight opcodes, one for each possible short-address upper 3-bit combination.

Operational Code	Mnemonic	Operand(s)	Bytes
00	NOP		1
01	AJMP	sadd	2
02	LJMP	ladd	3
03	RR	A	1
04	INC	A	1
05	INC	add	2
06	INC	@R0	1
07	INC	@R1	1
08	INC	R0	1
09	INC	R1	1
0A	INC	R2	1
0B	INC	R3	1
0C	INC	R4	1
0D	INC	R5	1
0E	INC	R6	1
0F	INC	R7	1
10	JBC	b,radd	3
11	ACALL	sadd	2
12	LCALL	ladd	3
13	RRC	A	1
14	DEC	A	1
15	DEC	add	1
16	DEC	@R0	1
17	DEC	@R1	1
18	DEC	R0	1
19	DEC	R1	1

Operational Code	Mnemonic	Operand(s)	Bytes	Operational Code	Mnemonic	Operand(s)	Bytes
1A	DEC	R2	1	47	ORL	A,@R1	1
1B	DEC	R3	1	48	ORL	A,R0	1
1C	DEC	R4	1	49	ORL	A,R1	1
1D	DEC	R5	1	4A	ORL	A,R2	1
1E	DEC	R6	1	4B	ORL	A,R3	1
1F	DEC	R7	1	4C	ORL	A,R4	1
20	JB	b,radd	3	4D	ORL	A,R5	1
21	AJMP	sadd	2	4E	ORL	A,R6	1
22	RET		1	4F	ORL	A,R7	1
23	RL	A	1	50	JNC	radd	2
24	ADD	A,#n	2	51	ACALL	sadd	2
25	ADD	A,add	2	52	ANL	add,A	2
26	ADD	A,@R0	1	53	ANL	add,#n	3
27	ADD	A,@R1	1	54	ANL	A,#n	2
28	ADD	A,R0	1	55	ANL	A,add	2
29	ADD	A,R1	1	56	ANL	A,@R0	1
2A	ADD	A,R2	1	57	ANL	A,@R1	1
2B	ADD	A,R3	1	58	ANL	A,R0	1
2C	ADD	A,R4	1	59	ANL	A,R1	1
2D	ADD	A,R5	1	5A	ANL	A,R2	1
2E	ADD	A,R6	1	5B	ANL	A,R3	1
2F	ADD	A,R7	1	5C	ANL	A,R4	1
30	JNB	b,radd	3	5D	ANL	A,R5	1
31	ACALL	sadd	2	5E	ANL	A,R6	1
32	RETI		1	5F	ANL	A,R7	1
33	RLC	A	1	60	JZ	radd	2
34	ADDC	A,#n	2	61	AJMP	sadd	2
35	ADDC	A,add	2	62	XRL	add,A	2
36	ADDC	@R0	1	63	XRL	add,#n	3
37	ADDC	@R1	1	64	XRL	A,#n	2
38	ADDC	A,R0	1	65	XRL	A,add	2
39	ADDC	A,R1	1	66	XRL	A,@R0	1
3A	ADDC	A,R2	1	67	XRL	A,@R1	1
3B	ADDC	A,R3	1	68	XRL	A,R0	1
3C	ADDC	A,R4	1	69	XRL	A,R1	1
3D	ADDC	A,R5	1	6A	XRL	A,R2	1
3E	ADDC	A,R6	1	6B	XRL	A,R3	1
3F	ADDC	A,R7	1	6C	XRL	A,R4	1
40	JC	radd	2	6D	XRL	A,R5	1
41	AJMP	sadd	2	6E	XRL	A,R6	1
42	ORL	add,A	2	6F	XRL	A,R7	1
43	ORL	add,#n	3	70	JNZ	radd	2
44	ORL	A,#n	2	71	ACALL	sadd	2
45	ORL	A,add	2	72	ORL	C,b	2
46	ORL	A,@R0	1	73	JMP	@A + DPTR	1

Operational Code	Mnemonic	Operand(s)	Bytes
74	MOV	A,#n	2
75	MOV	add,#n	3
76	MOV	@R0,#n	2
77	MOV	@R1,#n	2
78	MOV	R0,#n	2
79	MOV	R1,#n	2
7A	MOV	R2,#n	2
7B	MOV	R3,#n	2
7C	MOV	R4,#n	2
7D	MOV	R5,#n	2
7E	MOV	R6,#n	2
7F	MOV	R7,#n	2
80	SJMP	radd	2
81	AJMP	sadd	2
82	ANL	C,b	2
83	MOVC	A,@A + PC	1
84	DIV	AB	1
85	MOV	add,add	3
86	MOV	add,@R0	2
87	MOV	add,@R1	2
88	MOV	add,R0	2
89	MOV	add,R1	2
8A	MOV	add,R2	2
8B	MOV	add,R3	2
8C	MOV	add,R4	2
8D	MOV	add,R5	2
8E	MOV	add,R6	2
8F	MOV	add,R7	2
90	MOV	DPTR,#nn	3
91	ACALL	sadd	2
92	MOV	b,C	2
93	MOVC	A,@A + DPTR	1
94	SUBB	A,#n	2
95	SUBB	A,add	2
96	SUBB	A,@R0	1
97	SUBB	A,@R1	1
98	SUBB	A,R0	1
99	SUBB	A,R1	1
9A	SUBB	A,R2	1
9B	SUBB	A,R3	1
9C	SUBB	A,R4	1
9D	SUBB	A,R5	1
9E	SUBB	A,R6	1
9F	SUBB	A,R7	1
A0	ORL	C,b	2

Operational Code	Mnemonic	Operand(s)	Bytes
A1	AJMP	sadd	2
A2	MOV	C,b	2
A3	INC	DPTR	1
A4	MUL	AB	1
A5	unused		
A6	MOV	@R0,add	2
A7	MOV	@R1,add	2
A8	MOV	R0,add	2
A9	MOV	R1,add	2
AA	MOV	R2,add	2
AB	MOV	R3,add	2
AC	MOV	R4,add	2
AD	MOV	R5,add	2
AE	MOV	R6,add	2
AF	MOV	R7,add	2
B0	ANL	C,b	2
B1	ACALL	sadd	2
B2	CPL	b	2
B3	CPL	C	1
B4	CJNE	A,#n,radd	3
B5	CJNE	A,add,radd	3
B6	CJNE	@R0,#n,radd	3
B7	CJNE	@R1,#n,radd	3
B8	CJNE	R0,#n,radd	3
B9	CJNE	R1,#n,radd	3
BA	CJNE	R2,#n,radd	3
BB	CJNE	R3,#n,radd	3
BC	CJNE	R4,#n,radd	3
BD	CJNE	R5,#n,radd	3
BE	CJNE	R6,#n,radd	3
BF	CJNE	R7,#n,radd	3
C0	PUSH	add	2
C1	AJMP	sadd	2
C2	CLR	b	2
C3	CLR	C	1
C4	SWAP	A	1
C5	XCH	A,add	2
C6	XCH	A,@R0	1
C7	XCH	A,@R1	1
C8	XCH	A,R0	1
C9	XCH	A,R1	1
CA	XCH	A,R2	1
CB	XCH	A,R3	1
CC	XCH	A,R4	1
CD	XCH	A,R5	1

Operational Code	Mnemonic	Operand(s)	Bytes	Operational Code	Mnemonic	Operand(s)	Bytes
CE	XCH	A,R6	1	E7	MOV	A,@r1	1
CF	XCH	A,R7	1	E8	MOV	A,R0	1
D0	POP	add	2	E9	MOV	A,R1	1
D1	ACALL	sadd	2	EA	MOV	A,R2	1
D2	SETB	b	2	EB	MOV	A,R3	1
D3	SETB	C	1	EC	MOV	A,R4	1
D4	DA	A	1	ED	MOV	A,R5	1
D5	DJNZ	add,radd	3	EE	MOV	A,R6	1
D6	XCHD	A,@R0	1	EF	MOV	A,R7	1
D7	XCHD	A,@R1	1	F0	MOVX	@DPTR,A	1
D8	DJNZ	R0,radd	2	F1	ACALL	sadd	2
D9	DJNZ	R1,radd	2	F2	MOVX	@R0,A	1
DA	DJNZ	R2,radd	2	F3	MOVX	@R1,A	1
DB	DJNZ	R3,radd	2	F4	CPL	A	1
DC	DJNZ	R4,radd	2	F5	MOV	add,A	2
DD	DJNZ	R5,radd	2	F6	MOV	@R0,A	1
DE	DJNZ	R6,radd	2	F7	MOV	@R1,A	1
DF	DJNZ	R7,radd	2	F8	MOV	R0,A	1
E0	MOVX	A,@DPTR	1	F9	MOV	R1,A	1
E1	AJMP	sadd	2	FA	MOV	R2,A	1
E2	MOVX	A,@R0	1	FB	MOV	R3,A	1
E3	MOVX	A,@R1	1	FC	MOV	R4,A	1
E4	CLR	A	1	FD	MOV	R5,A	1
E5	MOV	A,add	2	FE	MOV	R6,A	1
E6	MOV	A,@R0	1	FF	MOV	R7,A	1

B

How to Use the A51 Assembler

B.0 Introduction

In the early days of digital computing, (the 1940s), computers were programmed in binary, resulting in programs that appeared as:

```
1111010111010110
0001011011110010
1100010101000101
.
.
.
```

and are generally unintelligible to anyone. Early in this process, programmers became tired of typing all those 1s and 0s, so a shorthand notation for binary (hexadecimal) was adopted to shorten the typing effort:

```
92AF
F5D6
16F2
C545
.
.
.
```

using 0 – 9 for binary 0000 to 1001, and A–F for binary 1010 to 1111. The result is still unintelligible, but more compact.

Each line of code is an instruction to the computer, and the programmers composed descriptions for the instructions that could be written as:

```
Load the accumulator with a number
Move the accumulator to register 1
Move memory location 3 to location 2
```

The programmers translated these descriptions to the equivalent hex codes using pencil and paper. Very soon these long descriptions were shortened to:

```
LODE A,NUM
MOVE 1,A
MOVE 2,3
.
.
.
```

Mnemonics were born to speed up the programming process by retaining the essence of the instruction. Finally, programs became so long, and computing so inexpensive, that programs that translated the mnemonics into their equivalent hex codes were written to facilitate the programming process.

The translation programs go by many names:

- *Interpreter:* Translates each line of the program independently to an abbreviated ASCII (non-hex) equivalent
- *Assembler:* Translates the entire program, as a whole, to hex
- *Compiler:* Converts an Interpreter translation to hex

Then there are the "cross" varieties, which assemble or compile code on computer A for use on computer B. This type of assembler is included with this book: A cross assembler for the 8051, which runs on PC type computers. The assembler is a file on the programming CD included with this book named A51.EXE.

B.1 Using the Assembler

The assembler included with this book is a version that has been supplied by Keil, Inc. The intent is to supply an assembler that is easy to use, enabling the student to get to the business of writing programs with a minimum of delay.

The Big Picture

An assembler is a translator machine. Computer programs, written using a defined set of rules (the syntax), are put into the assembler and hex code pops out (if the syntax has been followed). The file must be an ASCII text file.

The assembler produces two output files:

1. A file with the same name as the input ASCII text file, which has the extension .LST, is the assembled file complete with line numbers, memory addresses, hex codes, mnemonics, and comments. Any *errors* found during assembly will be noted in the .LST file, at the point in the program where they occur.
2. A file with the same name as the input ASCII text file, which has the extension .OBJ, is the hex format file that can be loaded into the debugger and run.

Example:
A small program that blinks LEDs on an 8051 system is edited and saved as an ASCII file named *try.a51.*

```
        org 0000h
loop:   mov 90h,#0ffh   ;LEDs off
        acall time      ;delay
        mov 90h,#7fh    ;turn on LED 1
        acall time
        mov 90h,#0bfh   ;turn on LED 2
        acall time
        mov 90h,#3fh    ;both LEDs on
        acall time
        sjmp loop
time:   mov r0,#03h
in1:    mov r1,#00h
in2:    mov r2,#00h
wait:   djnz r2,wait
        djnz r1,in2
        djnz r0,in1
        ret
        end
```

The .LST file, which is produced by the assembler, has these features:

Address	Object	Line	Source		Comment
0000		1		org 0000h	
0000	7590FF	2	loop:	mov 90h,#0ffh	;LEDs off
0003	1116	3		acall time	;delay
0005	75907F	4		mov 90h,#7fh	;turn on LED 1
0008	1116	5		acall time	
000A	7590BF	6		mov 90h,#0bfh	;turn on LED 2
000D	1116	7		acall time	
000F	75903F	8		mov 90h,#3fh	;both LEDs on
0012	1116	9		acall time	
0014	80EA	10		sjmp loop	
0016	7803	11	time:	mov r0,#03h	
0018	7900	12	In1:	mov r1,#00h	
001A	7A00	13	In2:	mov r2,#00h	
001C	DAFE	14	wait:	djnz r2,wait	
001E	D9FA	15		djnz r1,In2	
0020	D8F6	16		djnz r0,In1	
0022	22	17		ret	

The .OBJ file contains the OBJ hex code from the .LST file.

The assembler will assemble your program and inform you of any errors that are found. You can type the .LST file to the computer screen or print the listing to a printer. All errors *in syntax* will be shown by the assembler in the .LST file. Keep in mind that a program that has been successfully assembled is not guaranteed to work; it is only grammatically correct. (Sentences in English can be written that are grammatically correct but make no sense; see any government form for an example). Re-edit your program until assembly is successful.

B.2 Assembler Directives

An assembler is a program and has instructions just as does any program. These are called *directives* or *pseudo operations* because they inform the assembler what to do with the mnemonics that it is to assemble. The pseudo ops are distinctly different from the mnemonics of the computer code being assembled so that they stand out in the program listing. For the A51 assembler, the most common are:

org xxxx ORiGinate the following code starting at address xxxx.

Example Program		Address	Hex
org 0400h	becomes:	0400	79
MOV r2,#00h		0401	00

The org pseudo op lets you put code and data anywhere in program memory you wish. Normally the program starts at 0000h using an org 0000h.

label equ xxxx EQUate the label name to the number xxxx

Example Program		Address	Hex
org 0000h	becomes:	0000	74
fred equ 12h		0001	12
mov a,#fred			

equ turns numbers into names; it makes the program much more readable because the name chosen for the label can have some meaning in the program, whereas the number will not.

db xx Define a Byte: place the 8-bit number xx next in memory

Example Program		Address	Hex
org 0100h	becomes:	0100	34
db 34h		0101	56
db 56h			

db "abc"

Example Program		Address	Hex
org 0200h	becomes:	0200	31
db "123"		0201	32
		0202	33

db xx takes the number xx (from 0 to 255d) and converts it to hex in the next memory location. db "abc" will convert *any* character that can be typed into the space between the quote marks into the equivalent ASCII (no parity) hex code for that character, and place multiple characters sequentially in memory. db permits the programmer to place any hex byte anywhere in memory.

dw xxxx   Define a Word: Place the 16-bit number xxxx in memory

Example Program		**Address**	**Hex**
org 0abcdh	becomes:	ABCD	12
dw 1234h		ABCE	34

dw is a 16 bit version of db.

end   The End: Tells the assembler to stop assembling

Other directives exist that are rarely used by student programmers. Refer to the assembler documentation contained in the CD Books folders.

B.3 Numbers

Numbers follow one simple rule: They must start with a number from 0 to 9. For example:

```
1234
0abcdh
0ffh
5aceh
```

Numbers in the program can be written in decimal or hex form as:

```
1234 = 1234 decimal
0xdd = DD hexadecimal
0ddh = DD hexadecimal
```

The first form of the hexadecimal number (0xdd) is a C language standard; the second form (0ddh) is a common assembly language standard.

B.4 Labels

Labels are names invented by the programmer that stand for a number in the program, such as a constant in the equ directive above, or a number that represents a memory location in the program. Labels used for memory locations follow two simple rules:

1. All labels must START with an alphabetic character and END with a : (colon).
2. A label can have no more than 128 characters: A(a)–Z(z), 0–9, _, and ?.

The following are examples:

```
fred:
m1:
p1234:
xyz:
```

The restriction that all numbers begin with a number is now apparent; hexadecimal numbers beginning with A to F would be mistaken by the assembler as a label and chaos would result.

◆ COMMENTS ◆

- *Anything* that follows a semicolon (;) in a line of a program is ignored by the assembler. Comments *must* start with a ;. For example,

  ```
  ;this is a comment and will be ignored by the assembler
  ```

- If you are assembling a program and get a LOT of syntax errors, you probably forgot to include a semicolon in your comments.
- A label may be used only once in a program.

Typing a Line

To make the program readable, it is recommended that you type all opcodes about 10 spaces or so to the right of the left margin of your text. Start all labels at the left margin of text, and place any comments to the right of the opcode entry. The finished line should appear as follows:

```
label:      opcode      ;comment
```

Inspection of the programs included with this book will provide many clues as to what syntax is acceptable to the assembler. Experiment with the assembler by writing short programs to get a clear understanding of what each output file contains.

B.5 Reserved Symbols Used as Labels

Any sort of a name used in a program is called a symbol. A label is a symbol with a colon after it. Certain symbols have already been predefined by the A51 programmers and *may not* be used by anyone else as a *label*. Predefined symbols are also called reserved symbols because they are reserved for A51. Using a reserved symbol as a label in a program will result in an assembly-time error and the message:

```
ATTEMPT TO DEFINE AN ALREADY DEFINED LABEL
```

Check the program to see if it contains a label that is also found in the following lists.

Keil Symbols Reserved for All Assemblers

Following is a list of Keil symbols reserved for all assemblers:

A	DA	INPAGE	MOD	REPT
AB	DATA	INSEG	MOV	RET
ACALL	DB	IRP	MOVC	RETI
ADD	DBIT	IRPC	MOVX	RL
ADDC	DEC	ISEG	MUL	RLC
AJMP	DIV	JB	NAME	RR
AND	DJNZ	JBC	NE	RRC
ANL	DPTR	JC	NOP	RSEG
AR0	DS	JE	NOT	SEG
AR1	DSEG	JG	NUL	SEGMENT
AR2	DW	JLE	NUMBER	SET
AR3	ELSE	JMP	OR	SETB
AR4	ELSEIF	JNB	ORG	SHL
AR5	END	JNC	ORL	SHR
AR6	ENDIF	JNE	OVERLAYABLE	SJMP
AR7	ENDM	JNZ	PAGE	SUB
BIT	ENDP	JSG	PC	SUBB
BITADDRESSABLE	EQ	JSGE	POP	SWAP
BLOCK	EQU	JSL	PUBLIC	UNIT
BSEG	EXITM	JSLE	PUSH	USING
C	EXTRN	JZ	R0	XCH
CALL	GE	LCALL	R1	XCHD
CJNE	GT	LE	R2	XDATA
CLR	HIGH	LJMP	R3	XOR
CMP	IDATA	LOCAL	R4	XRL
CODE	IF	LOW	R5	XSEG
CPL	INBLOCK	LT	R6	
CSEG	INC	MACRO	R7	

8051 Reserved Symbols

Following is a list of 8051 symbols reserved for all assemblers:

AC	ES	IE0	P	PT0	RI	SM1	TF1	TO
ACC	ET0	IE1	P0	PT1	RS0	SM2	THO	TRO
B	ET1}	INT0	P1	PX0	RS1	SP	TH1	TR1
CY	EX0	INT1	P2	PX1	RXD	T1	TI	TXD
DPH	EX1	IT0	P3	RB8	SBUP	TB8	TLO	WR
DPL	F0	IT1	PS	RD	SCON	TCON	TL1	
EA	IE	OV	PSW	REN	SM0	TF0	TMOD	

Register Byte and Bit Symbols

The symbols in Table B.1 may be used in instructions for registers that are both *byte* and *bit* addressable. For instance, using P0 in an instruction involves 1 byte of P0 data. Using P0.1 in an instruction will affect only bit 1 of the byte in P0.

TABLE B.1

Symbol	Address	Symbol	Address	Symbol	Address	Symbol	Address	Symbol	Address
AC	=00D6	IE	=00A8	P1.1	=0091	PSW.6	=00D6	T2CON.3	=00CB
ACC	=00E0	IE.0	=00A8	P1.2	=0092	PSW.7	=00D7	T2CON.4	=00CC
ACC.0	=00E0	IE.1	=00A9	P1.3	=0093	PTO	=00B9	T2CON.5	=00CD
ACC.1	=00E1	IE.2	=00AA	P1.4	=0094	PT1	=00BB	T2CON.6	=00CE
ACC.2	=00E2	IE.3	=00AB	P1.5	=0095	PT2	=00BD	T2CON.7	=00CF
ACC.3	=00E3	IE.4	=00AC	P1.6	=0096	PXO	=00B8	TB8	=009B
ACC.4	=00E4	IE.5	=00AD	P1.7	=0097	PX1	=00BA	TCLK	=00CC
ACC.5	=00E5	IE.7	=00AF	P2	=00A0	RB8	=009A	TCON	=0088
ACC.6	=00E6	IE0	=0089	P2.0	=00A0	RCAP2H	=00CB	TCON.0	=0088
ACC.7	=00E7	IE1	=008B	P2.1	=00A1	RCAP2L	=00CA	TCON.1	=0089
B	=00F0	INT0	=00B2	P2.2	=00A2	RCLK	=00CD	TCON.2	=008A
B.0	=00F0	INT1	=00B3	P2.3	=00A3	REN	=009C	TCON.3	=008B
B.1	=00F1	IP	=00B8	P2.4	=00A4	RI	=0098	TCON.4	=008C
B.2	=00F2	IP.0	=00B8	P2.5	=00A5	RS0	=00D3	TCON.5	=008D
B.3	=00F3	IP.1	=00B9	P2.6	=00A6	RS1	=00D4	TCON.6	=008E
B.4	=00F4	IP.2	=00BA	P2.7	=00A7	RXD	=00B0	TCON.7	=008F
B.5	=00F5	IP.3	=00BB	P3	=00B0	SBUF	=0099	TF0	=008D
B.6	=00F6	IP.4	=00BC	P3.0	=00B0	SCON	=0098	TF1	=008F
B.7	=00F7	IP.5	=00BD	P3.1	=00B1	SCON.0	=0098	TF2	=00CF
CPRL2	=00C8	IT0	=0088	P3.2	=00B2	SCON.1	=0099	TH0	=008C
CT2	=00C9	IT1	=008A	P3.3	=00B3	SCON.2	=009A	TH1	=008D
CY	=00D7	OV	=00D2	P3.4	=00B4	SCON.3	=009B	TH2	=00CD
DPH	=0083	P	=00D0	P3.5	=00B5	SCON.4	=009C	TI	=0099
DPL	=0082	P0	=0080	P3.6	=00B6	SCON.5	=009D	TL0	=008A
EA	=00AF	P0.0	=0080	P3.7	=00B7	SCON.6	=009E	TL1	=008B
ES	=00AC	P0.1	=0081	PCON	=0087	SCON.7	=009F	TL2	=000C
ET0	=00A9	P0.2	=0082	PS	=00BC	SM0	=009F	TMOD	=0089
ET1	=00AB	P0.3	=0083	PSW	=00D0	SM1	=009E	TR0	=008C
ET2	=00AD	P0.4	=0084	PSW.0	=00D0	SM2	=009D	TR1	=008E
EX0	=00A8	P0.5	=0085	PSW.1	=0001	SP	=0081	TR2	=00CA
EX1	=00AA	P0.6	=0086	PSW.2	=00D2	T2CON	=00C8	TXD	=00B1
EXEN2	=00CB	P0.7	=0087	PSW.3	=00D3	T2CON.0	=00C8		
EXF2	=00CE	P1	=0090	PSW.4	=00D4	T2CON.1	=00C9		
F0	=00D5	P1.0	=0090	PSW.5	=00D5	T2CON.2	=00CA		

B.6 Upper- and Lowercase

A51 is *not* case sensitive. Symbols may be typed in any mixture of upper- and lowercase the programmer desires. For instance, mOv, MoV, MOV, and mov are all the same to A51. In this book, uppercase was used for opcodes in explanatory chapters simply to reinforce the opcode mnemonics. Later chapters use lowercase, and this is the practice for most real-world programming.

C

Using µVision2

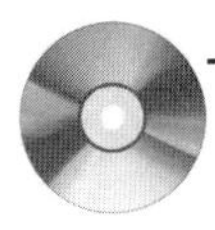

C.0 Installing µVision2 Development Software

The CD-ROM provided with this book contains an evaluation version of Keil Software Incorporated's µVision2 Integrated Development Environment (IDE) program. Keil has generously allowed the author to include µVision2 and its screen images with this book. The evaluation version is similar to Keil's professional version, except that the debugger will not load programs containing more than 2K bytes of code. No program in this book (including answers to problems) will exceed 2K bytes unless the programmer makes a mistake.

◆ COMMENT ◆

The debugger is limited to programs containing no more than 2048 bytes of code. This does *not* mean that program counter addresses are limited to 07FFh and below. Programs may be placed anywhere in the 64K address space of the 8051, as long as they do not exceed a total of 2K.

µVision2 may be used on any Pentium class PC-compatible computer running under a Windows operating system from 95 onward. The computer must have the following minimum hardware features:

20 MB free space on a hard drive used as the C: drive

16 MB memory

A floppy drive used as drive A[1]

To install µVision2 on a computer, perform the following operations:

1. Insert the CD-ROM in the computer CD drive and find the folder on it named Keil.
2. Copy the CD folder named Keil to the C: drive. The Keil folder contains the folders UV2, C51, Floppy Setup, and the files TOOLS and Using Other Drives.
3. Copy (Send To) the µVision2 shortcut in the Keil folder on the CD to the desktop screen.

µVision2 is now installed on the computer in the folder C:\Keil, and may be started by clicking on the shortcut icon. Before starting µVision2, however, some file organization has to be done.

[1]µVision2 is set up for this book assuming that student work will be done using public computers and that homework may be submitted for evaluation on removable media, most commonly floppy disks.

Configuring µVision2 to Use Any Drive

µVision2 places all files in the drive:/folder specified by the user when a New Project is defined. The drive used must be one that can be written to. Identify or create a new folder on the chosen drive and start µVision2.

1. Open the Projects menu and select New Project. A Create New Project window will appear with a Save in the box where the project files are to be stored. Browse until the chosen disk and folder are found and type a project name into the File name box. A "project name.UV2" project file has now been created in the folder, along with a "project name.PLG" log file.
2. The Files window will now show a "Target 1" folder, and a "Select Device for Target 'Target 1'" window will appear. Browse to the Intel folder, open it, and select 8051AH (unless some other microcontroller is to be used). Click OK and then click No to the "Copy Standard 8051 startup code . . ." question in the pop-up µVision2 box.
3. Expanding the "Target 1" folder in the Files window shows that it contains a subfolder named "Source Group 1." Both the "Target 1" and "Source Group 1" folders may be renamed to suit the user as follows:

   Open the Project menu and click the Targets, Groups, Files . . . menu choice.

   Type a new target name into the Target to Add box and Add it.

   Click on the new target name in the Available Targets box and Set as Current Target.

   Click on the "Target 1" name and Remove Target. Select the Group/Add Files tab.

   Type a new group name in the Group to Add box and Add it.

   Click on the "Source Group 1" name in the Available Groups box and Remove Group.

   Click on the new group name and click Add Files to Group button.

   Type in the name of the new source file in the File name box, using the name.a51 format.

   Select Asm Source files from the Files of type box. Click the Add button.

   Close all windows and note that all names have changed to the new ones selected.
4. Unless an assembly code source program with the same name as the one typed into the File name box just used above exists, it must be created. (A red "X" next to the source file name indicates µVision2 cannot find it.)
   To create a source file, go to the File drop-down menu and select New. An empty Text file window will appear in the file space. Type in an assembly code source program and Save As the file using the same name as the new

source file used in the sequence above. Close the project and then Open it again. The red X should be gone and the new source file ready to be assembled and debugged.

Generating Source Program Listing Files

µVision2 will not unconditionally generate a listing file that contains assembler results unless conditional listing for the project is disabled. To enable listing, go to the Project menu and select Options for Target "name." From the Options window. Select the Listing tab and clear the check from the Conditional box found in the Assembler Listing area.

EPROM Programming Using µVision2 HEX Output Files

If EPROMs are to be programmed, a project.HEX file must be created by µVision2. Go to the Project menu and select Options for Target "name." From the Options window, select the Output tab and check the Create HEX File box before closing the window.

µVision2 Disk Files

µVision2 creates seven project-named files (eight if a HEX file is made) in the drive/folder specified when a project is assembled and debugged. Two source program-named files (name.a51} are also produced.

All project files created by µVision2 have the root name given to the project. For example, the book project is named "8051 Microcontroller Projects." µVision2 makes the following project-specific files:

8051 Microcontroller Projects.UV2—used to load the project on start-up

8051 Microcontroller Projects.OPT—options chosen for the project

8051 Microcontroller Projects_Uv2.BAK—project file backup

8051 Microcontroller Projects_Uv2_Opt.BAK—options file backup

8051 Microcontroller Projects.HEX—an Intel hex file for burning EPROMs

8051 Microcontroller Projects.PLGE—project activity log file

8051 Microcontroller Projects.M51—linker map file

8051 Microcontroller Projects—µVision2 information file

The two files generated from the source assembly code file are used in debugging. For example, the source file used in the book is Student.a51. Files generated from this source file are:

Student.OBJ—a binary file used by the debugger to test the program

Student.LST—summary of assembler operation; very useful in finding errors

Using µVision2 with a Floppy Disk

µVision2 organizes all user input programs under the heading of a *Project*. Each Project holds a *Target* folder (from the target microcontroller under development), and each Target folder contains a *Group* folder, which holds *Source* program files. The first two folders exist only inside the structure of µVision2 and do not appear on a user's disk. The source files, however, are very real and are provided by the user on a disk. Due to the concept of a project, µVision2 must load a project file that contains information about each target, group, and source file(s) before any work can begin.

For use with this book, a folder named "Floppy Drive Setup" has been included in the Keil folder. Inside the folder are eleven files: eight 8051 Microcontroller Projects files and three Student files.

1. Insert a floppy in the A: drive and copy the 11 files in the Floppy Drive Setup folder (either from the CD-ROM or from the C: drive Keil folder) to the A: drive. The files are:

   8051 Microcontroller Projects.UV2—Used to load the project on start-up

   8051 Microcontroller Projects.OPT—Options chosen for the project

   8051 Microcontroller Projects_Uv2.BAK—Project file backup

   8051 Microcontroller Projects_Uv2_Opt.BAK—Options file backup

   8051 Microcontroller Projects.HEX—An Intel hex file for burning EPROMs

   8051 Microcontroller Projects.PLGE—Project activity log file

   8051 Microcontroller Projects.M51—Linker map file

   8051 Microcontroller Projects—µVision2 information file

   Student.a51—A user supplied assembly code source file.

   Student.OBJ—a binary file used by the debugger to test the program

   Student.LST—summary of assembler operation; very useful in finding errors

2. Go to the A: drive and select the first file. Right-click on the file icon and select "Properties" from the menu. Remove any check mark found in the "Read-only" box found at the bottom of the (filename) Properties General tab screen. Repeat this procedure for the remaining files on the A: drive.

◆ CAUTION ◆

Windows versions, with the exception of XP, will copy files from the CD-ROM as "Read-only" and "Archive." Failure to remove the read-only property of the A: drive files will result in the inability to change any of them, rendering µVision2 useless.

C.1 Starting µVision2

µVision2 may now be started by clicking on the shortcut icon. A screen similar to that shown in Figure C.1 should appear after clicking the shortcut.

The first thing to know about µVision2 is that it was intended for use by an experienced programmer, often one of many programmers working on the same 8051 project. These programmers may have used any number of assembler and debugger programs prior to using µVision2, and they already know what an IDE program should do. A beginning programmer, however, usually does not have any idea of what to expect from an IDE. The beginner is now faced with two challenges: (1) how to write programs for the 8051 and (2) how to use an IDE to test those programs.

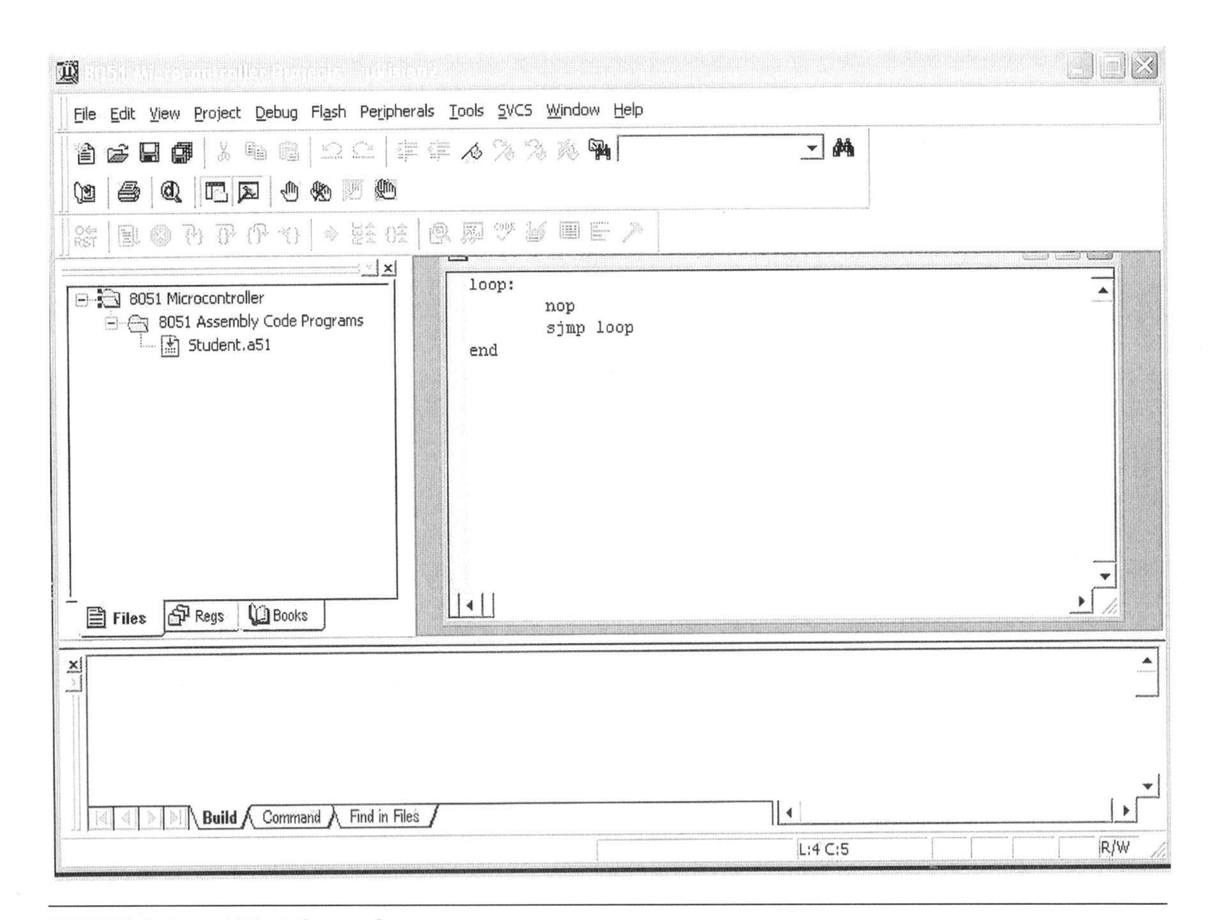

FIGURE C.1 ◆ µVision2 Startup Screen

This section seeks to help beginners get quickly started using Keil's µVision2 IDE. To do so, many of the more sophisticated and advanced features of µVision2 have been intentionally ignored. Those who consider themselves experienced programmers may take full advantage of µVision2 by reading the "µVision2 Getting Started" user guide (in PDF format) found in the Books window and listed under the Help menu.

The second thing to know about µVision2 is that every IDE program needed to write, edit, assemble, list, and debug a working program can be run from a screen menu item or icon shown in Figure C.1. In addition to mouse selection, certain function keys can easily invoke the most common IDE operations.

Loading a Project into µVision2

The µVision2 startup screen shown in Figure C.1 has a Menu bar at the top, an array of command buttons, and three main windows areas, known as the Project, Work Space, and Output windows:

The Project window displays Target, Group, and Source file path.

The Work Space window displays the file currently opened, Student.a51 in this example.

The Output window displays Source file assembly progress.

The first step in developing a source program is to start µVision2 and load a Project file using the Menu bar at the top of the µVision2 screen. Click on the Projects Menu and select Open Project from the drop-down menu. Browse to the A: drive and open the 8051 Microcontroller Projects file. µVision2 will load that 8051 Microcontroller Projects file and display, as shown in Figure C.1, that the path to the Student.a51 assembly code program is:

```
8051 Microcontroller (Target name)
    8051 Assembly Code Programs (Group name)
        Student.a51 (an 8051 assembly code program)
```

Student.a51 is not a folder, but an 8051 assembly code source file. Clicking on Student.a51 will load it into the Work Space window for editing, assembling, and debugging.

◆ CAUTION ◆

Closing µVision2 with a project loaded will make it attempt to load the same project, from the same drive, the next time it is started. Clearly, this may be a problem if the previous user did not use the "book-standard" file names and drive. When in doubt, load a project file using the instructions give above.

C.2 Editing, Assembling, and Debugging a Sample Student.a51 Program

Editing and Assembling

Replace the small startup program shown in the Work Space window of Figure C.1 with the following program. The Work Space window shows the last file opened (Student.a51) and is in edit mode whenever Student.a51 is loaded.

```
            big equ 010h            ;use names for numbers
            small equ 00h
            org 0000h               ;assembler directive: PC = 0000h
loop:                               ;main program loop label
            mov r0,#small           ;move 00h hex into r0
countup:                            ;loop here until r0 = 0ffh
            inc r0                  ;increment r0
            cjne r0,#big,countup    ;until r0 = 0ffh
            mov r1,#big             ;move 0ffh into r1
countdown:                          ;loop here until r1 = 00h
            djnz r1,countdown       ;decrement r1 to 00h
            sjmp loop               ;jump back to the beginning
end                                 ;assembler directive to stop assembling
```

◆ COMMENT ◆

Note that the Work Space window shows source file opcodes and directives in blue, if they are spelled correctly.

To assemble Student.a51, press the F7 function key (or select Build target from the Project menu). As Student.a51 is being assembled, the Output window will show what the assembler is up to:

```
Build target '8051 Microcontroller'
assembling student.a51
linking . . .
Program Size: (varies by program)
creating hex file from "8051 Microcontroller Projects" . . .
"8051 Microcontroller Projects"—0 Error(s), 0 Warnings).
```

The last line is *crucially important*; if the assembler discovers any errors in the source file, it will not finish assembling, but end with the dreaded "Target not created" message. Each line of the program that has an error will be listed to guide the programmer. The best way to find the error(s) is to open Student.LST on the A: drive and scroll down until the offending line number(s) is found. The assembler will post an error message below the line and point to the problem with a ^ symbol.

Debugging a Sample Student.OBJ

Successful assembly produces a file named Student.OBJ, a binary *object* file that can be debugged.

◆ CAUTION ◆

Getting a file to assemble just means that all the rules of syntax have been followed. It does not mean the program is going to work. Careful debugging proves that it works.

To debug Student.a51, click on the debug button (a red "d" inside a magnifying glass) or select Stop/Start Debug Session from the Debug menu. As soon as the debugging session begins, the µVision2 screen will change to that shown in Figure C.2. (Click off the window that reminds us we are running a 2K evaluation version of µVision2.)

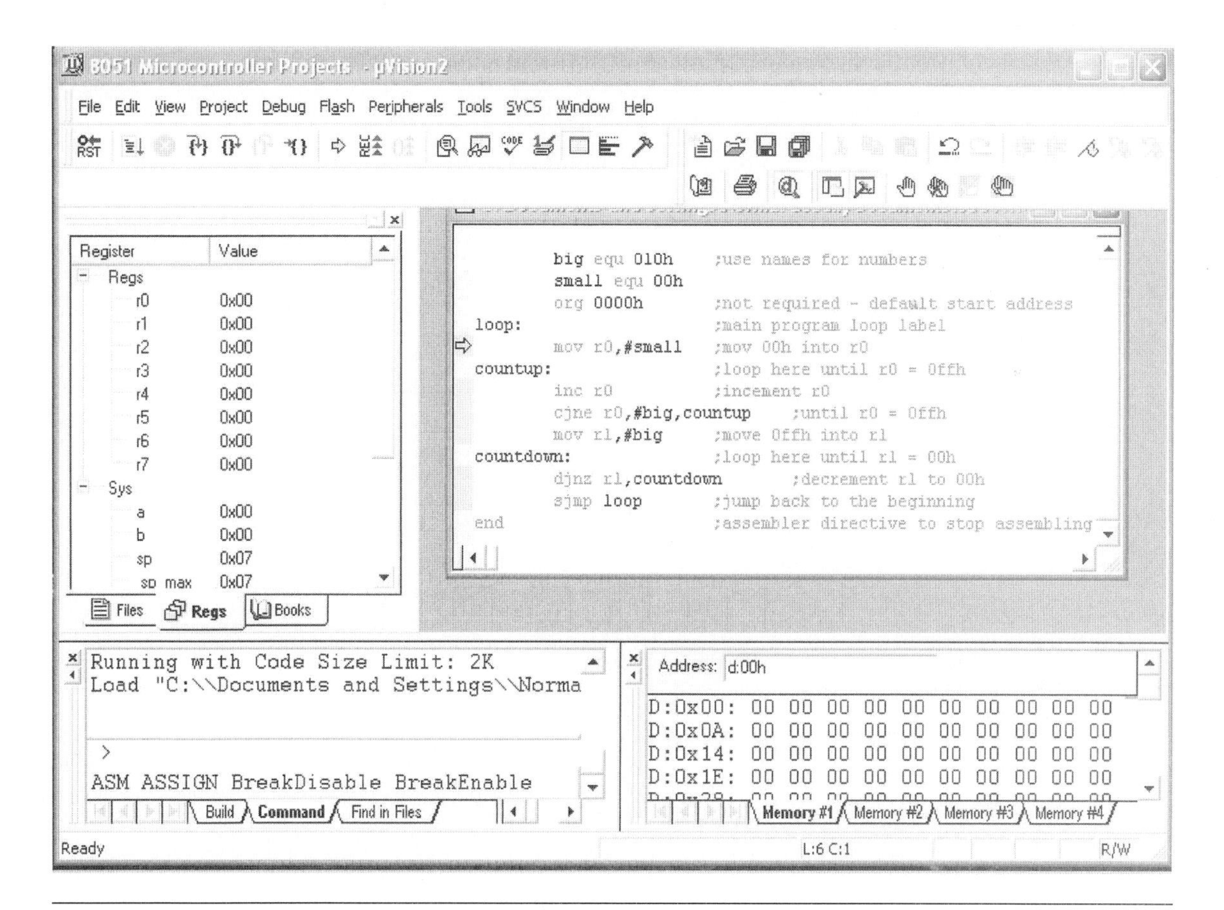

FIGURE C.2 ◆ µVision2 Debugging Screen

◆ CAUTION ◆

Once in the Debug mode, μVision2 will not respond to most commands except debugger commands. Stop debugging before trying, for example, to edit or re-assemble the program under test. The debugger will not take any notice of changes made to the program while debugging.

Figure C.2 shows several changes from Figure C.1. The Project window has switched over to show all of the internal registers of the 8051, while the Output window has added a Memory window. The Work Space window now shows Student.OBJ ready for debugging.

To debug the sample program, perform the following steps:

1. Place the cursor next to the yellow debug arrow that points to the next instruction to be done.
2. Press the F11 function key to execute (Step) each instruction.
3. Watch r0, and then r1, in the Project window change as the program loops.
4. Verify that r0 and r1 are performing the expected operation.

The last two steps are the most important. Stop debugging by again clicking the debug button.

C.3 Debugger Windows

To see what the program is doing, the programmer must be able to inspect the contents of every register and memory location in an 8051-based system. To test the program, it may also be necessary to change the contents of those locations.

Register Display

Every 8051 register is displayed in the Project window during a debugging session, if the "Regs" tab is chosen (the default tab when debugging starts). These include:

r0 to r7 of the *currently selected* bank

a, b, sp, sp_max (how big the stack ever gets), dptr, and PC

states used as the program executes

seconds of program execution from the start

psw flags (expand to see p, fl, ov, rs (from 0 to 3), f0, ac, and cy)

If all of the register contents cannot be seen, drag the Projects window to enlarge it.

Changing Register Contents

Click on any register *contents* in the Projects window and then *slowly* click again. Enter a new value in the contents box that appears and press the return key.

Memory Display

The Memory window shows the contents of any area of memory type entered into the Address box. Valid memory address types are:

Internal data from d:00h to d:0ffh (SFR above 7Fh)

Code data beginning at c:0000h

External data beginning at x:0000h

Up to four Memory windows may be defined, from Memory #1 to Memory #4, on the tabs shown.

Changing Memory Contents

Right-click any memory cell in the Memory window and choose "Modify Memory at (address)," which brings up an "Enter Bytes(s) at (address)" window. Memory change choices include:

A single byte from 00h to 0ffh

The ASCII equivalent of a single character inside quotation marks, for instance, 'a'

A string of ASCII bytes for a string of characters inside quotation marks, for instance, "hello"

The debugger will ignore any erroneous change to register or memory contents.

Peripheral Display

8051 peripheral windows are selected from the Peripherals menu. These include:

Interrupt Register

I/O Ports P0 to P3

Serial UART

Timers T0 and T1

Each peripheral window may be dragged to a convenient spot on the screen.

Serial Data Output and Input

In addition to the peripheral windows, two serial windows may be chosen from the View menu, Serial Window #1 and Serial Window #2. Serial data transmitted from the UART may be observed in Serial Window #1. Data typed in Serial Window #1, after placing the cursor in the window, will be received by the UART.

Serial data transmitted to Serial Window #1 may be displayed in ASCII mode or Hex(adecimal) mode. Choose the display mode by right-clicking the mouse when the cursor is placed inside Serial Window #1. A menu will appear, and the desired mode will be chosen. The window may also be cleared from the same menu.

Only use Serial Window #1 for 8051 programs.

Data typed into the Serial Window #1 does not appear on the window.

Changing Peripheral Contents

Each peripheral window has a set of byte or bit boxes, unique to each peripheral, that may be changed. Data that may be changed is signified by a box with a white interior. Information-only data is in a gray box. Change values as follows:

> Byte values: Place the cursor in a byte box and type in new data followed by the return key.
>
> Bit values: A check mark indicates a 1; no check indicates a 0. Click a check on or off to select a 1 or 0 bit value.

The only exception to the preceding paragraph is found in the Interrupt System window. To change the priority of an interrupt:

> Click on the interrupt type to get a blue highlight.
>
> Go to the Pri box and type in a 1 or a 0 priority level.
>
> Click on the interrupt type again to change its priority.

C.4 Debugging Fundamentals

Debugging demands two skills: attention to detail and patience. The whole object of a debugging session is to *make the program show you that it works.* To show you that it works, you must be able to:

1. See the contents of every register, memory location, and peripheral used by the program.
2. Change program input data to test a range of program possibilities.

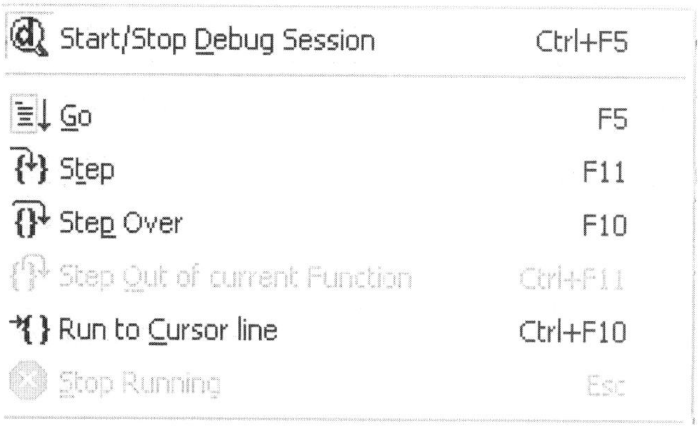

FIGURE C.3 ◆ Debugger Icons

3. Step through each program instruction until you know it works as intended.
4. Run the program to the end of repetitively looping code sections.
5. Run the program continuously.

Section C.3 provides information on how to inspect and change data in every program memory address, register, or peripheral, as well as how to view input serial data. This section covers how to accomplish the last three steps outlined above.

Step, Run, Reset, and Halt Debugger Toolbar Buttons

The debugger toolbar has button icons that may be clicked to perform debugger actions. Figure C.3 shows these icons.

Step Each Instruction

Step (execute) the instruction opposite the yellow debugger arrow: Place the cursor at the tip of the yellow debug arrow and press the F11 function key, or click on the Step button on the debug toolbar or Debug menu. The instruction opposite the yellow debugger arrow is executed, and the arrow/cursor combination points to the next instruction.

◆ COMMENT ◆

Registers that are affected by an instruction will be highlighted in blue in the Regs window.

Every instruction (except an NOP) should produce a result in some register, memory address, or peripheral. Check to see if the result is what you expected and where you expected it.

Run-to-Cursor Line

Looping code in a program, such as incrementing or decrementing a register, may take a lot of debugging time if done one step at a time until the loop conditions are satisfied. A good way to debug loops is to step through the loop a few times until it appears to work. Then place the cursor at the instruction line that immediately follows the loop and run the program to the instruction at the cursor location. Press the Ctrl-F10 key or click on the run-to-cursor button on the debug toolbar or Debug menu.

Do not place the cursor at a label that is on a line without an opcode. The cursor must be placed somewhere on an 8051 instruction opcode line.

Run the Program

Most programs in this book, except those that involve real-time port pin or serial data interaction, may be tested using single-step and run-to-cursor techniques. Real-time programs, after having been thoroughly tested by step or run-to-cursor techniques, may be run continuously by pressing the F5 function key or by clicking on the Run button on the debug toolbar or Debug menu. Stop Running by clicking the Halt button on the debug toolbar or Debug menu.

Reset the CPU

To reset the program, Halt the simulation and click on the Reset (RST) button on the debug toolbar. To completely restore all memory (and most register values) to zero, stop debugging and then re-start debugging.

Change the Crystal Frequency

The default crystal frequency for the debugger is 12 MHz. To change the crystal frequency, go to the Project menu and choose Options for Target (assuming the Target folder is highlighted) and click the Target tab on the pop-up options window. Type in the desired frequency in the Xtl(MHz) window provided.

D

The 8255 Programmable I/O Port

D◆0 Introduction

Eight-bit microprocessor families include peripheral chips that are used with the CPU to provide many of the I/O functions that are now found integrated inside a microcontroller. As the 8-bit microprocessor fades into obsolescence, these peripheral chips are finding a new life in augmenting microcontroller I/O capability.

These peripheral chips include serial and parallel I/O as well as interrupt controllers and dynamic RAM controllers. The 8051 loses two parallel I/O ports when used with external memory, and part of a third to serial data communication and interrupt functions. To make up for this loss, a programmable parallel port chip, the 8255, is often added to an 8051 system, as discussed in Chapter 9. This appendix describes how to use the 8255 as a basic parallel I/O port. The 8255 is capable of many sophisticated I/O functions, including interrupts and handshaking. Refer to the manufacturers' literature for a complete description of 8255 capabilities and programming.

D◆1 Functional Description

The 8255 is packaged in a 40-pin DIP. The function of each pin is shown in Table D.1. Figure D.1 shows the 8255 pinout and command byte structure. The 8255 features three 8-bit programmable parallel I/O ports named A, B, and C. Port C can be used as two separate ports of 4 bits each if properly programmed.

Addressing the 8255

The 8255 contains one Control register and three port registers, as shown in Figure D.1. The 8255 registers are addressed by address bits A0 and A1 on pins 9 and 8. The addresses for each of the four internal registers in the 8255 are shown in Table D.2.

Note, for mode 0, that when a port register is written, the data byte written is *latched* into the port for *output*. When a port register is read, the port pin logic levels are read from the port pins at that *instant*. Modes 1 and 2 latch both output and input data. Note also that the Control register should *only* be written, not read.

Before any port can be used, the 8255 must be programmed by writing the proper control bits to the Control register. The three ports may then be accessed by the 8051 program. The 8255 uses the address lines A0 and A1 to access the Control register and the three ports. The $\overline{RD}$, $\overline{WR}$, and $\overline{CS}$ lines are enabled by

TABLE D.1

Pin	Name	Dir	Function	Pin	Name	Dir	Function
			I/O Port A				***Upper Nibble***
4	PA0	B	Programmable Port A Bit 0	13	PC4	B	Programmable Port C Bit 4
3	PA1	B	Programmable Port A Bit 1	12	PC5	B	Programmable Port C Bit 5
2	PA2	B	Programmable Port A Bit 2	11	PC6	B	Programmable Port C Bit 6
1	PA3	B	Programmable Port A Bit 3	10	PC7	B	Programmable Port C Bit 7
40	PA4	B	Programmable Port A Bit 4				
39	PA5	B	Programmable Port A Bit 5				**Control Inputs**
38	PA6	B	Programmable Port A Bit 6	9	A0	I	Internal registers address bit 0
37	PA7	B	Programmable Port A Bit 7	8	A1	I	Internal registers address bit 1
				6	/CS	I	Chip select, active-low
			I/O Port B	5	/RD	I	Read internal register, active-low
18	PB0	B	Programmable Port B Bit 0	36	/WR	I	Write internal register, active-low
19	PB1	B	Programmable Port B Bit 1	35	RESET	I	Reset, active-high
20	PB2	B	Programmable Port B Bit 2				
21	PB3	B	Programmable Port B Bit 3				**Data**
22	PB4	B	Programmable Port B Bit 4	34	D0	B	Data Bit 0
23	PB5	B	Programmable Port B Bit 5	33	D1	B	Data Bit 1
24	PB6	B	Programmable Port B Bit 6	32	D2	B	Data Bit 2
25	PB7	B	Programmable Port B Bit 7	31	D3	B	Data Bit 3
				30	D4	B	Data Bit 4
			I/O Port C	29	D5	B	Data Bit 5
			Lower Nibble	28	D6	B	Data Bit 6
14	PC0	B	Programmable Port C Bit 0	27	D7	B	Data Bit 7
15	PC1	B	Programmable Port C Bit 1	26	Vcc	I	Power supply voltage, +5V
16	PC2	B	Programmable Port C Bit 2	7	GND	I	Power supply ground
17	PC3	B	Programmable Port C Bit 3				

Note that the 8255 does *not* use pins 40 and 20 for Vcc and ground.

the particular decoding scheme used by the 8051 system designer. The resulting control and address states yield the following actions:

A1	A0	$\overline{RD}$	$\overline{WR}$	$\overline{CS}$	Action
0	0	L	H	L	Read the contents of port A
0	1	L	H	L	Read the contents of port B
1	0	L	H	L	Read the contents of port C
0	0	H	L	L	Write to the port A latch
0	1	H	L	L	Write to the port B latch
1	0	H	L	L	Write to the port C latch
1	1	H	L	L	Write to the Control register
X	X	X	X	H	Data bus to high impedance

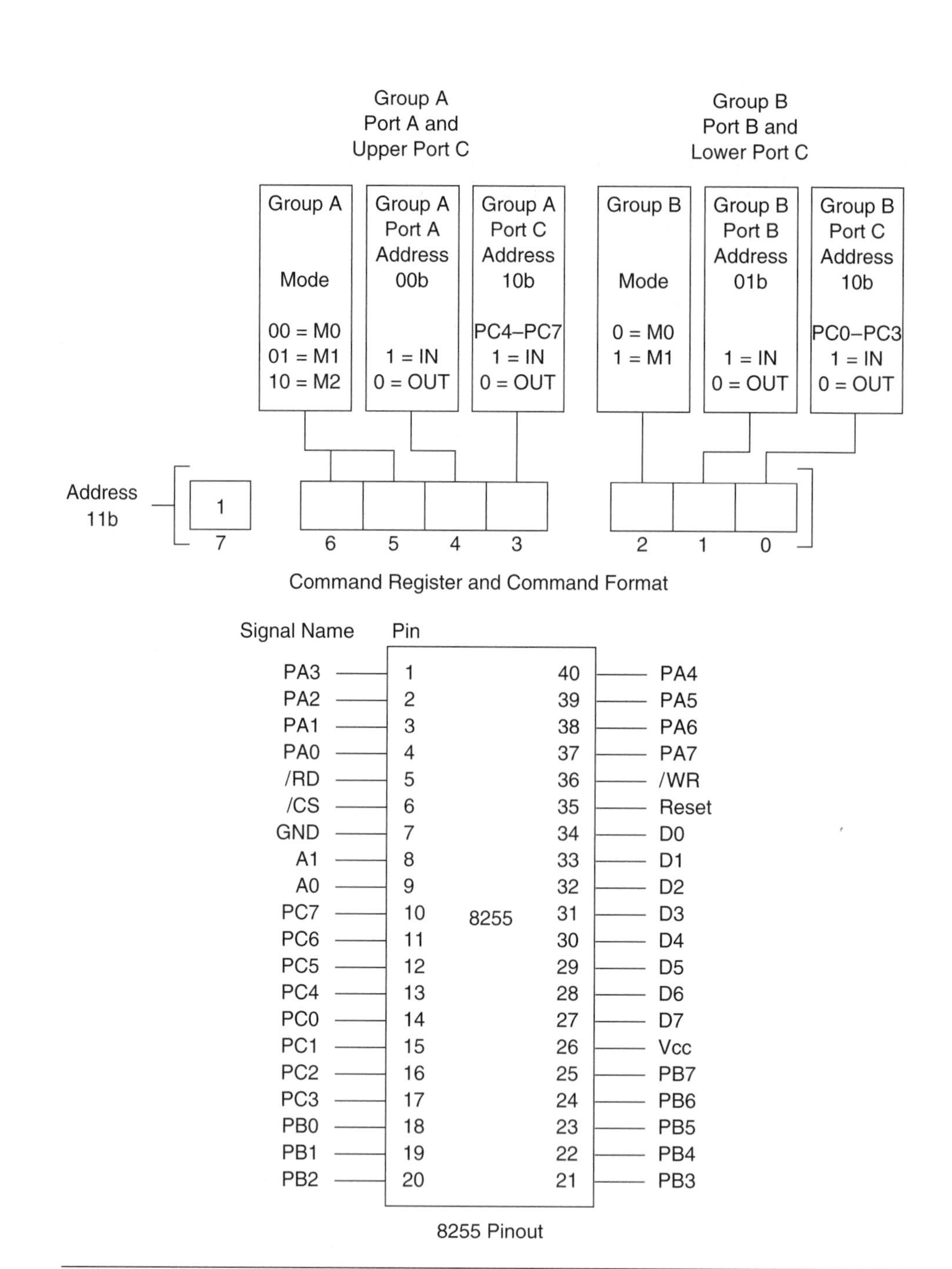

FIGURE D.1 ◆ 8255 PPI

TABLE D.2

8255 Register Address		Addressed Register
A1	A0	
0	0	Port A
0	1	Port B
1	0	Port C
1	1	Control register

The 8255 appears much like an internal port of the 8051 once it has been programmed.

D.2 Programming the 8255

Control bytes written to the Control register use each bit of the byte to program some feature of the 8255:

Bit	State	Result
7	1	Program ports for mode and input or output
7	0	Set/reset individual bits of port C

When bit 7 is a 1 then the ports are programmed as:

Bit	State	Result
6,5	00	Set port A and C4–C7 in I/O mode 0
6,5	01	Set port A and C4–C7 in I/O mode 1
6,5	10	Set port A and C4–C7 in I/O mode 2
4	0	Set port A as an output port
4	1	Set port A as an input port
3	0	Set C4–C7 as an output port
3	1	Set C4–C7 as an input port
2	0	Set port B and C0–C3 in I/O mode 0
2	1	Set port B and C0–C3 in I/O mode 1
1	0	Set port B as an output port
1	1	Set port B as an input port
0	0	Set C0–C3 as an output port
0	1	Set C0–C3 as an input port

8255 I/O Modes

Port A and the high part of port C may be programmed in one of three modes; port B and the lower part of port C may be programmed in one of two modes. The modes are:

Mode 0—Basic I/O: Data written to the port is latched; data read from the port is read from the input pins. (This mode is identical to 8051 port operation.)

Mode 1—Strobed I/O: This handshaking mode uses ports A and B as I/O and port C to generate handshaking signals to the devices connected to ports A and B and an interrupt signal to the host microcontroller.

Mode 2—Strobed bi-directional I/O: This mode is similar to mode 1 with the ability to use port A as a bi-directional data bus.

Modes 1 and 2 require setting interrupt enable bits in the port C Data register. These modes are intended to be used with intelligent peripherals such as printers.

Reset Condition

Upon reset all the port data latches and the control register contents are cleared to 00. The ports are all in the input mode.

E

The Rest of the Family

E◆0 Introduction

As mentioned briefly in Chapter 3, the 8051 microcontroller is the first of a large family of designs that build upon the basic 8051 architecture. New 8051 family members continue to be introduced.

Intel Corporation, knowing that few customers wished to have only one source for parts, licensed other semiconductor manufacturers to build and continue the development of the 8051 family. One of the most active developers of new derivatives from the original 8051 architecture is Philips Semiconductors, the giant Dutch semiconductor company. The following summary is derived from Philips's Web site, a must-see for anyone involved in 8051 applications.

E◆1 The 8052

One of the first 8051 modifications was the 8052, and the associated 8032 ROMless and 8752 EPROM parts. The 8052 adds 128 bytes of internal RAM to the original 128 bytes of the 8051, increases ROM from 4K to 8K, and provides an additional timer, T2, with an associated control register, T2CON. Figure E.1 shows the additional hardware associated with timer T2. Figure E.2 shows T2CON.

Timer T2 has the timing or event counting features of T0 and T1. Timing is done using the internal clock, whereas counting uses an external signal on port 1.0. Timer T2 also can provide an alternate baud rate clock to the UART receiver and/or transmitter in modes 1 and 3. Using two baud clock sources for the UART transmitter and receiver makes it possible to transmit at one frequency (using T1 for instance) and receive at another frequency using T2. T2 may also function in a totally new manner: as a *capture timer*.

Capture Timers

A capture timer is a timer that can store its current value in a set of registers by hardware rather than software means. The 8051 can read the value of timers T0 or T1 and store the timer value, at any instant, using software instructions. The 8052, however, can read, or "capture," the current T2 timer value using an external signal to trigger when the T2 values are to be stored. New special function registers RCAP2H and RCAP2L are used to store the captured values. The external signal named T2EX, which controls when T2 values are captured, is connected to port 1.1. T2EX is a negative *edge* active signal; no action occurs until T2EX goes high-to-low.

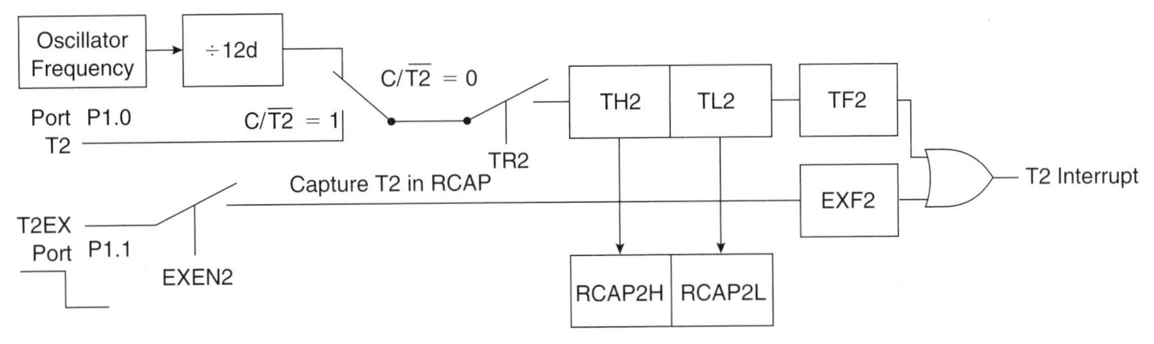

Timer 2 Configured to Capture Count

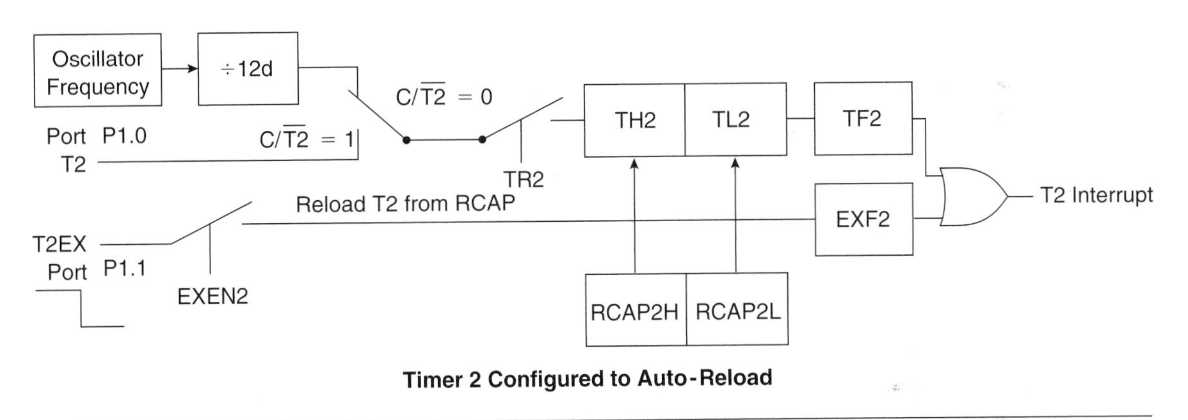

Timer 2 Configured to Auto-Reload

FIGURE E.1 ◆ Timer T2 Capture/Auto-Reload Configurations

The T2 counter may also reverse the operation of a capture timer, that is, it may also be auto-reloaded from the capture registers under the control of T2EX. The capture registers may be read from or written to under program control.

Timer T2 is equipped with the expected Overflow flag, TF2, and an additional flag, EXF2, that can be set by the external control signal T2EX. The new flags are part of the T2CON Special function register.

T2 Software Additions

The addition of the T2 (TH2 and TL2), T2CON, RCAP2H, and RCAP2L special function registers does *not* add to the 8051 instruction set. The mnemonics for moving data to and from the new registers, as well as setting and clearing control and flag bits, are part of the original 8051 instruction set. What *is* new are

TF2	EXF2	RCLK	TCLK	EXEN2	TR2	C/$\overline{\text{T2}}$	CP/$\overline{\text{RL2}}$
Bit 7	6	5	4	3	2	1	0

Bit	Symbol	Function
7	TF2	Timer 2 Overflow flag. Set to 1 when T2 overflows from maximum count to 0. Is not reset when program vectors to interrupt address 002Bh.
6	EXF2	External input 2 Interrupt flag. Set to 1 if EXEN2 bit is set to 1, and a negative edge signal occurs on pin T2EX. Is not reset when program vectors to interrupt address 002Bh.
5	RCLK	UART receiver clock bit. UART receiver uses T2 overflow pulses when set to 1. UART receiver uses T1 overflow pulses when cleared to 0. UART must be in modes 1 or 3 to use timer pulses for receiver timing.
4	TCLK	UART transmitter clock bit. UART transmitter uses T2 overflow pulses when set to 1. UART transmitter uses T1 overflow pulses when cleared to 0. UART must be in modes 1 or 3 to user timer pulses for transmitter timing.
3	EXEN2	External interrupt 2 enable bit. Enables capture or auto-reload by T2EX signal when set to 1. Prevents T2EX action when cleared to 0.
2	TR2	Timer 2 run control bit. Timer 2 will count or time when set to 1. Timer 2 disabled when cleared to 0.
1	C/$\overline{\text{T2}}$	Timer 2 pulse source bit. T2 counts external pulses from port pin 1.0 when set to 1. T2 counts oscillator pulses when set to 0.
0	CP/$\overline{\text{RL2}}$	Capture/reload bit. T2 will operate in capture mode when set to 1. T2 will operate in reload mode when cleared to 0. This bit is ignored if RCLK or TCLK bit is set to 1. When RCLK or TCLK is set to 1, T2 is forced into auto-reload mode on T2 overflow.

FIGURE E.2 ◆ T2CON Timer 2 Control Register

the addresses and operand names of the T2 special function registers, which are listed below:

Special Function Register	Address (hex)	Bit Addressable
T2H	CD	No
T2L	CC	No
RCAP2H	CB	No
RCAP2L	CA	No
T2CON	C8	Yes

Thus, the instruction to move data to T2H, mov t2h,#34h, or to run T2, set tr2, is used exactly as might be used for T0 or T1.

Timer T2 Interrupt

The only software complication introduced by T2 is the fact that only one interrupt is generated whenever TF2 or EXF2 is set. A timer 2 interrupt is enabled by setting the ET2 bit in the IE register. Timer 2 interrupts, whether caused by the TF2 Overflow flag or the EXF2 external signal, will cause an interrupt to location 002Bh in program code. The program may have to determine which of the timer 2 interrupt flags caused an interrupt by inspecting the flags in T2CON. T2 interrupt flags are *not* reset when the interrupt address at 002Bh is vectored to; the program must reset the flags.

The T2 interrupt has the lowest priority of the six interrupt sources. Interrupt priority (unless modified by setting bits in the IP register) is as follows:

IE0 (Highest Priority)
TF0
IE1
TF1
RI/SI
TF2/EXF2 (Lowest Priority)

Priority bit PT2 of the IP register may be used to adjust the priority of the T2 interrupts.

Uses for Timer 2

The addition of timer 2 enables the 8052 to perform tasks not possible using the 8051. Most notable of these is the capture feature. The 8051 *can* capture a value in T1 or T0, when stimulated by an external signal on $\overline{\text{INT0}}$ or $\overline{\text{INT1}}$. There will be some *inaccuracy* however, due to the time it takes to vector to the external interrupt program location, and read the timer. The 8052 can capture the *exact* count whenever T2EX is active.

Using T2 to generate baud rates for the UART frees timer T1 for other uses. Dual-baud rate applications (one rate for transmission and another for reception), while possible, are not common.

E.2 A/D and D/A Equipped Family Members

Chapter 10 shows how the 8051 may be used with external A/D and D/A converters. Analog-to-digital conversions are such common and useful applications that it is no surprise to find family members with converters incorporated on the chip. A survey of Philips family members includes:

Part Number	A/D (bits)	D/A (bits)	A/D MUX	Other Features
P87C552	10	PWM	10 Channel	Watchdog Timer
P87C557	10	PWM	8 Channel	48 I/O
P83CE598	10	PWM	8 Channel	CAN-Controller

Watchdog Timer

A watchdog timer (WDT) is an admission that programmers make mistakes, and that nature is not kind. The function of the timer is to reset the CPU should the timer ever count up to its limit and overflow. The programmer may enable the WDT and periodically, as the program loops, reset the WDT. Should the program not reset the WDT as a result of becoming "trapped" in an endless loop, then the WDT will time out and reset the CPU.

Other factors may cause the program to become trapped. Electrical noise may cause a malfunction inside the CPU circuitry that could lead to program interruption. The WDT is virtually noise-proof and should time out and reset the CPU.

Pulse Width Modulation (PWM)

Pulse width modulation is a digital-to-analog conversion method that changes the duty cycle of an output pulse from 0% to 100%. The output pulse is provided on a single output pin. A binary 00, for instance, may result in a 0% duty cycle output (always low), whereas a binary FFh value may result in a 100% duty cycle output (always high). PWM outputs are particularly useful when they are used to drive DC motors. A DC motor's speed is typically a function of the DC *average* voltage fed to its armature. The DC value of the PWM output ranges from zero to whatever maximum value is used to power the motor. The speed of the motor then varies from zero to some maximum value.

Analog Comparators

Analog comparators compare two analog voltages. One voltage is typically a reference voltage, and the other the unknown value. When the unknown voltage exceeds the reference, the output of the comparator switches from one binary state to another. The comparator output may be used to interrupt the program or cause some other program change.

Inter-Integrated Circuit (I2C) Serial Bus

The I2C bus is a popular industrial control local serial data bus that uses two wires, one for data and the other for a reference clock, to interchange data between devices. The concept of I2C is similar to the serial data UART in modes 1, 2, or 3, with the important distinction that data interchange is synchronous, not asynchronous. The I2C bus is identical to UART mode 0, with the important distinction that many functions that would have to be done using software in UART mode 0 are automatically done using I2C bus hardware.

Controller Area Network (CAN)

CAN is an industry-standard network protocol that permits many processors to be connected together for data interchange. Data may be interchanged at rates up to 1 megabit per second.

E.3 Faster and Smaller

Faster

The maximum clock rate of the original 8051 family is typically 12 to 16 MHz. Newer members of the family now run at speeds up to 33 Mhz. For instance, the following *full-featured* Philips parts sport the indicated clock rates:

Part Number	Clock Frequency (MHz)
P87C51MC2	24
P87C51RD2	33

Full-featured parts are those that offer most of the 8051 family ports, timers, RAM, ROM, and so on.

Fastest

A review of Chapter 3 shows that the original 8051 design uses 12 clock cycles, divided into six states, to generate 1 machine cycle. Intel introduced an 8051 derivative, numbered as the 80C151, that requires only 2 clock cycles per machine cycle. The reduction in clock cycles yields an execution rate up to six times that of the original 8051. Another feature of the 80C151 is an instruction *pipeline.* A pipeline is an internal code storage area where code bytes may be fetched and stored *before they are executed.*

The original 8051 design would fetch, execute, fetch, execute, and so on. No new code could be fetched until the last instruction had executed. A pipeline design, however, stores several instructions in the pipeline at the same time the CPU is executing previous instructions. The overlap between fetching and executing speeds up overall CPU execution of a program. The 80C151 has now been replaced by the 80251.

Smaller

Less is sometimes more. Frequently an application demands a limited subset of features offered by full-featured 8051 family members. To meet this need, Philips, and others, have developed 8051 family members that have fewer pins and features. The following summary lists several of the smaller (and cheaper) Philips parts:

Part Number	Pins	Port Pins	Timers	UART	RAM	ROM	Other
7LPC760	14	12	2	Yes	128	1K	I2C
7LPC761	16	14	2	Yes	128	2K	WDT
7LPC768	20	18	2	Yes	128	4K	PWM, WDT
7LPC769	20	18	2	Yes	128	4K	A/D, D/A, WDT

E.4 Bigger

In this age of 32- and soon-to-be 64-bit microprocessors, an 8-bit microcontroller would seem to be lacking in power. Of course, as pointed out in Chapter 1, many industrial control applications are well served by an 8-bit microcontroller. But program size in the 8051 family is generally limited to 64K, a problem in some high-level embedded applications. Additional address and data lines both increase memory size and decrease program execution time. As data path widths are increased, overall program execution time is decreased because more bytes can be moved in one memory cycle.

The XA Family

Philips, and others, have developed an *extended architecture* version of the 8051 family, the PXA. The overall objective of the XA architecture is to retain the attractive features of the 8051, particularly the bit-level operations, while increasing memory and data size.

External Features

The XA family is housed in a 44-pin LQFP (Low-profile Quad Flat Package) and a PLCC (Plastic Leaded Chip Carrier). Four 8-bit ports, a 16-bit data path, a 20-bit address bus (24-bit addresses are supported), and two UARTs are featured. As is the case with the 8051, most pins have several functions. Initial clock speed is limited to 30 MHz. Data and code memory retain the Harvard architecture arrangement of 1M of code and data space. Data path size, 8- or 16-bit, may be selected by the user.

RAM

On-chip RAM may be as large as 2K. External RAM may be as large as 16M.

ROM

The G37 has 32K of (EP)ROM. External code space may be as large as 16M.

24 Memory Address Bits and 8-Bit Segment Registers

An address in data space is formed by concatenating (placing to the left, as the most significant byte) an 8-bit segment register with a 16-bit offset. The offset may be a register, a number, or some combination of register and number. Three segment registers are available to the programmer: the Data Segment (DS), Extra Segment (ES), and Code Segment (CS) registers. Each segment register may specify 1 of 256 segments in memory, each of which may be 64K in size, for a total of 16M of code or data memory.

Working Registers

The working registers are sized at 16 bits (1 word). The registers are divided into two 8-bit registers. Each 8-bit-register half of the word-sized register is addressable. Thus, word register R7 is made up of R7H (high byte) and R7L (low byte).

There are four banks of four word-size registers, each bank containing registers R0 to R3. Registers R4 to R7 exist as single units; R7 is the stack pointer.

All of the registers may be used for arithmetic and logical operations, thus avoiding having to use only register A for these operations. The working registers may also be used as pointers to memory.

Special function registers remain at 8 bits.

Operating System Hardware

An 8-bit microcontroller typically is dedicated to executing a single program. An XA family application, however, may be designed such that several different programs may be run, in a *multi-tasking* mode. Multi-tasking implies that some master program schedule *when* each individual (application) program is granted access to the CPU resources. The master program is usually called an *operating system,* after such famous types such as DOS, UNIX, OS/2, Windows, and so on.

Operating systems generally require unique hardware assets in order to easily control the application programs. The XA family is equipped with two stack pointers: the system stack pointer (SSP) is reserved for the operating system; the user stack pointer (USP) is used by the application programs. Other protection features are incorporated to keep an application program from "trashing" the operating system program.

Other Features

A watchdog timer, 8052 style T2 timer, and PWM capability are included in the design. A modified pipeline is included, which allows simultaneous instruction execution and code fetching, greatly increasing overall program execution speed.

Instruction additions include 15 conditional jumps, a 16×16 bit multiply and 32/16 divide, sign extend, arithmetic shifts, load effective address, breakpoint set, and a software reset.

Interrupt capability has been expanded to 9 event (hardware) interrupts, 21 exception and trap (operating system) interrupts, and 7 software (program) interrupts.

Clearly the XA family of 8051s are not intended for simple controller operation. To cover the capabilities of the family, particularly the operating system features, would require an entire textbook.

Software

Although the XA certainly may be programmed in assembly code, it has been designed to take advantage of high-level languages. The following is a list of XA IDE, mentioned on the Philips Web site:

Altium (www.altium.com) provides extensive software support for the XA, including a macro assembler, linker/locator, C/C++ compilers, C libraries, 8051 to XA translator, debugger, and hardware monitor.

Raisonance (www.raisonance.com) offers a complete development "suite" of both software IDE and hardware prototype boards for the XA family. The IDE includes a macro assembler, linker, C compiler, and real-time operating system (RTOS) program.

The 80251

In 1996, Intel introduced the 80251 as an updated and enhanced successor to the 8051. At a minimum, the 80251 can replace the 8051 and use any program originally developed for the 8051. The main advantage to using the 80251 as a drop-in replacement for the 8051 is that the 80251 executes code six times faster than the 8051.

A better hardware reason to migrate to the 80251 is to take full advantage of expanded data size, additional instructions, a programmable counter array, and two UARTS. A good software reason to switch to the 80251 is that sufficient code space exists to hold large C programs. Other highlights of the 80251 architecture include:

Data in byte (8 bits), word (16 bits), and double-word (32 bits) size

16 MB address space for external code and data bytes

1K of internal RAM; 16K of internal code

Five compare/capture PCA modules

A complete set of conditional jump instructions

16-bit arithmetic and logic instructions

The 80251 is also produced by Atmel and Sanyo. Keil provides an 80251 IDE similar to the one used in this book.

E.5 The Future

A Web search by the author showed that total worldwide sales of all microcontrollers in 2002 was over 10 billion dollars, with 8-bit models making up *half* that amount. 8051-derivative microcontrollers accounted for about 20% of the 8-bit market, or close to 1 billion dollars. The 8051 family is manufactured by Intel, Philips, Siemens, OKI, Temic, and other semiconductor companies. Billions of dollars and millions of programmer-hours are invested in 8051 manufacturing, development systems, software, and resultant 8051-based products. Clearly, the 8051 architecture has become an industry standard, and we may anticipate that it will remain so for many years to come.

F

Control Registers

F.0 Introduction

For the convenience of the programmer, the control special-function register figures from Chapter 3 are shown here for easy reference. A listing of all byte and bit RAM and SFR addresses is given next, followed by an ASCII table.

F.1 Control Special-Function Register Figures

TCON and TMOD Function Registers

7	6	5	4	3	2	1	0
TF1	TR1	TF0	TR0	IE1	IT1	IE0	IT0

7	6	5	4	3	2	1	0
Gate	$C/\overline{T}$	M1	M0	Gate	$C/\overline{T}$	M1	M0
[Timer 1]				[Timer 0]			

SCON and PCON Function Registers

7	6	5	4	3	2	1	0
SM0	SM1	SM2	REN	TB8	RB8	TI	RI

7	6	5	4	3	2	1	0
SMOD	—	—	—	GF1	GF0	PD	IDL

IE and IP Function Registers

7	6	5	4	3	2	1	0
EA	—	ET2	ES	ET1	EX1	ET0	EX0

7	6	5	4	3	2	1	0
—	—	PT2	PS	PT1	PX1	PT0	PX0

F.2 Internal RAM and SFR Addresses

The following table lists internal RAM address by hex number, name, and function. Bit addresses, where appropriate, are also listed. Numbers in **bold** are the SFR contents on reset.

RAM Addresses (hex)

Byte	Bit 7	6	5	4	3	2	1	0	Name	Function
00									R0	Register bank 0
01									R1	
02									R2	Default register bank on reset
03									R3	
04									R4	
05									R5	
06									R6	
07									R7	SP is set to 07h on reset
08									R0	Register bank 1
09									R1	
0A									R2	
0B									R3	
0C									R4	
0D									R5	
0E									R6	
0F									R7	
10									R0	Register bank 2
11									R1	
12									R2	
13									R3	
14									R4	
15									R5	
16									R6	
17									R7	
18									R0	Register bank 3
19									R1	
1A									R2	
1B									R3	
1C									R4	
1D									R5	
1E									R6	
1F									R7	
20	07	06	05	04	03	02	01	00		Bit-addressable RAM
21	0F	0E	0D	0C	0B	0A	09	08		May be addressed as entire bytes, or each bit may be addressed as shown
22	17	16	15	14	13	12	11	10		
23	1F	1E	1D	1C	1B	1A	19	18		
24	27	26	25	24	23	22	21	20		

RAM Addresses (hex)

Byte	Bit								Name	Function	
25	2F	2E	2D	2C	2B	2A	29	28			
26	37	36	35	34	33	32	31	30		Note that the following bit mnemonics are used when addressing bits:	
27	3F	3E	3D	3C	3B	3A	39	38			
28	47	46	45	44	43	42	41	40			
29	4F	4E	4D	4C	4B	4A	49	48			
2A	57	56	55	54	53	52	51	50		ANL C,b	
2B	5F	5E	5D	5C	5B	5A	59	58		ANL C,/b	
2C	67	66	65	64	63	62	61	60		ORL C,b	
2D	6F	6E	6D	6C	6B	6A	69	68		ORL, C,/b	
2E	77	76	75	74	73	72	71	70		CPL b	
2F	7F	7E	7D	7C	7B	7A	79	78		CLR b	
										MOV C,b	
										MOV b,c	
										SETB b	
										JB b,radd	
										JNB b,radd	
										JBC b,radd	
30–7F										General Purpose RAM	
80	87	86	85	84	83	82	81	80	P0.7–P0.0	Port 0 pins 7–0	**FFh**
81									SP	Stack Pointer	**07h**
82									DPL	Data Pointer low byte	**00h**
83									DPH	Data pointer high byte	**00h**
87									PCON	Power Mode Control SFR	**00h**
88	8F	8E	8D	8C	8B	8A	89	88	TCON	Timer Control SFR	**00h**
89									TMOD	Timer Mode Control SFR	**00h**
8A									TL0	Timer 0 low byte	**00h**
8B									TL1	Timer 1 low byte	**00h**
8C									TH0	Timer 0 high byte	**00h**
8D									TH1	Timer 1 high byte	**00h**
90	97	96	95	94	93	92	91	90	P1.7–P1.0	Port 1 pins 7–0	**FFh**
98	9F	9E	9D	9C	9B	9A	99	98	SCON	Serial Port Control SFR	**00h**
99									SBUF	Serial Data Buffer	**XXh**
A0	A7	A6	A5	A4	A3	A2	A1	A0	P2.7—P2.0	Port 2 pins 7—0	**FFh**
A8	AF			AC	AB	AA	A9	A8	IE.7—IE.0	Interrupt Enable SFR	**00h**
B0	B7	B6	B5	B4	B3	B2	B1	B0	P3.7–P3.0	Port 3 pins 7–0	**FFh**
B8				BD	BC	BA	B9	B8	IP.7–IP.0	Interrupt Priority SFR	**00h**
D0	D7	D6	D5	D4	D3	D2	D1	D0	PSW.7–PSW.0	Program status word	**00h**
E0	E7	E6	E5	E4	E3	E2	E1	E0	ACC.7–ACC.0	Accumulator A	**00h**
F0	F7	F6	F5	F4	F3	F2	F1	F0	B.7–B.0	B register	**00h**

◆ COMMENT ◆

- Addresses that are not listed do not *physically* exist.
- Byte addresses are also the bit address of bit 0 for the byte. For example, port 3 byte address is B0h. Bit 0 of port 3 is also addressed as B0h. Opcodes used for bit operations, however, are unique.
- The SFR addressable bits may be addressed using *absolute* hex addresses, by *SFR.bit* names, or by *functional* names that are shown for each bit of the control registers in the previous section. For example, bit 7 of the IE SFR may be addressed at hex address *AFh,* by the SFR bit name of *IE.7,* or by the functional name *EA.*

F.3 ASCII Codes for Text and Control Characters—No Parity

Hex	Character	Hex	Character	Hex	Character	Hex	Character	Hex	Character
00	NUL	1A	SUB	34	4	4E	N	68	h
01	SOH	1B	ESC	35	5	4F	O	69	i
02	STX	1C	FS	36	6	50	P	6A	j
03	ETX	1D	GS	37	7	51	Q	6B	k
04	EOT	1E	RS	38	8	52	R	6C	l
05	ENQ	1F	US	39	9	53	S	6D	m
06	ACK	20	(space)	3A	:	54	T	6E	n
07	BEL	21	!	3B	;	55	U	6F	o
08	BS	22	"	3C	<	56	V	70	p
09	HT	23	#	3D	=	57	W	71	q
0A	LF	24	$	3E	>	58	X	72	r
0B	VT	25	%	3F	?	59	Y	73	s
0C	FF	26	&	40	@	5A	Z	74	t
0D	CR	27	'	41	A	5B	[	75	u
0E	SO	28	(	42	B	5C	\	76	v
0F	SI	29	)	43	C	5D	]	77	w
10	DLE	2A	*	44	D	5E	^	78	x
11	DC1	2B	+	45	E	5F	—	79	y
12	DC2	2C	,	46	F	60	'	7A	z
13	DC3	2D	–	47	G	61	a	7B	{
14	DC4	2E	.	48	H	62	b	7C	|
15	NAK	2F	/	49	I	63	c	7D	}
16	SYN	30	0	4A	J	64	d	7E	~
17	ETB	31	1	4B	K	65	e	7F	(del)
18	CAN	32	2	4C	L	66	f		
19	EM	33	3	4D	M	67	g		

Index